The Philosophy of Ecology

The Philosophy of Ecology

From Science to Synthesis

EDITED BY DAVID R. KELLER
AND FRANK B. GOLLEY

The University of Georgia Press / *Athens and London*

© 2000 by the University of Georgia Press
Athens, Georgia 30602
All rights reserved
Designed by Sandra Strother Hudson
Set in Minion by G&S Typesetters

Printed digitally in the United States of America

Library of Congress Cataloging-in-Publication Data

The philosophy of ecology : from science to synthesis /
edited by David R. Keller and Frank B. Golley.
xiii, 366 p. : ill. ; 24 cm.
Includes bibliographical references (p. 323–359) and index.
ISBN 0-8203-2219-9 (alk. paper) —
ISBN 0-8203-2220-2 (pbk. : alk. paper)
1. Ecology — Philosophy. I. Keller, David R., 1962– .
II. Golley, Frank B.
QH540.5 .P49 2000
577'.01 — dc21   00-023419

Paperback ISBN-13: 978-0-8203-2220-9

British Library Cataloging-in-Publication Data available

www.ugapress.org

Dedicated to Our Wives

Anina Merrill

and

Priscilla Golley

CONTENTS

LIST OF FIGURES AND TABLES

PREFACE

Friedrich Nietzsche (1989: 16) says our ideas and actions grow out of unique contexts, unique *soils*. We met in the Environmental Ethics Certificate Program at the University of Georgia. This book grew out of those experiences and experiences that followed. In terms of age, physical size, and disciplinary interests we are different, but our joint interest in nature and ethical concerns about the human/nature dynamic created a bridge across the divisions, and we became friends and colleagues. We enjoy walking and talking: walking through the piedmont forests of the Southeast and the red-rock canyons of southern Utah, peering into rotted stumps at the microcosmic life-worlds inside, marveling at luxuriant moss clinging to the surface of wet sandstone, and talking about the human relationship to nonhuman nature throughout the history of the Occidental tradition.

These formative experiences were followed by thousands of hours of reading books and journal articles, staring at computer screens, making scores of telephone calls, and sending dozens of letters and faxes and hundreds of e-mail messages. As we examined the connections between ecology and philosophy, we were impressed by the opportunity to bring together ideas and points of view that are possibly marginal in the core of these disciplines yet assume a major significance in terms of human well-being. The result is this introductory reader in the philosophy of ecology, which lays the groundwork for a second book on the possibility of deriving ethical norms from ecological science.

We chose as our first task the development of a reader that would provide students and interested professionals with some background in the philosophy of ecology. Little has been written about the philosophy of ecology, although the subject has been of continual interest to ecologists from the time of Frederic Clements and before. Most authors have focused on epistemology and the practice of ecology (see, e.g., Peters 1991; Shrader-Frechette and McCoy 1993; and Pickett et al. 1994). But the subject poses other metaphysical, epistemic, and moral problems about the *nature* of nature, if we can understand it, and the place of humans in our earthly home. These issues, central to the concerns of the philosopher and the ecologist, underlie myriad social and environmental policy challenges. The goal of this book is to stimulate discussion. We hope it is followed by many better books, new conversations, and new experiences.

The following editorial adjustments have been made to previously published texts:

Some paragraph breaks have been adjusted.

Individual lists of works cited have been combined into one and citation styles have been regularized.

Typographical errors have been corrected.

Some spellings have been modernized, and British spellings have been changed to U.S. spellings.

Mechanical and stylistic consistency have been imposed to the extent possible without altering the authors' intent.

ACKNOWLEDGMENTS

This book could not have been completed without the constant encouragement and support of our families and friends, and Barbara Ras at the University of Georgia Press. Also thanks to Charlie Brummer, Elaine Englehardt, Frederick Ferré, Mark Hunter, Anina Merrill, David Rothenberg, and Robert Ulanowicz for their constructively critical comments. We are grateful to Tina Black, Kathyrn Henrie, Lynne Hetzel, and Coby Nilsson for help with the text. We appreciate Mary Brown and Micheal Gillilan for their assistance in the production of the figures and tables, and Scott Abbott for the German-to-English translation in chapter 2.

Permissions granted as follows: Figure 1 courtesy of the National Library of Medicine, Bethesda, Maryland. Chapter 2 reprinted by permission of *American Midland Naturalist* and the University of Notre Dame. Chapter 3 reprinted courtesy of *Ecology*. Chapter 5 reprinted with permission from Elsevier Science. Chapter 7 © 1975 by the University of Chicago, Chapters 17 and 20 © 1982 by the University of Chicago, and Chapter 21 © 1988 by the University of Chicago. Reprinted by permission. Chapters 4, 16, and 22 reprinted by permission of Kluwer Academic Publishers. Chapter 9 reprinted by permission of Academic Press. Chapter 10 reprinted by permission of Prentice-Hall and the Estate of Karl Popper. Chapters 11 and 13 reprinted by permission of *American Zoologist*. Chapter 14 © 1977 American Association for the Advancement of Science. Reprinted with permission. Chapter 15 reprinted courtesy of *Ludus Vitalis*. Chapter 18 reprinted by permission of Cambridge University Press. Chapter 19 reprinted courtesy of The Royal Society.

Also thanks to the faculty and staff at our respective academic institutions, Utah Valley State College and the University of Georgia, for their patronage and unflagging confidence in our project.

The Philosophy of Ecology

Ecology as a Science of Synthesis

The approach of a new century—especially a new millennium—encourages reflection. As we enter the twenty-first century, ecology is proving to be a timely and captivating subject. Ecology is timely because it is being enthusiastically heralded as a powerful and needed corrective for the malaise of contemporary industrial civilization (e.g., Capra 1988; Merchant 1990; Roszak et al. 1995; Ferré 1996). And ecology is captivating due to the sheer comprehensiveness of its scope and complexity of its subject matter; ecology addresses everything from the genetics, physiology, and ethology of animals (including humans) to watersheds, the atmosphere, geologic processes, and influences of solar radiation and meteor impacts—in short, the totality of nature.[1]

What is nature? Is nature the aggregate of physical processes and phenomena that carry on independently of human cognizance, and are these processes capable of being discovered by scientific methods? Or is nature simply the subjective projection of socially inculcated worldviews onto a corporeal backdrop? The former answer, known as scientific (or metaphysical-epistemic) realism, contends that nature has an objective existence independent of any perception of it, and that observation can disclose the laws governing natural systems. The latter answer—that of social contextualism or constructivism—contends that what a scientist "discovers" in nature is actually an interpretation contingent upon that scientist's culturally rooted biases and prejudices, a view popularized by "poststructuralist" thinkers.

Is only one of these conceptions of nature correct? While scientists unanimously maintain some form of metaphysical realism, a metaphysics of realism ineluctably raises epistemic questions, such as: *How* do we know nature? Are there general principles directing natural processes, or are events singular? Constructivists unanimously reject realism, instead asking: How are various conceptions of "nature" informed by different historical and cultural perspectives? A perspective is an amalgamation of factors: biological, social, religious. The power and influence of these factors are not to be underestimated in their effect on a person's conception of reality. The Western scientist is no exception, constructivists contend.

Few twentieth-century scientists, and no postmodernists, believe human beings are capable of epistemic objectivity—that is, *sub specie aeternitatis* (a "view from eternity"). Human knowing is temporally contingent (always a "view from here"). But this does not mean, on the other hand, that *all* knowledge varies from person to person. Because statements about the world are third-personal (e.g., "the limestone ledge has moss on it") rather than first-personal, some convergence of belief is possible. Agreement is possible because individual observers share similar experiences. Therefore, objectivity (not in the strong sense of godlike but in a weak sense of *extrasubjective*) in natural science is feasible (Williams 1985).

Looking at the world scientifically is like looking at the world through sunglasses: the color of the lens, like the cultural filter through which each scientist sees the world, adds a quality to the world that is not inherent *in* the world. The world is not green, yellow, or rose. The lenses may make fuzzy distant objects look sharp or fuzzy close objects look sharp. Yet part of the image is of the world as it exists independently of human observation. Trees extend from the ground upward, rivers flow downhill within the confines of their banks, and the sun traverses the sky the same way, day to day, year after year. Some elements of the complete image are first-personal, and other elements of the image are third-personal. Thoroughgoing scientific realism is not humanly possible, but a mitigated realism is (we will return to this topic below).

The focus of this book is on the area where philosophy informs ecological science and ecological science informs philosophy. Debates in the philosophy of science are directly relevant to ecology, and the discoveries of ecological science are relevant to environmental ethics and public policy.

In understanding the philosophy of ecology, it will be useful to begin with a consideration of the general features of ecology.

Some Meanings of *Ecology*

The word *ecology* connotes "ecological worldview."[2] An ecological worldview emphasizes interaction and connectedness. This theme can be developed several ways:

1. All living and nonliving things are integral parts of the biospherical web (ontological interconnectedness).

2. The essence or identity of a living thing is an expression of connections and context (internal relations).

3. To understand the makeup of the biosphere, connections and relations between parts must be considered, not just the parts themselves (holism).

4. All life-forms—including *Homo sapiens*—result from the same processes (naturalism).

5. Given the affinities between humans and nonhumans, nonhuman nature has value above and beyond instrumental, resource utility for human beings (nonanthropocentrism).

6. Humans have caused serious negative impacts (pollution, anthropogenic extinction) on the earth, leading to the need for environmental ethics.

In a much narrower sense, *ecology* denotes the *science* of ecology, which has a specific history consisting of individual scientists practicing ecology within definite social and institutional milieus (Golley 1993: 167). Although the etymological pedigree of *ecology* is scientific, the term's broad current usage reflects little of this origin. Since scientific ecology has been associated with social events in the public mind—in the United States, notably with the conservation movement of the early twentieth century and more recently with the environmental movement beginning in the 1960s and 1970s—the original meaning has been usurped by the wider popular connotation. Ecology has become vaguely synonymous with "environmentalism,"[3] even though for decades it was used exclusively to refer to a formal scientific pursuit (Egerton 1983: 260).

The broader use has led to confusion between connotations of scientific ecology proper and ecological-like thinking. For example, the eminent American historian Leo Marx refers to geographer and conservationist George Perkins Marsh (1801–1882) as the "father of American ecology" (1970: 946) because Marsh recognized the deleterious effects human activity has had on the environment. The proliferation of such comments has prompted one lexicographer (Voorhees 1983) to see the necessity of coining the term *protoecologist* to describe thinkers who exhibited an ecological outlook before ecology was established as a science.

Surprisingly, the broad sense of ecology emphasized by the literati may or may not have anything to do with scientific ecology (Brennan 1988: 31–42). Having an ecological outlook does not mandate embracing the lessons of scientific ecology; nor do scientific ecologists necessarily have "ecological worldviews." The ecological themes emphasized in nonscientific literature are, quintessentially, reactions against the mechanistic model of nature that has dominated Western science since the Renaissance, while scientific ecology in significant ways is a *continuation* of mechanism. Consequently, many of the themes promulgated by writers and social activists are not well represented within the domain of scientific ecology.

Not surprisingly, the strain of ecological thinking that emanates from the humanities frequently incorporates decidedly nonscientific religious or spiritual themes. This coupling of literary ecology to scientific ecology has caused chagrin for many scientists who have struggled to distinguish ecology as a legitimate scientific pursuit. One ecologist laments that until scientific ecologists shake the mystical associations of romanticized ecology, "the theological parallels will continue to be strongly associated with our work" (Dayton 1979: 15; see also Sagoff 1997). One effective, if impractical, solution would be for scientists to give up the word altogether. Tom Fenchel (1989: 641) expresses the view of many scientists when he suggests that *ecology* be given to the environmentalists and each subdivision be referred to by its special name, e.g., biogeochemistry, limnology, evolutionary ecology, and so on.

This is not to say that there are *no* connections between nonscientific and scientific forms of ecology. In the sense that ecology is about how humans recognize their relatedness to nature, ecological-like awareness predates Western science itself, extending back to our earliest roots as nomadic hunter-gatherers (Oelschlaeger 1991).[4]

It is useful to further clarify the distinction between scientific and nonscientific ecology by sketching in sweeping strokes some prominent figures in the history of ecological-like thinking. We do this with an eye to clarifying the differences of nonscientific forms of ecology such as romantic ecology and political ecology from scientific ecology.[5]

Romantic (Literary) Ecology

Romantic ecology springs from the writers who revolted against the radically anthropocentric and thoroughly mechanistic worldview of modernity.[6] As the modern period evolved in the nineteenth century, writers such as William Wordsworth (1770–1850), Ralph Waldo Emerson (1803–1882), Henry David Thoreau (1817–1862), Walt Whitman (1819–1892), and John Muir (1838–1914) presented a pastoral or arcadian contrast to the urban, industrial disutopia of Europe and America.

According to the metaphysics of mechanism widely espoused by scientists, philosophers, and theologians, the material world is a magnificent machine; nature itself is devoid of any teleological direction, value, or purpose. Nature ticks on precisely and predictably according to the strict, deterministic, causal laws of physics; all natural phenomena can be described in terms of inert matter in motion.[7] Johannes Kepler (1571–1630) described this vision well at the beginning of the seventeenth century when he said: "I am much occupied with the investigation of physical causes. My aim is to show that the celestial machine is to be likened not to a divine organism but rather to a clockwork."[8]

The metaphysics of mechanism was a crucial cornerstone in a constellation of factors that led to an unprecedented subjugation and instrumentalization of nature during the Industrial Revolution. Consistent with God's injunction to "fill the earth and subdue it; and have dominion over the fish of the sea and over the birds of the air and over every living thing that moves upon the earth" (Genesis 1:28), industrialists valued nonhuman nature as a resource (White 1967). Nature, they believed, has no value in and of itself, but acquires value according to its use by humankind (Locke 1992).

Romantic writers perceived the instrumentalization and devaluation of nature along with the separation of humans from the rest of nature. Mechanism made nature "a dull affair, soundless, scentless, colourless; merely the hurrying of material, endlessly, meaninglessly" (Whitehead 1967: 54). Nature for the romantics was not dull but awe-inspiring. In England, Wordsworth praised the sublime quality of his native Lakes District. In America, Thoreau had begun "to regard man as an inhabitant, or part and parcel of nature, rather than a member of society" (1980b: 93). And in contrast to Plato's anthropocentrism, which profoundly affected Western thinking—summed up in Socrates' declaration: "I'm a lover of learning, and trees and open country won't teach me anything, whereas men in the town do" (1982: 479, 230d)—Whitman extolled the epistemic and spiritual value of nature:

Now I see the secret of the making of the best persons,
It is to grow in the open air and to eat and sleep with the earth . . .
Now I re-examine philosophies and religions,
They may prove well in lecture-rooms, yet not prove at all under the spacious clouds and along the landscape and flowing currents. (1965: 152)

Muir was a pantheist who considered nature God (Cohen 1984), not the insensate machine of modern science. During his 1867 trek across the southeastern United States, he lamented the instrumentalization of nature: "The World, we are told, was made especially for man—a presumption not supported by the facts. A numerous class of men are painfully astonished whenever they find anything, living or dead, in all God's universe, which they cannot eat or render in some way what they call useful to themselves" (1992: 136).

The themes of romantic ecology are well represented by a variety of authors, Robinson Jeffers (1938), Edward Abbey (1988), Gary Snyder (1974), Barry Lopez (1990), and Terry Tempest Williams (1995), to name a few.

Political Ecology

The romantic or literary strain of ecological thinking reaches into the realm of politics, as in the case of Muir. Insofar as ecological thinking questions fundamental assumptions of the mechanistic worldview, ecology is a "subversive" subject (Sears 1964).

Although Aldo Leopold (1886–1948) and Rachel Carson (1907–1964)—two pillars of American environmentalism—were scientists, both are remembered for stepping beyond science into the realm of social commentary. Leopold (1987) argued that moral considerability, which historically had expanded to include previously excluded groups (e.g., blacks and women), should be *further* expanded to include the land as a whole. Ethics, Leopold said, should include entire biotic communities, not just human individuals. Carson argued that using pesticides was an example of anthropocentrism based on the misguided belief that humans can control nature: "The 'control of nature' is a phrase conceived in arrogance . . . when it was supposed that nature exists for the convenience of man. . . . It is our alarming misfortune that so primitive a science has armed itself with the most modern and terrible weapons, and that in turning them against the insects it has also turned them against the earth" (1994: 297).

Political ecology has now flowered into a variety of environmental ethics. "Deep ecology," a phrase coined by the Norwegian philosopher and mountaineer Arne Naess (1973), currently enjoys wide popularity. Deep ecology, states Naess (1993), is the psychological awareness that the individualistic, egoistic self is inextricably linked to a larger, biospherical self. Naess and other deep ecologists suggest that the environmental ills associated with modernism stem from the failure to acknowledge this primordial relationship.

In a totally different vein, the "social ecology" of Marxian thinker Murray Bookchin (1991) maintains that the exploitation of nature is the result of unjust social frameworks (i.e., capitalism), and that a just social framework (i.e., anarchy) will also be an ecologically sound social framework.

"Ecological feminism" alleges that patriarchy—men's domination of women—and anthropocentrism—human domination of nonhuman nature—are based on the same logic of domination: bifurcation and prioritization. A category is cut in two (man/woman, human/nature), and one subcategory is placed over the other (man over woman, human over nature). Because patriarchy and anthropocentrism are validated by the same conceptual logic, ecofeminists contend that any feminist ethic will also be an environmental ethic, and vice versa (Warren 1990). Dismantling the ubiquitous logic of domination will help weaken both patriarchy and anthropocentrism.

Scientific Ecology

For the purposes of historical exegesis, scientific ecology can be broken down into two periods demarcated by the point at which ecology was formally identified as a specific area of scientific inquiry: protoecology and ecology.

Protoecology

In large part, the genealogy of protoecological science can be traced through the legacy of natural history beginning with the ancient Greek philosopher Theophrastus of Eresus (372–288 B.C.E.). Theophrastus was Aristotle's most distinguished student and successor at the Lyceum, and he is considered by some to be the founder of botany (Locy 1925: 34). Theophrastus retained Aristotle's naturalism—as opposed to Plato's supernaturalism—and continued many of Aristotle's natural history studies of flora and fauna.

Theophrastus's work is an early example of ecological-type thinking, and in this area he surpassed his mentor. Theophrastus recognized that biota actively adapt to their surroundings (Zeller 1963: 200), that is, adaptation is more than external compulsion; adaptation is an organism's internal response to environmental conditions.[9] For example, Theophrastus noticed that the chameleon and octopus change color to blend in with their surroundings. In the case of the latter, Plutarch (c. 46–120) interpreted Theophrastus as arguing that "the change [of color] of the octopus is something that it does, rather than something that happens to it; for it changes [color] deliberately, using this as a device both to hide from [the creatures] it fears and to capture [those] on which it feeds" (Fortenbaugh et al. 1992: 171).

These investigations of the interrelationships of biota and environmental conditions were continued by Roman authors, for example, Pliny the Elder (23–79) in his thirty-seven-volume *Historia Naturalis*. Yet the Greek spirit of natural history lay relatively fallow until it was rediscovered in the Renaissance (Allee et al. 1949: 16). The Swedish botanist Carl von Linné (1707–1778)—also known as Carolus Linnaeus—explained in his famous 1749 essay "The Oeconomy of Nature" that all interactions between organisms and the environment are controlled, with mechanical precision, by the hydrological cycle. French biologists Georges de Buffon (1707–1788) and Jean-Baptiste Lamarck (1744–1829) speculated that species are not static categories but in fact evolve. In his 1749 *Histoire Naturelle*, Buffon recognized that organisms are affected by environmental conditions; in his 1809 *Philosophie Zoologique*, Lamarck contended that species change by the inheritance of traits determined by their ancestors' use or disuse of organs. In his 1798 *Essay on the Principle of Population*, the British economist Thomas Malthus (1766–1834) emphasized the relationship between the environment and populations by arguing that food production increases arithmetically while population growth increases geometrically, invariably resulting in overpopulation, famine, and wide-ranging death. Charles Darwin (1809–1882), in his 1859 landmark work, *On the Origin of Species by Means of Natural Selection, or The Preservation of Favored Races in the Struggle for Life*, crystallized the theory of evolution by validating his conviction that there is an intimate connection between the makeup of organisms and environmental conditions. He called this connection "natural selection."

Ecology

The establishment of ecology as a formal science occurred in 1866 when the German zoologist Ernst Haeckel (1834–1919) coined the neologism *ecology* in a textbook (1866b) on the morphology, taxonomy, and evolution of animals (see figure 1). *Ecology* is a combination of the ancient Greek words οικοσ (*oikos*) and λογοζ (*logos*). *Oikos* means "house, not only of built houses,

Figure 1. Ernst Haeckel (1834–1919), inventor of the word ecology *and passionate supporter of Darwin in Germany*

but of any dwelling place," "domicile of a planet" (Liddell and Scott 1968: 1204, 1205; emphasis omitted). *Logos* means "ultimate truth"—the answer, origin, root of things, reason, computation, reckoning, explanation, rule, principle, law (ibid.: 1057–59). In late English and German, *logos* connotes scientific knowledge, for example the *logy* of geo*logy*, bio*logy*, and psycho*logy*. Hence *ecology* means "the scientific study of the earthly dwelling place" or "home." A synonymous term used by Russian scientists is *biogeocoenosis*, which means "life and earth functioning together" (E. P. Odum 1993: 27).

A social reformer and interpreter of Darwin, Haeckel's motivation for coining this new word was to draw attention to the inclusive study of organisms *in the environment*, in contradistinction to the narrower study of organisms in the laboratory (the province of physiology). To emphasize the investigation of organisms in their natural setting and the operations of natural selection, Haeckel wanted to distinguish ecology from biology in the narrow sense of dealing with the structure and classification of organisms themselves.[10]

In January 1869, Haeckel gave his most eloquent definition of ecology in an inaugural lecture at the University of Jena (Stauffer 1957: 141):

> By ecology we mean the body of knowledge concerning the economy of nature— the investigation of the total relations of the animal both to its inorganic and to its organic environment; including, above all, its friendly and inimical relations with those animals and plants with which it comes directly or indirectly into contact—in a word, ecology is the study of all those complex interrelations referred to by Darwin as the conditions of the struggle for existence. This science of ecology, often inaccurately referred to as "biology" in a narrow sense, has thus far formed the principal component of what is commonly referred to as "Natural History." (Haeckel 1879)[11]

The idea of ecology caught on. In 1885 Hanns Reiter used *ecology* in the title of a book for the first time (Egerton 1977: 195), and interest in ecology developed rapidly. In 1893, John Burden-Sanderson, president of the Royal Society of Great Britain, commented that the future of ecology was especially bright because it dealt with organisms in their environments (Burden-Sanderson 1893). Universities began to offer ecology courses, and in 1913 the first professional society, the British Ecological Society, was established (Sheail 1987). Thus the founders and organizers of the science were alive and active until the middle of the twentieth century.

Defining Contemporary Ecology

Ecological science is a complex subject because it concerns the study of all forms of life over the expanse of time that life has existed on earth, and all the environmental relationships in which life is present. Ecology relies on physics, chemistry, geology, molecular biology, and the biology of the organism to explain certain phenomena. It is mainly concerned with organisms, groups of organisms, interactions between organisms and between their environments, and complexes made up of the physical, chemical, and biological components called *ecosystems.*

Ecology differs from other natural sciences in its emphasis on the primacy of direct observation. The discovery of patterns in nature does not necessarily require instruments or machines. Nature can be directly experienced. Ecologists use machines, of course (computers have enormously aided theoretical modeling of populations and systems), but in the field machinery is not required for discovery to proceed. For example, Thoreau (1980a) made his remarkable discovery of the pattern of succession simply by walking through the fields and forests of Concord, Massachusetts, and observing squirrels burying hickory nuts in stands of hemlocks. Of course, physics, chemistry, and molecular biology concern natural systems, but the features of physical, chemical, or biological systems of interest require elaborate machinery and abstract concepts and language to describe their structure and function.

It is difficult to find fundamental or simple patterns in nature. Ecological science, like other modern sciences, aspired to identify general laws that would explain its observations. However, as generalizations have been proposed and even applied, inevitably exceptions have been found. The specific replaces the general. Currently ecological science exists in a schizophrenic state: ecologists want to find generalizations, but increasingly ecology has become a science of case studies.

Scientific ecology is grounded in several fundamental principles, including the principles of *system* and *evolution.* System is concerned with the question: How does it work? Evolution is concerned with the question: How did this system come to be this way? In both cases the answers are expressed mechanistically.

A system is an organized pattern in which interconnected subsystems create a whole that has its own properties, which are not predictable from the parts or their connections. An ecosystem has at least two parts: organisms and an environment. Evolution involves the generation of variety so that a number of similar entities are produced, each with often-minute differences. As these entities interact with their specific environments, some survive and others perish. Evolution is the generation of variety and selection from that variety. It is a creative phenomenon that shapes ecological systems because it produces change.

The modern sciences that derive from ultimate concepts such as system and evolution are continually forming and reforming. To make ecology comprehensible, one is tempted to create a series of dualities—a common human practice!—and then to contrast them; for example, arcadian versus imperial ecology (Worster 1994) or systems versus population ecology (Hagen 1992).

One helpful way of conceptually organizing scientific ecology has been suggested by G. Evelyn Hutchinson (1978: 215). According to Hutchinson's heuristic, ecology can be divided into two schools based on merological and holological paradigms.[12] *Mero* comes from the ancient Greek word *méros* meaning
"part" or "portion" (Liddell and Scott 1968: 1104); *holo* comes from the ancient
Greek word *hólos* meaning "whole" or "entire" (ibid.: 1218). Thus, in the most
simplistic terms, merology is the study of parts, and holology is the study of
wholes.

Merological ecology (or autecology) focuses on the analysis of the components of ecological entities. Holological ecology (or synecology) focuses on the
relationships between entities in an ecological system, rather than on the entities themselves. The basic assumption of merological ecology is that ecological
structure is largely determined by populations of organisms, and populations
are determined by the properties of the individual organisms that comprise
them. The basic assumption of holological ecology is that ecological structure
is determined by flows of materials and energy. Accordingly, population ecology is founded on the merological paradigm, whereas ecosystem ecology is
founded on the holological paradigm.

Certainly this division of ecology into two approaches is an oversimplifi-
cation. Even so, Hutchinson's heuristic is useful in capturing the flavor of
twentieth-century ecology, especially during the 1980s and 1990s, when the differences between these approaches became especially acute.

This schism of paradigms has resulted in widespread calls for integration in
ecology (Jones and Lawton 1994; Pickett et al. 1994). Pickett et al. argue that
"unboxing" organisms in the holological paradigm provides a promising route
of overcoming the breach between population and ecosystem ecology (8).

Basic Philosophy of Ecology
While the emphasis on discrete organisms, on the one hand, and
materials and energy flux, on the other, has kept ecology from achieving conceptual unity, ecologists generally agree on three philosophical issues: (1) naturalism, (2) mitigated scientific realism, and (3) the comprehensive scope of
ecology.

Before looking at each of these three items in turn, let us consider the word
nature itself, a word so ubiquitous yet laden with so many various connotations
that it often hinders clear thinking (Ferré 1988: 28). As philosopher Frederick
Ferré points out, *nature* has two different primary meanings. In the first, arti-
facts of human manipulation (art, cities, dioxin) are not considered part of
nature per se; *nature* in this sense is everything apart from the artificial (na-
ture₁). In the second, *nature* connotes everything apart from the supernatural,
including the human in all its manifestations—art, culture, civilization, and

the products of human manipulation (nature$_2$). It is important to be clear about which definition of nature is used in a given instance.

Confusingly, ecologists employ both connotations simultaneously. On the one hand, ecologists consider humans to be part of nature (nature$_2$). On the other hand, a system or population is considered "natural" if it has not been manipulated to serve a human purpose or been perturbed by human activity (nature$_1$). While the present environment is influenced by human activity (humans have affected every system and every flow in the biosphere in some way; McKibben 1989), the field study of organisms that are relatively unperturbed by human activity is the point of departure for many ecological investigations.

NATURALISM. Ecology shares with the other natural sciences its methodology and philosophical foundations. One of these foundations is *naturalism,* the position that there is but one system of reality—nature$_2$—and this system does not depend on supernatural factors. Naturalism does not necessarily preclude supernatural existence, but understanding the natural world does not require understanding the supernatural world. Naturalism thus challenges positions that posit the cause and regulation of the universe as prior to or ontologically distinct from nature itself.

Naturalism asserts that the totality of nature can be explained using the same methods and terminology. There is no part of reality to which these methods are not applicable, because all entities that pass in and out of existence are the result of spatiotemporal processes. A pebble, a passing squall, a bacterium, and a human being all result from operations of nature—and hence all exist in the same system of space and time.

It is crucial to notice that the defining characteristic of naturalism is not metaphysical but methodological (Danto 1972: 448). Ecologists, as philosophical naturalists, agree that all real things are discoverable by the same methods and are describable in the same language, but disagree on the ultimate constitution of nature itself (a central theme of this book). For example, is there a "balance of nature," or is nature inherently chaotic?

MITIGATED SCIENTIFIC REALISM. Another philosophical foundation ecology shares with other natural sciences is metaphysical-epistemic realism, the position that there are patterns or regularities in nature existing independently of human perceivers, and these patterns, to some extent, are objectively knowable by humans. Ecology attempts to discern these patterns. Together with geology and physical geography, ecology begins in the field and studies nature directly as it presents itself to humans. The observation of patterns becomes the basis for the development of hypotheses that will be tested through the methods of normal science. While ecologists disagree on what exactly the essence of nature *is,* they agree that there are patterns in nature that are more than conceptual categories created by the human imagination.

Our theory is that a thoroughgoing social constructivism is wrong in holding that any *meaning* ascribed to nature is socially produced. While much or most of the meanings we find in nature—pastoral Eden or dismal wasteland in which life is "solitary, poor, nasty, brutish, and short" (Hobbes 1985)—is socially based, we believe there is some extrasubjective, transcultural meaning in nature which humans can discern. Therefore, scientific ecology, to a certain degree, is at odds with social constructivism. Social constructivists deny that the words *nature* and *wilderness* have any objective, transcultural meaning. It is important to note that constructivists are not Berkeleian idealists; they have no problem with the notion that corporeal processes exist independently of human perception. However, in their view the semantics of the words *nature* and *wilderness* (e.g., pastoral or threatening, balanced or chaotic) is socially determined.

Constructivism has been developed both in terms of the Critical Theory of Frankfurt school scholars such as Lukács, Horkheimer, Adorno, and Habermas (e.g., Vogel 1996), as well as the "postmodernism" or "poststructuralism" of French philosophers such as Jacques Derrida, Jean-François Lyotard, Gilles Deleuze, and Michel Foucault. Poststructuralism is heir to Friedrich Nietzsche's (1844–1900) critique of modernism, particularly Cartesian philosophy of self and science.

The influence of René Descartes' (1596–1650) declaration: "I am really distinct from my body, and can exist without it" (1989: 115) cannot be overestimated. In making this claim, Descartes provided a solid foundation for thoroughgoing objectivity in science. The scientist, as a disembodied mental entity separated from the flux of the corporeal world, can observe phenomena as a detached and unaffected observer surveying external phenomena in their fundamental essence. Nietzsche alleged that this concept of selfhood so central to modernity—and its attached notion of objectivity—is a fiction (e.g., 1968: 267, §481; 269, §485); rather, all statements about the world are culturally based or individually created *interpretations*.

The second camp of social constructivists, following Nietzsche, assert that whatever we think we know about nature as an objective and independent entity is just an interpretation or projection of a particular point of view (Bird 1987; King 1990; Evernden 1992; Cronon 1996). In Neil Evernden's words, nature, "the entity which we take for granted as an objective reality has, in fact, a complex origin as a social creation" (1992: 109).

Ecologists, according to the constructivist account, cannot learn objective "truths" about natural systems because they are themselves inescapably rooted in a social framework or worldview. "Ecological knowledge" is actually a reflection of a particular cultural perspective. There is no fundamental "essence" in nature to discern; we cannot even agree on what nature *is*, claim constructivists. All scientific knowledge, they point out, reflects a certain political agenda, a certain worldview.

It is true that the practice of science always has a subjective component. One lesson of the twentieth century has been that science cannot be exhaustively objective in the Cartesian sense, as Nobel Prize–winning physicists Albert Einstein (1879–1955) and Werner Heisenberg (1901–1976) demonstrated. Einstein confirmed that the Newtonian absolutes of space and time are relative to an observer (Reichenbach 1951), and Heisenberg showed how a singular phenomenon like photons may appear to be waves or particles depending on the observer's choice of instruments (Heisenberg 1927).[13]

Our thesis is that the developments of the twentieth century imply a *mitigated* scientific realism. While many ecologists do not claim that ecology is an entirely objective pursuit free of cultural prejudices, they are nevertheless unwilling to give up the idea that there are essential aspects of nature that are accessible to human inquisitors. This idea implies neither that nature is accessible in *all* of its attributes nor that it is simple and deterministic. Ecological research demonstrates the complexity and indeterminacy of natural systems.

Thus, according to a metaphysics and epistemology of mitigated scientific realism, ecology is informed not just through empirical investigation of material processes, but also by the aesthetic, spiritual, and social filters through which we all inevitably experience the world. Neo-Nietzschean social constructivists are right to bring attention to the inescapable temporal contingency of perspective and the central role perspective has on each person's interpretation of the world, scientists included.

While convergence of beliefs can earn the status of extrasubjective truth, scientific knowledge can never earn the status of complete objectivity, because all individuals are rooted in continuums of space and time. We view the cosmos from *within* the evolutionary processes that gave rise to us; we cannot step outside these processes and survey them with a godlike gaze. We can only interpret the processes to which we owe our existence from inside the processes (Oelschlaeger 1991)—the ultimate hermeneutic circle.

THE COMPREHENSIVE SCOPE OF ECOLOGY. Explicit in Haeckel's definition of ecology is the claim that ecology studies all of nature. The "economy of nature" to which Linnaeus referred involves all interactions and complex relationships between organisms and environments. If we agree, with Haeckel, that humans are animals (nature$_2$), then the ways human beings act in nature fall under the purview of ecological investigation. These ways include the creation of culturally based worldviews (cf. social constructivism) and the manufacture of objects such as machines, cities, and paintings.

How can we describe a human ecology that includes the study of technology, cities, and art as environmental phenomena? Since humans are part of nature, "human ecosystems" are natural ecosystems (Kormondy 1974). A city is a human ecosystem; it is a human technological strategy of living on earth. The point is that since ecology is a subject that begins in the natural sciences

(*Naturwissenschaften*) but ends up crossing into the human sciences (*Geisteswissenschaften*)—sociology, anthropology, psychology, religion—it is the study of everything, of total reality. The inclusive reach of ecology is manifest in the psychiatrist Leonard Duhl's definition: "*Ecology* may be defined as that inter-intra confrontation of biological, social and historical factors that embrace one's family, school, neighborhood, and the many overlapping communities that teach values, defenses, and offenses, the meaning of oneself and one's existence" (Anderson 1966). Ecology, as Duhl defines it, is truly an architectonic science—a *science of synthesis.*

The seduction of synthesis in ecology is clear. Its comprehensiveness requires the ecologist to be not merely a specialist but also an integrator. However, considering the magnitude of subject matter germane for ecological study and the limitations of the human mind in understanding the "totality of reality," a science of everything is impractical. Therefore, while the potential for synthesis has attracted many to ecology as an integrative subject, ecologists have seldom been overcome by such exalted ambitions. Rather, they have tended to subdivide nature into parts and to confine their study to those parts, applying Newtonian analysis to go deeper and deeper into the details of dynamic behavior and adaptation. Nevertheless, the alternative of synthesis is always present and generates an attraction for integrative work. The concomitant pull between analysis and synthesis is ever present in ecological inquiry.

Goals

Ecology shares with the sciences generally and the natural sciences specifically a set of fundamental assumptions. The task of elucidating these assumptions or "first principles" is the *philosophy of ecology.* Its aim is to identify those initial premises or presuppositions on which ecological theories are founded yet which cannot themselves be axiomatically demonstrated (Whitehead 1966: 105–7).[14] To illustrate one such foundational assumption, mechanism assumes there is only material and efficient causation (to borrow Aristotle's terminology from the *Metaphysics* 1013a24–1013b4 (1979: 74)). That there is no formal or final causation in nature cannot itself be logically shown.

Clarifying the fundamental assumptions of ecological science is useful in dispelling the popular myth that ecologists are unified in their vision of the essence of nature. The public looks to ecologists to provide a description of the inner workings of the biosphere, yet ecologists are far from arriving at any consensus. For example, Alston Chase (1995) claims that the concept of the ecosystem is based on the resurrected dogmas of design, balance, and equilibrium in nature. The truth is that these concepts are all hotly debated within the community of scientific ecology (vide Egerton 1973; Ehrenfeld 1993: 139–46).

Philosophy can help with many of the issues concerning fundamental assumptions that have beleaguered ecologists. Ecology is a complex and weakly organized subject that finds its foundations not in logical thought but from patterns of nature observed through cultural filters. As a result, many ecologists have not been clear about the variety of tacit presuppositions on which research has been based. The utility of philosophy for ecology lies in the clarification and critique of first principles. Philosophy of ecology, as part of philosophy of science, is both descriptive and prescriptive (normative): it seeks not only to uncover the initial presuppositions ecologists have used in their work, but also to suggest the initial presuppositions ecologists *should* use in their work.

Ecology also contributes to philosophy. Any ethic presupposes an ontology of selfhood (Keller 1995, 1997a). Environmental ethics presupposes a conception of human and nonhuman beings, and the relationship of the two. Ecological science can help inform the philosopher about the relationships upon which environmental ethics rests.

Topic Areas and Methodology

This book is divided into five parts that are concerned with foundational philosophical issues in the science of ecology. As with any complex and polymorphic subject, clean divisions are problematic, and these foundational issues are intimately entwined. For this reason, the path of our inquiry will proceed in a "spiral" fashion so that one issue may be touched on several times in different contexts. Many of the themes of part 1 resurface in part 4, for instance, because metaphysics affects method.

Part 1 examines the metaphysical character of ecological entities and processes. Ecologists, of course, have objects of study: individual organisms, symbiotic complexes, populations, guilds, species, communities, ecosystems, landscapes, biomes, etc. But the exact delineation of these objects of study proves to be extremely difficult. Are the objects of ecological study unambiguously bounded in space and time? Are they really stages in dynamic flows of energy and matter driven by solar energy? Further, how are the entities and processes of ecological study to be organized in order to make them comprehensible?

In part 2 we turn to four selected ecological concepts: community, niche, biological diversity, and stability. These four concepts are central to ecological science, yet they pose fundamental definitional problems. As becomes apparent in part 1, some ecologists think "biotic communities" exist more in the minds of ecologists than in nature. Interesting conversations have also taken place about exactly what a niche is. The concepts of diversity and stability have deep roots in the Occidental idea of design in nature, which has been severely criticized, especially during the twentieth century.

Parts 3 and 4 address the methods of ecology. Part 3 concerns rationalism and empiricism, a dualism illustrated in ecology by two stereotypes: the field naturalist and the systems modeler. The first type of ecologist answers the question, What is the best way to learn about ecological entities and process? by getting out into the marsh or mountains and making observations. The second type of ecologist would rather create mathematical computer models of ecosystems in the research laboratory. Ecologists agree that both methods are necessary but disagree on the relative weight each method should be given.

Part 4 focuses on the issue of reductionism versus holism, a quarrel that has been endemic to ecology for many years. Do ecological entities have "emergent" properties, or can the properties of ecological entities be understood by an analysis of the parts? What are the alternatives to reductionism and holism proper? Some of the most interesting dialogues about the methods of ecology have been aimed at resolving these issues.

The connection between evolutionary law and ecological science is treated in part 5. Ecology is concerned with organisms and the environment. The behavior of organisms is a function of their genetic constitution, which provides the information required for the organism to develop internally and to respond to outside influence. The genome and the environment interact through evolution and natural selection. For this reason ecology is deeply interested in evolutionary law, even though the focus of research in this field lies within the science of genetics. Geneticists and ecologists interact closely in evolutionary ecology and biology. The major problem in applying evolutionary precepts to ecology is that natural organisms in populations are subject to numerous environmental interactions. It is not possible to examine these interactions simultaneously, but the study of single cause-and-effect events does not reveal the total synergy of the evolutionary process. Can these methodological problems be overcome? We will review several proposals, particularly within the context of the debate about the role of adaptation in evolution.

Our method in this volume is to introduce an issue in a topic area and then let scientists and philosophers speak to that issue through published essays and edited excerpts. Conversations between philosophers and ecologists have not been prevalent, and this artificial device might well stimulate a desire to have more dialogues face-to-face.

Conclusion

Ecology endeavors to discover pattern and process in natural networks. These networks or systems comprise organic and inorganic elements in dynamic relationship. As an Occidental science, ecology has traditionally looked at nature as a grand and intricate machine functioning according to the deterministic Cartesian-Newtonian laws of physics. (Even the "organismic" metaphor of Clementsian ecology[15] treats the organism mechanistically.) While the machine metaphor has gone a long way toward increasing our understanding of nature, we believe there are additional sources of meaning that the mechanistic approach does not exhaust: the aesthetic, spiritual, and social meanings.

While some ecology focuses mainly on the mechanistic element, our thesis is that ecology can justifiably be informed by nonmechanistic sources of meaning. These sources are the processes inherent in nature that have somehow generated the beauty and complexity of living systems. But does the enterprise of ecology end with discovering *how* the "machinery" of nature works? We think not. The ecologist, knowing something about how such beauty is generated, faces questions about the extent to which these processes should be affirmed, inhibited, or manipulated. Ecology cannot escape making value judgments about nature. *Ecology entails ethics.*

Notes

1. Notoriously, *nature* has many meanings. Its polysemy is an undergirding theme of this book. Here we simply mean the aggregate of physical processes.

2. While our focus is on the Occidental tradition, it must be noted that ecological worldviews are well represented in Eastern thinking (see Callicott 1994).

3. There are hundreds (possibly thousands) of examples. Even within academia, where semantic precision is venerated, confusion between ecology qua worldview and ecology qua science is prevalent. For example, *Environmental Ethics* (a journal that has largely defined the field), from its inception in spring 1979 to summer 1996, published thirty papers with *ecology, ecological,* or *ecosystem* in the title (even excluding articles about the environmental philosophies deep ecology, social ecology, and ecological feminism), but only about one-third of these even tangentially touched on the topic of *scientific* ecology. Most articles use *ecology* to refer to various interpretations of what the author or authors take to be metaphysical implications of ecological science. In another example, Anna Bramwell's *Ecology in the Twentieth Century: A History* (1989) is a study of a particular political movement in Europe and has little to do with the science of ecology.

4. Soulé (1995: 733) makes a similar point about the roots of conservation biology.

5. For worthy narratives of the history of ecological ideas, see McIntosh 1985; Kingsland 1991; and Worster 1994. Allee et al. 1949: 13–42 focuses on protoecological science; Hagen 1992 and Golley 1993 focus on systems ecology in ecological science; Kingsland 1985 considers population ecology.

6. *Modernity, modern,* and *modernism* refer to the period of the Western tradition beginning with the Renaissance, continuing through the Enlightenment, and extending (roughly) to the present. Key concepts of modernism are the possibility of epistemic objectivity, mind-body dualism, and determinism of the material world.

7. The well-known natural theology of Archdeacon William Paley (1825) construed

nature as a corporeal clockwork designed by God, its every operation determined by divine will. Nature does not have an intrinsic impetus; any purposive phenomenon such as life comes from outside nature, from God.

8. Letter to Herwart von Hohenburg, February 10, 1605. Quoted in Oelschlaeger 1991: 77.

9. Here Theophrastus echoes Aristotle's criterion for voluntary action in the *Nicomachean Ethics* 1110a15–19 (1985: 54).

10. Curiously, seventy-two years after Haeckel coined the word *ecology*, the eminent philosopher of science Rudolp Carnap argued for the necessity of a special field of biology that "deals with the behavior of individual organisms and groups of organisms within their environment, with the dispositions to such behavior" (1938: 47), subsequently claiming (certainly to the astonishment of many ecologists): "There is no name in common use for this second field" (ibid., 48).

11. Translation from Allee et al. 1949: frontispiece. Stauffer remarks: "Although this translation . . . is a free rather than a literal rendering, I consider it very faithful to Haeckel's meaning" (1957: 141).

12. As Hagen (1989: 434) points out, Hutchinson retained this distinction throughout his career, though under different names. Earlier, Hutchinson referred to the merological paradigm as a "biosociological" or "biodemographic" approach, and to the holological paradigm as a "biogeochemical" approach (ibid.: 452, n. 1).

13. While both Einstein and Heisenberg transcended Newtonian thought, the two split on the issue of indeterminacy: Heisenberg asserted indeterminacy against Einstein's belief that "God does not play dice with the universe."

14. Mill (1974: 254) makes the same point in relation to ethics when he says what is ultimately "good" cannot be logically deduced. For example, that "health" or "pleasure" is good cannot be incontestably proven but must be assumed.

15. "Clementsian ecology" refers to the writings of Frederic Clements (1874–1945), who developed a form of quantitative plant ecology that viewed vegetation complexes as "superorganisms." See especially chapters 1 and 2, although the topic of Clementsian ecology is woven throughout the fabric of the entire book.

Entities and Process in Ecology

When ecologists enter a natural setting and begin their observations, they recognize a variety of entities and patterns. "Natural setting," used in this sense, consists of entities, objects, or things that appear to be distinct and bounded against a background matrix. "Boundedness" involves a recognizable difference between the properties of an entity and those of the matrix in which it is located. The observer perceives the difference and distinguishes the entity as separable from its background. For example, we encounter trees in the forest and we begin to give them technical names and note their size and condition. A bird flies in front of us and we do the same, except the bird's mobility leaves us uncertain if we saw a flash of white on the tail feathers as it flew away.

Entities are never static; they come into being and are destroyed. Some move quickly, like the bird, and others move slowly, like continental plates. So ecology is not simply concerned with discerning entities against an environmental backdrop, but also with discerning patterns of change. Nature's intrinsic dynamism makes the work of the ecologist all the more complicated and challenging.

Part 1 examines the metaphysical character of ecological entities and processes. Surprisingly little work has been done on this basic but complex topic in the philosophy of ecology.

Entities: Preliminary Metaphysical Considerations

An entity is something that exists as a discrete unit—that is, something that is distinct and bounded. The conditions that make the entity a discrete unit, discernible from its environment, differ: both the marmot and the granite boulder it sits upon we recognize as entities, but radically different types of entities. The *way* the entity is bounded—its internal structure—makes a difference.

The intentionality of the observer also makes a difference. When we recognize things in the world, we always do so from a temporal, or subjective, perspective. A tree may be seen as an excellent center beam for a house or a source of shade in the hot summer. As Kant (1965) points out, all knowledge arises from experience, but knowledge is not comprised only of experience: subjects experience the world in physiologically similar ways (Kant refers to these as "categories of the mind").

Moreover, subjectivity involves more than physiological features of the perceiving being. Subjectivity also involves socially inculcated interests and purposes. Philosopher Frederick Ferré (1996) remarks: "What is 'essential' or 'accidental' for entities is a matter of interests and purposes interwoven with the facts. The actual attributes, relations and functions of something are not irrelevant to the decisions we make. They provide the basis for our decisions. Entities 'are' the joint product of what we find and what we make" (325). The role of intentions in defining entities means that the scientific enterprise always involves an element of subjectivity. In terms of the long-running debate about objectivity in science (Harré 1967), the hope for pure objectivity is unrealizable. This suggests an epistemology of *mitigated* scientific realism as an alternative to a thoroughgoing scientific realism. Some part of knowledge is contingent upon the intentionality of the knower—that is, the subjectivity of the observer influences the observation.

Keeping in mind the role of the value judgments we make in analyzing the world, Ferré identifies six types of entities: (1) aggregate entities, (2) systematic entities, (3) organic entities, (4) formal entities, (5) compound entities, and (6) fundamental entities.

Aggregate entities, such as granite boulders, mountains, lakes, and glaciers, are characterized by external relations among the parts. Even a huge disturbance, such as the explosion of Mount Saint Helens in the Cascade Range of the northwestern United States (which blew away the top of the mountain), is insufficient for us to alter our recognition of the entity. Mount Saint Helens is radically changed, but it remains Mount Saint Helens. Aggregate entities provide a background for ecology, serving as the stage upon which the ecologic play is acted.

Systematic entities include ecosystems, which are characterized by feedback loops. Ecosystems retain coherence even under intense stress until the pressure overwhelms them and they collapse. The term *systematic* refers to this capacity to maintain structure and function under continually changing conditions.

Living organisms are examples of *organic entities.* Living organisms are made up of parts that are internally related. The whole organism is governed by this internal system of relationships that maintain homeostasis. Further, organic processes are creative in generating unique, new forms of life.

Formal entities are based on the subjective intentionality of the observer, for example, definitions. In biology, a "species" is a formal entity. A species is real as a group of related biota capable of interbreeding, but what makes a species "real" is that it was invented as a way of classifying organisms.

Ferré completes his taxonomy of entities with two final categories: *com-*

pound and *fundamental entities*. Compound entities have strong internal relations but are without an apparent internal system dynamics, for example, inorganic molecules. Finally, fundamental entities constitute the deep structure beneath entities in general—that is to say, they are the basic ontological units of nature.

All entities relevant to ecology are the joint product of what we observe and the context in which we understand our observation. It is the knowledge added by the knower that creates both a richness in the diversity of ecological entities and the endless quarrels among ecologists about the validity of entities and relationships.

Ecological Entities

Systematic entities, organic entities, and formal entities are closest to the immediate concerns of scientific ecologists. Formal entities are conceptual constructions, and systemic and organic entities are the direct objects of ecological investigation. Systemic and organic entities have noticeable boundaries, are identifiable against a matrix of space/time flux, and have some kind of internal structure.

Even so, given ontological interconnectedness, the boundaries of ecological entities are imprecise, in part because of the entities' *openness* (porousness or permeability). A closed entity would be isolated from its environment; no closed entities exist in nature. An organism continually exchanges matter and energy with its environment, and an ecosystem exchanges matter and energy with the larger system of which it is a part. These flows couple the system to the physical environment (we return to this point in the discussion of Tansley below). Linkages tend to blur the distinctness of ecological objects. The selection of a boundary is always arbitrary because boundaries vary over space and time.

Ecological boundaries also vary spatially. For example, in the center of the United States a great grassland borders an eastern deciduous forest. The boundary between these two regions of the country is made up of a mixture of trees and grasslands in a patchy, savanna-like system. If we examined a satellite photograph we would observe that the two broad regions of the country are distinct and easily recognized. At this scale there is a boundary, and we treat each region as separate and distinct.

Boundaries also may vary over time, as a stream margin varies between flood and drought. In this case the boundary of the stream, based on the presence or absence of water, may move upward and downward and laterally in or out of the floodplain. The boundary of the stream becomes an important environmental factor to organisms living in the wetlands bordering the stream.

The stream margin is similar to the "fuzzy boundaries" Zadeh (1965) uses to represent the imprecision of language. Fuzzy boundaries are more common than distinct ones in nature.

Despite the indeterminacy of boundaries, an ecological entity can be distinguished from its environmental matrix in terms of internal versus external processes: an entity is characterized by internal processes of connection that are stronger than the external linkages of the entity to other entities and the matrix. It is these strong internal connections that create the difference between "inside" and "outside" and permit us to distinguish entities from the broader environment.

Ecological Entities: Three Ontologies

The focus of ecology is the interaction of organisms with each other and with the inorganic environment. The constellation of these interactions forms the basic unit of ecological inquiry. Interestingly—but not surprisingly—ecologists have not agreed on the metaphysical status of the primary ecological entity. We will consider three prominent ontologies recognized by twentieth-century English-speaking ecologists: (1) the biotic community, (2) the individual organism, and (3) the ecosystem.

The Biotic Community

Occidental philosophers, scientists, and theologians have long seen grand design in nature. Along these lines, the American ecologist Frederic Clements speculated that an entire community of organisms—or "biotic community"—has a specific structure of internally related parts, like an organism itself. For this reason, Clements referred to a biotic community as a "superorganism."

In his preface to *Plant Succession: An Analysis of the Development of Vegetation* (1916), reprinted here as chapter 1, Clements asserts that the "developmental study of vegetation necessarily rests upon the assumption that the unit or climax formation [i.e., biotic community] is an organic entity." Through a process of development (succession), each plant association matures predictably according to a final, ultimate identity. Treating the plant community "as a complex organism with a characteristic development and structure in harmony with a particular habitat . . . represents the only complete and adequate view of vegetation[;] in short, . . . every climax formation has its phylogeny as well as its ontogeny."

Clements began forming his theory of the community as a University of Nebraska student at the end of the nineteenth century. He and fellow student Roscoe Pound (who would later become a famous jurist) built square quadrats in order to sample the prairie vegetation. They found that repeated observation of the prairie plants within the square yielded data on the species of plants present, the numbers of individuals of each species, and the patchiness or sociability of the individuals within species (Pound and Clements 1897). The patterns produced by repeated samples from quadrats gave quite different conclusions from those obtained by the traditional botanical observer who walked over the prairie listing species and their abundance.

By applying this methodology to observations across the western United States over a lifetime of study, Clements was able to form a synoptic geographical view of the vegetation. Flora form a community that will appear repeatedly across its ecological range of environments. By correlating climate (mainly temperature and precipitation), plant species distribution, and abundance, Clements and other ecologists working with similar methods created a regional plant community geography.

Clements also had before him abundant evidence of disturbance to vegetation. Fire, plowed land, grazed land, and abandoned agricultural land were commonplace. Clements noted that over time plants invaded the disturbed area and then replaced themselves in patterns of development that led ultimately to the vegetation that was present under similar conditions in different locations. Clements named this endpoint the "climatic climax" of the process of plant succession.

Combining the spatial and temporal descriptions of vegetation, Clements then made his creative leap. He postulated that the plant community was, by analogy, an individual superorganism. The superorganism was born on an abandoned field with the plant invaders present on the site; development took place, and eventually maturity was achieved. Because the mature state was set by the regional climate, all the sites undergoing succession eventually converged to a single state. Clements and Victor Shelford teamed up in the late 1930s and added animals to Clements's conception of vegetation.

The Individual Organism

Perhaps, contrary to the mainstream current of Western thought, grand design in nature is an illusion. Perhaps what appears to the human observer to be teleologically ordered is really the accidental association of various parts.

This is the essence of Henry Gleason's "individualistic hypothesis" of plant association. In "The Individualistic Concept of the Plant Association" (1939), reprinted here as chapter 2, Gleason argues that plant communities are not organized associations; rather, they are random assemblages of individual organisms. Gleason came to this conclusion when he applied the quadrat method to savanna prairie vegetation in Illinois and reached diametrically different conclusions from Clements. Gleason, like Clements, found that a few species were abundant in the squares and most species were uncommon or rare. The specific species patterns were best treated probabilistically. The conditions of the environment for plant growth differ on a microscale; in one place individuals of a species will be common and in another place individuals of the same species will be rare. Not all species had equal chance of appearing in every quadrat. Thus the species actually present were there due to the chance of dispersal and their ability to invade and colonize and then compete for resources, grow, and reproduce.

Gleason concluded that the species composition of a site is indeterministic. Plant associations are accidental assemblages: "Are we not justified in coming to the general conclusion, far removed from the prevailing opinion, that an association is not an organism, scarcely even a vegetation unit, but merely a *coincidence?*" (1926: 16; emphasis in original).

The Ecosystem

A third ontology is the *ecosystem*. Precipitated by South African ecologist John Phillips's defense of Clements (Phillips 1931, 1934, 1935a, 1935b), English botanist Arthur Tansley (1935 [reproduced in this volume]) argued that the organismic analogy between plant development and ecological succession is a poor one. Biotic communities, he said, are more like machines—ecosystems.

"The Use and Abuse of Vegetational Concepts and Terms" (1935), reprinted here as chapter 3, is Tansley's refutation of Clementsian ontology, published in honor of a major explorer of plant succession, the Chicago ecologist Henry Cowles. In this paper, Tansley defines the ecosystem as "the whole *system* (in the sense of physics), including not only the organism-complex, but also the whole complex of physical factors forming what we call the environment of the biome—the habitat factors in the widest sense. . . . It is the systems so formed which, from the point of view of the ecologists, are the basic units of nature on the face of the earth." In this definition Tansley distances ecology from biology and its embarrassing arguments about vitalism and entelechies and identifies ecology with physics. In doing so, Tansley follows the tradition of mechanistic materialism. Physics was considered the most fundamental science because it was believed that ultimately all knowledge would be explained by physical principles. Physics had made brilliant progress in the Cavendish Laboratory at Cambridge University, where Tansley was a lecturer in botany. Tansley also connects the ecosystem concept with the definition of ecology by Ernst Haeckel (1879), emphasizing the interactions of organisms and environment. The ecosystem, Tansley claims, is the basic unit of ecological entity.

Tansley's concept of the ecosystem is a totally different ecological entity from those made up exclusively of biological entities. The ecosystem involves the physical, chemical, and informational features of the environment characteristic of a space/time continuum, which are closely and reciprocally interacting with the biotic community. The ecosystem of interest is a subsystem nested in another more extensive ecosystem, which serves as the environment of the system of interest. Thus, from this perspective there are two kinds of environments: those of a particular habitat that react with the biotic community to form a whole system, and the environment outside the system that affects it, provides it with resources, and receives its outputs.

Tansley's contentions are problematic from our perspective. His choice of physics in his claim that ecosystems are one level of a hierarchy of physical systems that range from the planet Earth to the atom conflicts with ecologists' view of systems. The physical concept of an entity is of an isolated, material object, which is explained through the structural interaction of its parts. Ecological systems are different: their processes are stochastic and interconnected with processes outside the system. They are not closed and their essence derives as much from their connections to the environment of the system as from the interactions among the parts of the system. These problems do not invalidate the ecosystem concept, but they change the emphasis in a fundamental way. Biology has developed in such a way that there is no need for identification with physics. Indeed, contemporary ecology is closely allied to biology in many of its subfields. Instead, the modern emphasis of the ecosystem concept is on a complex of stochastic interactions that make up the actual systems we encounter in the field.

Ecologists critical of the ecosystem claim that it is idealistic and subjective because it erects a concept that is rooted neither in natural history nor in evolutionary theory, the two other primary sources of inspiration for ecological science. Because ecosystem boundaries are fuzzy, critics call the existence of the entity into question. Because the dynamic behavior of ecosystems is usually described in terms of the flows of matter, energy, and information, and not in terms of evolution and natural selection, or behavior, competition, and cooperation, evolutionary ecologists disparage it as physical, chemical, and mechanistic. Other authors have even claimed that it is "fascistic" because it is holistic (Chase 1995) and undervalues the individual vis-à-vis the organic whole.[1]

While all of these claims can be shown to be incorrect, it is interesting how negatively ecologists have reacted to the ecosystem concept. Apparently, it represents a serious alternative entity to those of the biological persuasion, and is thus threatening. It is doubtful if any other recent ecological concept has attracted such widespread and vituperative comment.

Paradigm Shifts

Clements's theory of vegetation dominated ecological thought in the United States for almost fifty years, notwithstanding the efforts of Gleason, Tansley, and others. Echoes of Clements still reverberate in ecological research projects. The Gaia hypothesis (e.g., Lovelock and Margulis 1974) is basically the concept of the Clementsian superorganism applied to the entire biosphere.

This elegantly teleological ontology was convincing to most ecologists. The theory corroborated the observations of ecologists, it provided a simple and deterministic scheme of organizing the observations, and it was predictive. The Clementsian paradigm was so dominant that Gleason's observations were discounted even though they were published three times (1917, 1926, 1939) during his lifetime.

It was only after the middle of the century that support for Clements's paradigm diminished when it was observed that the prairie did not respond to the drought of the 1930s as the biotic community model predicted. After seven years of drought (1933–1940), thousands of acres of mixed prairie had been destroyed and replaced by short-grass prairie, 20 percent of the soil was covered by cactus, and the recovery capacity of the grassland was compromised. As Ronald Tobey points out, "The grassland formation that Clements and [John] Weaver had once described as the terminal climatic climax, in perfect harmony with the environment, was destroyed and replaced by a different set of dominants" (1981: 201).

Ecologists such as Robert Whittaker (1953) and John Curtis (1959) showed how individual plant species responded to environmental factors. These new viewpoints led to a shift in the perspective of ecologists, a decrease in interest in Clements's paradigm, and growing recognition of the validity of Gleason's observations (McIntosh 1975). Today the term *association* refers to a collection of plant species at a site that is part of a set of sites all with roughly the same species composition and the same environmental conditions. Usually, these associations are named after the dominant species. Clearly, the shift from Clements's to Gleason's ontology was a scientific revolution in the sense of Thomas Kuhn's (1970) theory of scientific progress.

In "A Succession of Paradigms in Ecology: Essentialism to Materialism and Probabilism" (1980), reprinted here as chapter 4, Daniel Simberloff interprets this paradigm shift in ecology as part of a broad revolution in science, namely, the rejection of "essentialism" in favor of materialism and probabilism. What Simberloff means by "essentialism" and "idealism" is the belief that nature has an elegantly teleological structure, and natural things have set, unchanging essences typified by Plato's metaphysics. Simberloff argues that geneticists and physicists rejected the deterministic, teleological model of nature in the early twentieth century; they were followed by ecologists in the 1940s and 1950s with the rethinking of the superorganism model.

Simberloff's paper has value to us as a historical comment on the philosophical development of ecological thought in general, as well as for the discussion it generated. In the 1980 volume of *Synthese* titled "Conceptual Issues in Ecology," where the paper first appeared, Marjorie Grene excoriates Simberloff for abandoning the "standards of accuracy that, at least in the layman's view, ought to govern their discourse as scientists" (1980: 41)—for example, by equating idealism with the ancient Greeks (omitting Fichte, Hegel, and Berkeley), Greek thought with idealism (omitting Democritus), and idealism with determinism (as Hobbes was a materialist determinist). Richard Levins and Richard Lewontin's "Dialectics and Reductionism in Ecology" (1980 [reproduced in this volume]) proposes "dialectal materialism" to resolve the "confusions" of Simberloff's interpretation by focusing on the resolution of unity with discord rather than their separation. In spite of equivocations and ambiguities, Simberloff's point is clear: the Western intellectual tradition, as a whole, has been characterized by a pervasive belief in order, design, and balance in nature.

Ecological Hierarchies

The entities recognized by ecologists are multifarious, depending on the intentionality of the ecologist. They include individual organisms, species, ecosystems, populations, metapopulations, guilds, breeding and feeding groups, ecotopes, landscapes, and biomes.

In order to represent the variety of ecological entities and relations, philosophers of ecology have created the concept of the *nested hierarchy* (vide chapter 17), a taxonomy of entities based on scale. Smaller entities are nested inside larger ones, somewhat like a Russian matryska doll, which opens to reveal another smaller doll. Accordingly, larger-scale ecological entities, such as a landscape, contain smaller-scale entities, such as ecotopes (figure 2). Biomes, landscapes, and ecotopes are all ecosystems of different sizes. Biomes are larger than landscapes and provide a matrix for landscape systems. Biotic communities, populations, and individual organisms represent another kind of hierarchy of scale, with the community being largest and the individual being the smallest.

Because smaller entities combine to form larger ones, larger entities are in-

Biosphere
Biome or Ecoregion
Landscape
Ecotope

Community
Population
Organism

Figure 2. Two nested ecological hierarchies

clusive of smaller ones. Think of a watershed as an ecosystem. The watershed of the Escalante River in southern Utah is made up of many smaller watersheds, such as Calf Creek, Boulder Creek, Harris Wash, Coyote Gulch, and so forth. The smaller riparian ecosystems are present in the larger Escalante ecosystem. (For heuristic purposes, we could classify the Escalante ecosystem as a landscape, and Calf Creek, Boulder Creek, Harris Wash, and Coyote Gulch as ecotopes. The entire Colorado Plateau could be classified as an ecoregion.)

The nested hierarchy differs from a control hierarchy, such as an army, in which members of one level, such as privates, do not appear at another level, such as generals. Following anthropologist Carole Crumley (1987), we could use the word *heterarchy* as a synonym for the nested hierarchy in order to distinguish ecological hierarchies from control hierarchies.

A nested hierarchy is constructed on the principle of similarity (O'Neill et al. 1986). Similar criteria should be used to classify the nested entities. In the Escalante River example, we organized watersheds across different scales. Using a geographical criterion on one level and a biological criterion on another is illegitimate. If one is concerned about the flow of water across the land surface, then the nested hierarchy is a hydrological order of watersheds, ranging from headwater streams to the river basin as a whole.

Obviously scale is central to ecological hierarchies. If we stay within one level of a hierarchy, we can observe many entities that associate and interact with each other. For instance, we may encounter many different patches of forest in a landscape. These patches will differ from one another in the landform, the species and age of trees, and the type of the undergrowth. However, in an aerial photograph, all patches will be characterized as forest. If we shift to an even larger scale by looking at a photograph covering more area, we will observe new entities interacting on a new matrix. In our example the images of forest patches will be much smaller and may even disappear as entities converge. The forested land unit becomes a new kind of entity at this higher level of scale.

Is the background matrix an entity too? Yes. But the matrix is an entity at a different level of scale than the entity of interest. The matrix contains the entity. If we shift scale again, the matrix itself may be observed as an entity within a yet larger matrix. Ecologically, this dimensional property of nature extends from the whole planet to the smallest organism; cosmologically, it extends from the universe to subatomic particles.

Ecological Processes

Philosophers are accustomed to speaking of metaphysics in terms of *being* or *becoming*. For some metaphysicians the essence of reality is motionless (perfect Being, in the verbal sense), while for others the essence of reality is change (in other words, continuous Becoming). In the Western tradition, the emphasis on stasis runs from the ancient Eleatic philosopher Parmenides through Plato, Augustine, Descartes, Newton, and others. The emphasis on flux runs from the ancient Milesian philosopher Heraclitus through Spinoza, Hegel, Alexander, Bergson, Nietzsche, Whitehead, and others.

Nature is so complex that both approaches seem relevant. Entities both persist and perish. However, as Robert Ulanowicz (1986) and other philosophers of ecology point out, ecology has been dominated by the entity (Being) approach at the expense of the process (Becoming) approach. As we have seen, Simberloff traces the dominance of the entity approach to the Platonic and Aristotelian metaphysics of essence. And on a practical level, it is easier to catalog persistence than to map patterns of change. Whatever the reasons, the hegemony of the entity approach is unfortunate because it is impossible to talk about nature without talking about process.

From the process perspective, entities are processes rapidly replicating form, creating the possibility that our sense organs can apprehend structure (Ferré, personal communication, May 1999). Energy and matter flow through networks at different rates within the system; objects are nodes in the network where flows of energy and matter are consumed, stored, and/or transformed. Sometimes we can physically observe the nodes qua entities. Less often can we observe the interactions, although the predator consuming the prey or the movement of the pollinating insect above the flower, its legs covered by yellow pollen, are vivid examples. Usually the interaction is interpreted as a consequence of a process.

In ecology, these processes are not just biological; they are also physical. For example, gravity causes a sediment-laden stream to deposit the heavier sand particles on the levee bordering the water and the lighter silt particles on the floodplain behind the levee. The process can be as clear to us as the flying insect. In this complex of interaction we often use the metaphor of a network, derived from systems science, to describe pathways of interactions and flows in space/time.

Thus ecological study is greatly complicated by the fact that entities are not static; they change, and they change at varying rates. The life cycle of some insects or microorganisms may be only days long. In contrast, some processes are so slow in terms of human life that we do not readily discern them. The uplifting and wearing down of mountain ranges goes on continually but takes millions of years to accomplish. From our perspective mountains are virtually eternal. The field ecologist must carefully observe organisms over long periods—as Frank Frasier Darling (1937), who stalked red deer in northern Scotland for more than two years—in order to see interactions that can be interpreted as a process. But there are relatively few such collections of intense, long-term observations, and generalizations made from natural history or experiments usually are inadequate to describe the connecting processes of the hundreds or thousands of kinds of organisms that occur in a typical community.

As we noted above in the discussion of the ecosystem ontology, the stochasticity of ecological process makes the deterministic model of mechanism inadequate. To convey the indeterminism of ecological process, Claudia Pahl-Wostl (1995) proposes a "macroscopic uncertainty principle"[2] roughly analogous to the "microscopic uncertainty principle" of quantum mechanics (Heisenberg 1927): "I conjecture that an uncertainty principle at the macroscopic level of living systems can be postulated especially when the global system as a whole is considered" (1995: 224). As Robert Ulanowicz (1999) in chapter 5, the science of ecology is in need of a new, post-Newtonian, postmechanistic, postmodern metaphysics.

The necessity of paying attention to process is illustrated by the challenges faced by evolutionary ecologists (see part 5). Evolution involves the selection of a genome containing a unique set of genetic characteristics by the environment. The complex processes associated with genetics and reproduction produce individual organisms or groups of siblings with unique genetic properties. These organisms interact with an environment that is made up of other organisms at the same scale, and with the matrix in which the organisms occur. Organisms that survive may reproduce and continue the genetic line. Organisms that do not survive mark the end of a genetic line.

Evolution is a process. The interaction of genome and environment Darwin called "natural selection." Using the criterion of natural selection that acts upon most, if not all, organisms, some ecologists focus on the individual organism as the fundamental entity in ecology. Other ecologists and geneticists argue that selection does not always operate at the scale of the organism. They claim that a group of organisms, such as a beehive, may also be selected as a unit because the group has a better chance of survival than does the individual. In their opinion, populations, communities, and even ecosystems might be the primary evolutionary entities. This is decidedly the less popular opinion, but it has adherents such as David Wilson (1980).

Conclusion

Entities may be considered in several different ways. The naturalist observes entities in the field that take a particular form, behave in repeatable patterns, and have a history. These entities are individuals in some cases. In other cases they represent collections of organisms combined physically, chemically, or biologically. Coral is an example in which individualism makes little sense. Lichen, made up of an alga and a fungus, is another. Field biologists group like-appearing and like-acting individuals into categories for the sake of speaking about them and recording observations. The genus/species taxonomic system, invented by Carolus Linnaeus, serves the field ecologist as a generally satisfactory system of organizing ecological information.

As the essence of nature is flux, it appears that there are no absolute criteria by which we can distinguish one entity from another in space/time continuums. The answer to the question, What is an entity? is "it depends." It depends upon how the properties are arranged according to the goals and purposes of the ecologist; how energy, matter, and information are received and exchanged; and so on.

The most important point in this discussion is that the ecologist is allowing natural organization and process to become visible through close, long-term observation and manipulation, taking advantage of natural experiments when the system of interest is stressed or affected by unusual events. The ecologist is not forcing ecological entities into a mechanical model of nature in which parts clunk across the stage of nature. The dynamism and stochasticity of ecological processes and the subjective intuition and creativity of the individual are reasons why Ferré (1996) claims that ecological studies are models for postmodern science.

Notes

1. This same argument has been made in the context of animal rights versus land ethics debates in environmental philosophy. For example, Tom Regan writes: "Like political fascism, where the individual is made to serve the interests of the larger political community, an unbridled ecological holism, where it is permissible to force the individual to serve the interests of the larger life community, is fascistic too" (1992: 138).

2. As Ulanowicz (personal communication, 1998) points out, *uncertainty* is epistemic, and the issue of determinism is metaphysical, so in this context *indeterminism* is a better word than *uncertainty*.

CHAPTER 1

Preface to *Plant Succession: An Analysis of the Development of Vegetation*

Frederic E. Clements

The present book constitutes the general part of a monograph on Rocky Mountain vegetation which has been under way since 1899. It is hoped that another volume dealing with the details of the development and structure of the climax formations of the Great Plains, Rocky Mountains, and Great Basin may appear subsequently. The general principles advanced here are an outgrowth of the treatment in the "Development and Structure of Vegetation" (1904) and "Research Methods in Ecology" (1905), in which an endeavor to organize the whole field of present-day succession was made for the first time. The studies of the past decade have confirmed and broadened the original concepts, and have led irresistibly to the conclusion that they are of universal application. The summer of 1913 and the spring and summer of 1914 were spent in testing both principles and processes throughout the vegetation of the western half of the continent. The area scrutinized extends from the Great Plains to the Pacific Coast and from the Canadian Rockies to the Mexican boundary. The great climax formations of this region were traversed repeatedly, and their development and relations subjected to critical analysis and comparison.

As a consequence, it is felt that the earlier concept of the formation as a complex organism with a characteristic development and structure in harmony with a particular habitat is not only fully justified, but that it also represents the only complete and adequate view of vegetation. This concept has been broadened and definitized by the recognition of the developmental unity of the habitat. As a result, formation and habitat are regarded as the two inseparable phases of a development which terminates in a climax controlled by climate. Hence, the basic climax community is taken to be the formation, which exhibits seral or developmental stages as well as climax units. It is hardly necessary to point out that this places the study of vegetation upon a purely developmental basis, which is as objective as it is definite.

From *Plant Succession: An Analysis of the Development of Vegetation.* Washington, D.C.: Carnegie Institution of Washington, Publication No. 242 (1916), pp. 1–7. Plates omitted.

The recognition of development as the cause and explanation of all existing climax formations forced the conclusion that all vegetation has been developmentally related; in short, that every climax formation has its phylogeny as well as its ontogeny. This led at once to the further assumption that the processes or functions of vegetation today must have been essentially those of the geological past, and that the successional principles and processes seen in existing seres hold equally well for the analysis of each eosere. As a consequence, it has been possible to sketch in bold outline the succession of plant populations in the various eras and periods, and to organize in tentative fashion the new field of paleoecology. . . .

Concept and Causes of Succession

The Formation an Organism

The developmental study of vegetation necessarily rests upon the assumption that the unit or climax formation is an organic entity (Clements 1905: 199). As an organism the formation arises, grows, matures, and dies. Its response to the habitat is shown in processes or functions and in structures which are the record as well as the result of these functions. Furthermore, each climax formation is able to reproduce itself, repeating with essential fidelity the stages of its development. The life history of a formation is a complex but definite process, comparable in its chief features with the life history of an individual plant.

Universal Occurrence of Succession

Succession is the universal process of formation development. It has occurred again and again in the history of every climax formation, and must recur whenever proper conditions arise. No climax area lacks frequent evidence of succession, and the greater number present it in bewildering abundance. The evidence is most obvious in active physiographic areas, dunes, strands, lakes, floodplains, badlands, etc., and in areas disturbed by man. But the most stable association is never in complete equilibrium, nor is it free from disturbed areas in which secondary succession is evident. An outcrop of rock, a projecting boulder, a change in soil or in exposure, an increase or decrease in the water content or the light intensity, a rabbit burrow, an ant heap, the furrow of a plow, or the tracks worn by wheels, all these and many others initiate successions, often short and minute, but always significant. Even where the final community seems most homogeneous and its factors uniform, quantitative study by quadrat and instrument reveals a swing of population and a variation in the controlling factors. Invisible as these are to the ordinary observer, they are often very considerable, and in all cases are essentially materials for the study of succession. In consequence, a floristic or physiognomic study of an association, especially in a restricted area, can furnish no trustworthy conclusions as to the prevalence of succession. The latter can be determined only by investigation which is intensive in method and extensive in scope.

Viewpoints of Succession

A complete understanding of succession is possible only from the consideration of various viewpoints. Its most striking feature lies in the movement of populations, the waves of invasion, which rise and fall through the habitat from initiation to climax. These are marked by a corresponding progression of vegetation forms or phyads, from lichens and mosses to the final trees. On the physical side, the fundamental view is that which deals with the forces which initiate succession and the reactions which maintain it. This leads to the consideration of the responsive processes or functions which characterize the development, and the resulting structures, communities, zones, alternes, and layers. Finally, all of these viewpoints are summed up in that which regards succession as the growth or development and the reproduction of a complex organism. In this larger aspect succession includes both the ontogeny and the phylogeny of climax formations.

Succession and Sere

In thorough analysis of succession it becomes evident that the use of the term in both a concrete and an abstract sense tends to inexactness and uncertainty. With the recognition of new kinds of succession it seems desirable to restrict the word more and more to the phenomenon itself and to employ a new term for concrete examples of it. In consequence, a word has been sought which would be significant, short, euphonic, and easy of combination. These advantages are combined in the word *sere*, from a root common to both Latin and Greek, and hence permitting ready composition in either. The root *ser-* shows its meaning in Latin *sero*, join, connect; *sertum*, wreath; *series*, joining or binding together, hence sequence, course, succession, lineage. In Greek, it occurs in ειρω, to fasten together in a row, and in σειρα, σηρα, rope, band, line, lineage. *Sere* is essentially identical with *series*, but possesses the great advantage of being distinctive and of combining much more readily, as in *cosere*, *geosere*, etc.

Sere and Cosere

A sere is a unit succession. It comprises the development of a formation from the appearance of the first pioneers through the final or climax stage. Its normal course is from nudation to stabilization. All concrete successions are seres, though they may differ greatly in development and thus make it necessary to recognize various kinds, as is shown later. On the other hand, a unit succession or sere may recur two or more times on the same spot. Classical examples of this are found in moors and dunes, and in forest burns. A series of unit successions results, in which the units or seres are identical or related in development. They consist normally of the same stages and terminate in the same climax, and hence typify the reproductive process in the formation. Such a series of unit successions, i.e., of seres, in the same spot constitutes an organic entity. For this, the term *consere* or *cosere* (*cum*, together, *sere; consero*, bind into a whole) is proposed, in recognition of the developmental bond between the individual seres. Thus, while the sere is the developmental unit, and is purely ontogenetic, the cosere is the sum of such units throughout the whole life history of the climax formation, and is hence phylogenetic in some degree. Coseres are likewise related in a developmental series, and thus may form larger groups, eoseres, etc., as indicated in the later discussion. . . .

Processes in Succession

The development of a climax formation consists of several essential processes or functions. Every sere must be initiated, and its life-forms and species selected. It must progress from one stage to another, and finally must terminate in the highest stage possible under the climatic conditions present. Thus, succession is readily analyzed into initiation, selection, continuation, and termination. A complete analysis, however, resolves these into the basic processes of which all but the first are functions of vegetation, namely, (1) nudation, (2) migration, (3) ecesis, (4) competition, (5) reaction, (6) stabilization. These may be successive or interacting. They are successive in initial stages, and they interact in most complex fashion in all later ones. In addition, there are certain cardinal points to be considered in every case. Such are the direction of movement, the stages involved, the vegetation forms or materials, the climax, and the structural units which result.

Relation of Causes

Since succession is a series of complex processes, it follows that there can be no single cause for a particular sere. One cause initiates succession by producing a bare area, another selects the population, a third determines the sequence of stages, and a fourth terminates the development. As already indicated, these four processes—initiating, selecting, continuing, and terminating—are essential to every example of succession. As a consequence, it is difficult to regard any one as paramount. Furthermore, it is hard to determine their relative importance, though their difference in role is obvious. It is especially necessary to recognize that the most evident or striking cause may not be the most important. In fact, while the cause or process which produces a bare habitat is the outstanding one to the eye, in any concrete case, it is rather less important if anything than the others. While the two existing classifications of successions (Clements 1904; Cowles 1911) have both used the initiating cause as a basis, it seems clear that this is less significant in the life history of a climax formation than are the others. [The] same sere may result from several initial causes.

Kinds of Causes

All of the causative processes of succession may best be distinguished as initiating or initial, continuing or ecesic, and stabilizing or climatic. At first thought, the latter seems not to be a cause at all but an effect. As is shown later, however, the character of a successional development depends more upon the nature of the climatic climax than upon anything else. The latter determines the population from beginning to end, the direction of development, the number and kind of stages, the reactions of the successive stages, etc. Initial causes are those which produce a new or denuded soil upon which invasion is possible. Such are the chief physiographic processes, deposition and erosion, biotic factors such as man and animals, and climatic forces in some degree. . . .

Ecesic causes are those which produce the essential character of vegetational development, namely, the successive waves of invasion leading to a final climax. They have to do with the interaction of population and habitat, and are directive in the highest degree. The primary processes involved are invasion and reaction. The former includes three closely related processes: migration, competition, and ecesis. The last is final and critical, however, and hence is used to designate the causes which continue the development.

Proximate and Remote Causes

In dealing with the causes of development, and especially with initial causes, it must be borne in mind that forces in nature are almost inextricably interwoven. In all cases the best scientific method in analysis seems to be to deal with the immediate cause first, and then to trace its origin just as far as it is possible or profitable. Throughout a climax formation, physiography usually produces a large or the larger number of developmental areas. The influence of physiography in this respect is controlled or limited by the climate, which in its turn is determined by major physiographic features such as mountain barriers or ocean currents. These are subordinate as causes to the general terrestrial climates, which are the outcome of the astronomical relations between the sun and the earth. As a consequence, physiography may well be considered the immediate initial cause of the majority of primary successions, just as the chresard is the controlling cause of vegetation structure, though it is dependent on the one hand upon soil structure, and this upon physiography, and on the other upon the rainfall, etc.

Apart from the gain in clearness of analysis, greater emphasis upon the proximate cause seems warranted by the fact that it is the chresard to which the plant responds, and not the soil texture or the physiography. In like manner, the invasion of a new area is a direct consequence of the action of the causative process and not of the remote forces behind it. The failure to consider the sequence of causes has produced confusion in the past . . . and will make more confusion in the future as the complex relations of vegetation and habitat come to be studied intensively. . . .

Essential Nature of Succession

Developmental Aspect

The essential nature of succession is indicated by its name. It is a series of invasions, a sequence of plant communities marked by the change from lower to higher life-forms. The essence of succession lies in the interaction of three factors, namely, habitat, life-forms, and species, in the progressive development of a formation. In this development, habitat and population act and react upon each other, alternating as cause and effect until a state of equilibrium is reached. The factors of the habitat are the causes of the responses or functions of the community, and these are the causes of growth and development, and hence of structure, essentially as in the individual. Succession must then be regarded as the development or life history of the climax formation. It is the basic organic process of vegetation, which results in the adult or final form of this complex organism. All the stages which precede the climax are stages of growth. They have the same essential relation to the final stable structure of the organism that seedling and growing plant have to the adult individual. Moreover, just as the adult plant repeats its development, i.e., reproduces itself, whenever conditions permit, so also does the climax formation. The parallel may be extended much further. The flowering plant may repeat itself completely, may undergo primary reproduction from an initial embryonic cell, or the reproduction may be secondary or partial from a shoot. In like fashion, a climax formation may repeat every one of its essential stages of growth in a primary area, or it may reproduce itself only in its later stages, as in secondary areas. In short, the process of organic development is essentially alike for the individual and the community. The correspondence is obvious when the necessary difference in the complexity of the two organisms is recognized.

Functional Aspect

The motive force in succession, i.e., in the development of the formation as an organism, is to be found in the responses or functions of the group of individuals, just as the power of growth in the individual lies in the responses or functions of various organs. In both individual and community the clue to development is function, as the record of development is structure. Thus, succession is preeminently a process the progress of which is expressed in certain initial and intermediate structures or stages, but is finally recorded in the structure of the climax formation. The process is complex and often obscure, and its component functions yield only to persistent investigation and experiment. In consequence, the student of succession must recognize clearly that developmental stages, like the climax, are only a record of what has already happened. Each stage is, temporarily at least, a stable structure, and the actual processes can be revealed only by following the development of one stage into the succeeding one. In short, succession can be studied properly only by tracing the rise and fall of each stage, and not by a floristic picture of the population at the crest of each invasion.

CHAPTER 2

The Individualistic Concept
of the Plant Association

Henry A. Gleason

The units of vegetation . . . were just as easily visible to primitive man as they are to us today. They were equally visible to the sages of Greece and Rome two thousand years ago, to our European ancestors a thousand years ago. Through all these long years they were neglected by scientists, whose thoughts were turned in different directions, but they were recognized by the laity. As a result the languages of Europe all contain a number of terms which refer, usually rather loosely, to vegetation.

Certainly every botanist who studied plants in their natural habitat during the eighteenth century was familiar with many types of vegetation, but to them he gave little thought, possibly considering that they were not proper subjects for scientific investigation. To see vegetation units is one thing; to take cognizance of their existence, to investigate their nature, to become aware that they have structure and behavior which may be analyzed, to formulate a philosophy in explanation of them, are entirely different matters.

For our purposes, and with the admission that many desultory references to vegetational units may be found in the works of earlier authors, we may say that the first definite, scientific discussion of the subject may be attributed to Grisebach, and we may repeat his often quoted words: "Ich möchte eine Gruppe von Pflanzen, die einen abgeschlossenen Charakter trägt, wie eine Wiese, ein[en] Wald, u.s.w., eine pflanzengeographische Formation nennen" [I would like to call a group of plants with a self-contained quality, a meadow or a forest, for example, a botanical-geographic formation]. This was in 1838, exactly a century ago, and we may congratulate ourselves that . . . we are celebrating the centenary of the association-concept.

During the next sixty years, the study of vegetational units was sporadic, mostly superficial, and usually purely descriptive in nature. True, we can sometimes see in the literature of this period statements which may be interpreted as applying to the underlying philosophy of the association, but in most cases

From *American Midland Naturalist* 21 (1939): 92–110, pp. 92–107.

such statements are casual or accidental and do not indicate that the authors had given deep thought to the fundamental nature of the vegetation which they described.

Not until the advent of the twentieth century did botanists turn their minds seriously to the consideration of underlying questions. Since that time we have made great progress. We have developed methods for the exact observational study of the association. We have recognized conditions and processes in their development, their existence, and their disappearance, and these conditions and processes are quite unlike anything in the life history of an individual plant or animal, so that our recognition of them has required the development of new habits of thought. We have described in modern terms vegetation from nearly all parts of the world. We have developed systems of classification, by which the units of vegetation may be orderly arranged. We have invented and brought into accepted use a new terminology by which these conditions, processes, structures, and concepts may be described and discussed.

There is, however, one important question which has not yet been settled to the satisfaction of all concerned. This *is* the fundamental question, basic to all our work: What is a plant association? Out of the thousands of pages of literature which have been used in expounding various views on the matter, three well marked theories may be chosen, and all others may be regarded as merely variants from them. These three are:

1. The association is an organism, or a quasi-organism, not composed of cells like an individual plant or animal, but rather made up of individual plants and animals held together by a close bond of interdependence; an organism, or a quasi-organism, with properties different from, but analogous to, the vital properties of an individual, including phenomena similar to birth, life, and death, as well as constant structural features comparable to the structures of the individual.

2. The association is not an organism, but is a series of separate similar units, variable in size but repeated in numerous examples. As such, it is comparable to a species, which is also composed of variable individuals. Under this view, the association is considered by some to be a concrete entity, merely divided into separate pieces, while by others the association as a whole is regarded as a mental concept, based on the common characters of all its separate pieces, and capable of typification by one or more of those pieces which most nearly approach the average or ideal condition.

3. The vegetation-unit is a temporary and fluctuating phenomenon, dependent, in its origin, its structure, and its disappearance, on the selective action of the environment and on the nature of the surrounding vegetation. Under this view, the association has no similarity to an organism and is scarcely comparable to a species.

In the original paper, in which this idea was presented to the botanical public in 1926, it was called the individualistic concept, a term which may well be continued. Whether fortunately or unfortunately, my own work during these twelve years has been wholly taxonomic. Nevertheless, observation of numerous plant associations during my field work in this period has merely intensified my own belief in the fundamental truth of the individualistic concept. The exposition of the concept which follows is merely a restatement of the subject in different terms; it is in no way different in principles or conclusions from the first presentation in 1926.

The argument for the individualistic concept rests on a series of theses, each of which is so obvious, so well known, so universally understood and accepted by all ecologists, that none of them requires prolonged discussion.

1. Every species of plant has reproductive powers in excess of its need. The land surface of the world is already fully occupied by plants. Room for additional plants is made available only by the death of plants now existing. If seed germination were always perfect, if there was no mortality among plants before reaching their reproductive stage, it would be necessary for each existing plant to produce only one seed in order to perpetuate its species and to maintain the existing number of individuals. On the contrary, every species of plants produces a considerable number of seeds or other propagating bodies, often yielding an annual crop over a long period of years. Since the world is full of plants, it is a fact that, on the average, only one of them comes to a state of full maturity, but huge numbers of seeds or other propagules are capable of growth and ready to grow if favorable conditions are offered. A well kept lawn, for example, may produce very few weeds during the course of the summer, but if the lawn is plowed, the same expanse will promptly develop an astonishing crop of weeds of many species. The bottoms of drained ponds, the first season after drainage, produce many plants of terrestrial species. These and other examples of the same sort are so well known and so conspicuous that we may safely state that the surface of the world is heavily planted with an excess of seeds, most of which never develop, but many of which will develop if a favorable opportunity is offered.

2. Every species of plant has some method of migration. The means of migration are well understood and require no discussion. The effectiveness of migration is often not appreciated. The distance to which some seeds may be carried by currents of water, by wind, or by birds is known to some degree, but the migration of other less favorably adapted species is also remarkable. Such seeds may be carried by rodents or ants or washed away by heavy rains. It is a demonstrable fact for all plants, without regard to their methods of migration, that more seeds are finally deposited near the parent plant than are carried to a great distance, the number decreasing, roughly speaking, inversely as the

square of the distance. But effectiveness of migration is increased by various kinds of accidents and also by the known longevity of seeds in many species. We may therefore conclude, and our conclusion is supported by direct evidence familiar to all of us by personal experience, not only that the world is heavily planted with seeds, as stated under our first thesis, but also that these seeds come not only from the existing plants of the immediate vicinity, but also from plants at some distance. In any unit of vegetation the potentiality of plant production includes not only the natural species of this unit, but numerous species not now found in it and derived from other vegetational units of different character.

3. The environment in any particular station is variable. Probably the simplest instance of variability which may be mentioned is light. Each day starts at midnight with light at or near zero. A little before sunrise the light curve begins to rise, reaches a theoretical maximum at noon, and drops to zero shortly after sunset. The length of the curve varies with the season, being longest in summer and shortest in winter; the amplitude of the curve varies in precisely the same way, reaching its maximum on June 22 and its minimum six months later. Any and all of the 365 curves which constitute a year's cycle may be subject to irregular reductions in amplitude by cloudy weather. Nor is the quality of the light constant, but varies somewhat with the altitude of the sun and the condition of the atmosphere. These are all relatively simple matters, but there are also variations due to sunspots, and there may have been and in fact may still be variations due to changes in the inclination of the earth's axis or the eccentricity of its orbit, or to other causes even more remote and far slower in their action. Locally, for many plants the light is changed again by the shade of taller plants, and this shade varies in its effectiveness from hour to hour, from season to season, and as the shading trees grow taller, from year to year.

The variation in temperature of the surrounding air shows similar irregular variations superposed on cyclic progressions in a way quite similar to the variation in light. Soil moisture is more irregular, with abrupt rises followed by longer periods of decrease, and with great variability in the amplitude of the curves. Available soil moisture is another question, based primarily on the total soil moisture, but complicated by matters of temperature, acidity, and other abstruse conditions. Similar conditions hold for every other factor of the environment taken individually and with all of them taken collectively: they are complex and variable to the last degree.

A second class of environmental variations may be called fluctuations. They are illustrated by our irregular alternation of cold and warm, of dry and wet years, of late and early seasons.

Still a third class of variations is important in its effect on plant life, and includes cumulative environmental changes which progress over a period of years or centuries or ages. Such, for example, are the silting up of a pond, the deepening and widening of a ravine by erosion, the exhaustion of soil fertility by percolation of rain water, the accumulation of humus, the increase in temperature following the retreat of a glacier, or the decrease in rainfall in the rain shadow of a mountain range during its elevation. Although excessively slow, the cumulative effect of these environmental changes is ultimately profound.

All three of these classes of environmental variation are in operation simultaneously in every situation. The first class is regular and predictable; the second class is irregular and unpredictable; the third class is slow and often immeasurable. The amplitude of the fluctuations in the second class is normally much greater than the steady progression of the third class. A single year of deficient rainfall may cause a greater change in the depth of water in a pond than fifty years of silting, or may have more effect on the crops of Montana than five thousand feet of additional elevation of the Rocky Mountains. If we assume that we are now approaching another advance of continental glaciers, the annual drop in temperature associated with it is far less than the fluctuation from a warm to a cold year. Nevertheless, the only fundamental difference between the three classes is the factor of time.

4. The development of a vegetational unit depends on one or the other of two conditions, the appearance of new ground or the disappearance of an existing association. The appearance of new ground is a matter of very little importance. More land may be added around the coasts of our continents by further elevation of the coastal plain or the building up of the shores by coastal deposits. More land may be added in supra-alpine regions by reduction of altitude through erosion. More land can not be added by increasing rainfall in deserts: our deserts are already fully occupied by plant life; some deserts merely support less plant life than others.

In the vast majority of cases, an association appears on the ground previously occupied by a different association. It makes no difference whether the earlier association is removed by the slow processes of ordinary succession or suddenly by some cataclysm. If the existing vegetation is destroyed by the axe, by fire, by a landslide, or by hot lava, some plants go first and some go last. If the vegetation is changed by a slow process of succession, some of the original inhabitants still go first and some linger. In both cases, some plants of the next association, the pioneers, appear first and others are delayed. The only essential difference between them is the factor of time, greatly shortened in the first case, often greatly prolonged in the second.

The various factors which collectively constitute the environment of a plant may be separately measured and diagrammed, although often with much difficulty, but no one has ever succeeded in reducing to a single statement or a single equation the total environment of any plant. For practical purposes the only measure of the environment is its result, as expressed in plant life. An oak tree may easily live to be three hundred years old. Let us see what this implies concerning its relation to the environment. It means that through this whole period, in spite of its extraordinary variability, the environment has never once exceeded the limits tolerated by the living protoplasm of the tree. The weather has never been too hot or too cold, too wet or too dry; the soil has never been too acid or too alkaline; no environmental factor has ever been too much or too little, or if the limits have sometimes been surpassed, it was for a period too short to be fatal.

The clearest example of temporary excess is the annual cold winter, during which the activities of the oak sink so low that we call the tree dormant. The oak is not only able to adapt itself to the variations in environment, but is able to vary its own life processes enormously to meet such a critical condition as winter cold. Not every plant meets the emergency in the same way. Some herbs actually die, but are fortunate in producing a seed which is not killed by cold and which produces another generation the following summer. There are still other plants whose seeds are also killed by the cold. These plants do not live with the oak, or, if they do, they must be reestablished every summer by fresh seeds from a more favorable climate. Such plants are familiar as cultivated in our gardens. The canna and dahlia may flourish not far from the oak, but they disappear forever if not removed by us to a more favorable climate for the winter and replaced the following spring.

Now leaving these trite illustrations, which have been used merely to recall to our minds what we all know from common experience, let us consider the question from a broader standpoint. We at once arrive at the general theorem, that each plant seizes and uses the particular time-period during which the environment is in a condition suitable to it. When the environment passes these limits, the plant dies. For the old oak, this limit has not been reached for the three hundred years of its life and, so far as we can imagine, not for still longer periods in the past when preceding generations of oaks were growing on the same spot, nor will it be surpassed for many years in the future, during which the descendants of the present tree will be living there. For the dahlia, the limit is reached before the plant has completed a single generation: it is cut off in October while it still bears young leaves and unopened buds. During the same summer, the Galinsoga which infests our gardens in this vicinity may produce several generations. If our lives were measured by days instead of years, we can imagine an ecologist saying, "I have seen three generations of Galinsoga on this one spot of ground. Evidently we are dealing with a stable environment and Galinsoga will live here forever." He would be wrong. With our knowledge of vegetational conditions actually extending over about three centuries, we now say, "Oaks have occupied this spot of ground for three hundred years. Evidently we have here a stable environment and the oak will live here forever." If our lives were seventy centuries instead of seventy years, would we not see that we were again wrong? Like the short-lived Galinsoga, which utilizes the time-period which we call summer, the long-lived oak is utilizing a longer time-period, and the time has been, and in the future will be again, when the race of oaks can no longer live on this particular spot.

During the time-period when a species may occupy a station, diverse effects are exerted upon the plants by the variable environment. Variations of the first type, which are regular, periodic, and uniform, such as night and day, winter and summer, act uniformly and unavoidably upon all plants. The physiological processes of every plant are adjusted to meet them. While their investigation forms an important part of plant physiology, their effect on plant distribution and their interest to the plant ecologist are negligible.

Variations or fluctuations of the second type, including such phenomena as cold and warm years, of dry and wet periods, of late and early seasons, have a pronounced ecological effect. This is evidenced in agriculture and horticulture by years of heavy or light yield, by the abundance or lack of pests and parasites, by the lateness or earliness of bloom or fruit, by the losses due to late frosts in spring or early frosts in autumn. If such fluctuation is maintained for even a few years, its results become so grave as to be of national importance, as shown recently by the effect of only four years of deficient rainfall in our Plains States.

The effect of these fluctuations on natural vegetation is precisely the same qualitatively, but less pronounced quantitatively, since crop plants are deliberately introduced into a region, while natural vegetation has been adjusted to the fluctuations through previous experience. In natural vegetation, the effect of fluctuation in the environment is seen in phenological phenomena, in the amount of annual growth, in the vigor of individuals, in the number of seeds produced, in the number of seedlings produced, and therefore in the relative number of individuals. These fluctuations rarely last very long, seldom affecting more than a single year; they therefore rarely cause the disappearance of a species or the appearance of a new one.

That fluctuations of environment affect the relative number of individuals of a species is a fact which has rarely been demonstrated in research in pure ecology, since accurate quantitative studies of vegetation have seldom been repeated in the same area over a series of years. I have personal records from the same area taken at varying intervals over twenty years and can assure you that they show conspicuous variations in the number of individuals of some species, and scarcely none for other species. We may assume that the physiological processes and environmental demands of the latter are so broadly adjusted that no fluctuations have ensued of sufficient magnitude to affect the plants. In the former group we may also assume that their physiological processes are more strictly defined or that the normal environment is already off the optimum for the species, so that fluctuations do interfere demonstrably with the number of individuals. Abundant records of this phenomenon are available through the careful observations on our grazing lands. Here the environmental fluctuation is largely in the number of animals feeding on the plants, but the effect is precisely the same. Differences in grazing lead immediately to the reduction of certain species and the multiplication of others.

The rare disappearance of a species because of environmental fluctuations [is] probably due to the fact that the fluctuations of any one year have been repeated at various times in the past, and species which would be exterminated by them have already been removed. Nevertheless, the long period of drought recently ended in the western states will probably have this effect in many instances, although I am not able to cite any single example at the present time.

The relatively rare appearance of a new species for the same cause, is due primarily to the time factor involved in plant migration. Environmental fluctuations are of relatively short duration; migration of plants is slow and before additional species can reach the spot, migration has been stopped by the return of the environment to normal. If next winter were to be omitted and summer temperatures should continue for another twelve months without interruption, we still could not expect to find mangroves along the shores of the north Atlantic. Sufficient time must always be allowed for the reaction of plants to any environmental change.

The variations of the third type, the slow-moving, long-continuing environmental changes caused by physiographic processes, geological developments, or climatic changes, cause no directly observable or measurable effect on vegetation, at least in most cases. In these days the effect of erosion during a single year can sometimes be measured or estimated but it has taken a century for it to make an impression on our minds. The silting of a pond with the resultant changes in plant life is also fairly rapid, so that it has sometimes been observed and recorded by a single person during his own lifetime. Even in these relatively rapid examples, and of course in the slower variations of climate, the effect of the change is always masked by the wider amplitude and quicker action of the fluctuations just discussed. Nevertheless, these fluctuations are based on a norm, and if the norm itself varies the amplitude of the fluctuations tend always to abate in one direction and to extend in another.

Suffice it then to repeat that on every spot of ground the environment is continually in a state of flux, and that the time-period in which a certain environmental complex is operative is seized on by the particular kinds of plants which can use it. The vegetation of every spot of ground is therefore also continually in a state of flux, showing constant variations in the kinds of species present, in the number of individuals of each, and in the vigor and reproductive capacity of the plants. . . .

In summary it may be stated that environment varies constantly in time and continuously in space; environment selects from all available immigrants those species which constitute the present vegetation, and as a result vegetation varies constantly in time and continuously in space. Those who disagree with the individualistic concept will very properly raise at this time certain questions, based on facts which at first thought seem to invalidate the whole concept. Before these questions are stated here, so that their obvious implications may be refuted, two general statements may be introduced.

First, an association, or better one of those detached pieces of vegetation which we may call a community, is a visible phenomenon. As such it has dimensions and area, and consequently boundary. While its area may be large, the community is nevertheless a very tangible thing, which may be mapped, surveyed, photographed, and analyzed. Over this area it maintains a remarkable degree of structural uniformity in its plant life. Homogeneity of structure, over a considerable extent, terminated by definite limits, are the three fundamental features on which the community is based. Without these three features, Grisebach would never have published his statement of a century ago; without them, all our studies of synecology would never have been developed. Also, besides its extent in space, every community has a duration in time. Uniformity, area, boundary, and duration are the essentials of a plant community.

Second, every community occupies a position in two series of environmental variation. In the space-series, as the community exists *here*, in this spot, it is part of a space-variation, and its environment differs from the adjacent communities. In the time-series, as the community exists *now*, at this time, it is part of a time-variation and in its environment differs from the communities which preceded it or will follow it.

The individualistic concept postulates a continuous variation in space and time. How can we reconcile this with the admitted uniformity in space and time?

In any community of reasonable extent, the variation of rainfall, temperature, length of day, and similar factors from one end to the other is extremely small. Not only is their effect proportionately small, but this effect is overshadowed by the much greater seasonal fluctuation. Soil, also is often uniform over the whole community, and when it [is] not uniform but varies significantly within small distances, we are prone to overlook its effects and to classify the variable vegetation in a single community. More important still, the dominant plants, which are distributed over the whole area of the community, exert such a uniform effect on the other species that discrepancies in the physical environment are more or less smoothed out or obliterated.

Nevertheless, it is difficult or impossible to find in any community two quadrats which are precisely similar. The community is a complex or mosaic of slight irregularities, all of which blend into an entirety of apparent homogeneity. We have all known of this lack of perfect uniformity, and have endeavored to evade it by developing the concept of a minimal area, but we have failed to realize its significance as indicating the general variability of vegetation.

Cumulative progressive changes in environment are generally so slow in their development that they make no pronounced effect from one year to the next, while the wider swings of fluctuating environment are so short in their duration that their full effect is not experienced. Also, comparatively few observers have kept careful records of vegetational change over a series of years. Furthermore, as in space variation, environment control by plants tends to overshadow environmental variation. Nevertheless, succession, which is merely vegetational change, is accepted by all as a fact; exact statistical records, when available, do show continuous variations in structure, and in many locations complete vegetational changes have occurred within the experience of a single observer.

The postulated uniformity of the community is therefore far from absolute. A community is uniform, either in space or in time, only to a reasonable degree. This uniformity is sufficient to enable us to recognize the community and to accept it as a unit of vegetation, while its variability, although slight, is sufficient to indicate the impossibility of considering any such area of vegetation as a definitely organized unit.

If vegetation varies continuously in space, how can we explain the abrupt transitions from one community to another, which are so conspicuous a feature of natural vegetation in many regions?

Abrupt transitions are in every case correlated with abrupt variations in the environment or with abrupt differences in the immigrating plant population. Some abrupt changes in environment are due to physical conditions, notably the soil, which may change notably within a short distance. The other changes are due to environmental control of the physical factors by the plant life itself. These account for most of the abrupt transitions in the eastern states, or in any other region where a dense vegetation is possible. The sharp demarcation of zones around a pond or bog, for example, is caused almost entirely by vegetational control.

Abrupt differences in immigration exist only in areas which have recently been disturbed, such as an abandoned field, a lake shore recently worked over by waves, a ballast heap, or a tract of newly filled ground. In such places the accidents of immigration often lead to the temporary establishment of distinct patches of vegetation, each characterized by one or a few species. Continued migration tends to smooth out these irregularities in a short time; environmental fluctuation favors certain species over others; denser growth leads to environmental control by certain species, and in the course of a few years such patchwork vegetation has blended to a relatively homogeneous community.

If vegetation varies continuously in space, how can we account for the repetition of the same vegetation in many separate communities? The answer to this question is simple. There is no exact repetition of the same vegetation from one community to the next. There is an approximate repetition only.

It is a fact that in any region several to many examples of vegetation may be found in which the differences are so slight that they are not observed, or if observed are considered as unimportant and negligible. In any community absolute homogeneity is impossible, and the observed heterogeneity may well be due to chance. But in a single community, if it is large enough, differences between two ends may be discovered. Every ecologist who has undertaken quantitative analysis of vegetation will probably agree with this statement. I once examined two adjacent sections, each an exact mile square, of virgin hardwood forest. The soil was uniform and level and there was no surface drainage system, nor any indication of wetter and drier parts. Careful quantitative studies in each section showed conspicuous differences, not important differences, to be sure, which could not be explained by any visible feature. At the Biological Station of the University of Michigan, the aspen association, with a single continuous community some six miles long, exhibits demonstrable variation from one end to the other, with no visible reason.

Between two different communities, not too far removed from each other, the observable differences in structure are of essentially the same degree of magnitude as these fluctuations within the same community, and one may easily tend to credit them also to the effect of chance. But part of them, possibly only a small part, is due to the space-variation in the environment, and another part, again possibly small, is due to a difference in the available plant population upon which environmental selection operates. If a series of communities are observed at successively greater distances, these differences cumulate, so that those at the ends of the series may be strikingly different, although connected by imperceptible or apparently negligible intermediates.

One shortcoming of our ecology has been that our field work has generally been confined to a small area. We have investigated and described all the associations in a small area, instead of trying to trace a single association over its whole extent. In any small area, environmental variation, essentially repeated in many spots, produces several well-marked types of environment, each characterized by a similar vegetation. We justifiably draw the conclusion, from this limited evidence, that association-types are definite. But as soon as we extend our observations, we begin to realize that each separate community is merely one minute part of a vast and ever-changing kaleidoscope of vegetation, a part which is restricted in its size, limited in its duration, never duplicated except in its present immediate vicinity, and there only as a coincidence, and rarely if ever repeated.

In other words, the similarities between adjacent communities, which have led to the views that the association is analogous to a species, or analogous to an organism, are not perfect similarities. They are caused by *nearly* similar environmental selection, intensified by *nearly* similar environmental control, from a nearly similar population. In addition to the imperfections of similarity caused by chance, and largely masked by them, are other variations of a cumulative nature. These, increasing in importance and in conspicuousness as more distant communities are considered, finally lead to vegetation of such unlike nature that they would never be classed in the same association-type.

In my original paper (1917) on this subject I mentioned as an example the alluvial forests along the Mississippi River and its tributaries over a stretch of about a thousand miles from its mouth to the northward. From mile to mile these forests show no considerable change; over a space of a hundred miles the changes may or may not be of ecological importance and the more important differences are apparently due to lack of time for sufficient migration to smooth out the variation. Yet these differences cumulate. One by one species disappear; one by one other species appear, and by the time one has reached, say, Indianapolis, there has been an almost complete change in the appearance and composition of the forest. Or, if the observer swings more to the west and travels up the Missouri and the Platte, he can see the disappearance of species one by one without corresponding replacement, until the forest is reduced to a fringe of willows and finally disappears completely in western Nebraska. Within the state of Michigan, the beech-maple climax forest, always considered to be a definite, well distinguished association-type, exhibits profound changes from one end of the state to the other.

I also venture to say, without personal experience to verify the opinion, that even more remarkable transitions might be discovered elsewhere. For example, the forests of the foothills of the Rocky Mountains in Colorado, composed there largely of *Pinus ponderosa*, might be traced northward with similar gradual variation, thence eastward along the northern boundary of the grassland in Canada, and again southward to the forests of Illinois, and lead us to the extraordinary conclusion that the *Pinus ponderosa* forests of Colorado represent the same association as the *Quercus velutina* forests of Illinois and the aspen groves of Manitoba.

Over such distances as the three I have mentioned the flora from which any area may be populated changes greatly. Such environmental factors as temperature, rainfall, and length of growing season also vary greatly and to an extent that completely surpasses the local fluctuations between any adjacent communities. With the vegetation determined by environmental selection from the available plant population, and with both of these underlying features altered, obviously the resultant vegetation must also be entirely changed.

It must be remembered that we admit the essential uniformity of vegetation within a single community, and the frequent striking uniformity between adjacent communities. But the fact that these small cumulative differences do exist is basically important in the consideration of the general concept of the plant-association. They indicate that each community, and for that matter each fraction of one, is the product of its own independent causative factors, that each community in what we now choose to call an association-type is independent of every other one, except as a possible source of immigrating species. With no genetic connection, with no dynamic connection, with only superficial or accidental similarity, how can we logically class such a series of communities into a definite association-type? Truly the plant community is an individualistic phenomenon.

Every species of plant and animal migrates, whether as a mature individual, as many species of animals, or as a reproductive body, as the vast majority of plants. Among animals, migration is sometimes selective in its direction and goal, as illustrated by birds which follow definite routes to definite established breeding grounds. With other animals and with all plants, migration is purely fortuitous. It progresses by various means, it brings the organisms into various places and to varying distances, but only those organisms which have reached a favorable environment are able to continue their life. Into this favorable environment other species also immigrate, and from all of the arrivals the environment selects those species which may live and dooms the others.

In this migration each migrating body acts for itself and moves by itself, almost always completely independent of other species. The idea of an association migrating *en masse* and later reproducing itself faithfully is entirely without foundation. Those cases in which there is a semblance of such a condition are caused by the proximity of the original association and the advantage which its species therefore have in migration. Even then, certain species always precede and certain others lag behind.

CHAPTER 3

The Use and Abuse of Vegetational Concepts and Terms

Arthur G. Tansley

It is now generally admitted by plant ecologists, not only that vegetation is constantly undergoing various kinds of change, but that the increasing habit of concentrating attention on these changes instead of studying plant communities as if they were static entities is leading to a far deeper insight into the nature of vegetation and the parts it plays in the world. A great part of vegetational change is generally known as *succession,* which has become a recognized technical term in ecology, though there still seems to be some difference of opinion as to the proper limits of its connotation; and it is the study of succession in the widest sense which has contributed and is contributing more than any other single line of investigation to the deeper knowledge alluded to. . . .

In 1920 and in 1926 I wrote general articles [published in 1920 and 1929, respectively] on this and some related topics. My return to the subject today is immediately stimulated by the appearance of Professor John Phillips's three articles in the *Journal of Ecology* (1934, 1935a, 1935b) which seem to me to call rather urgently for comment and criticism. At the same time I shall take the opportunity of trying to clarify some of the logical foundations of modern vegetational theory.

If some of my comments are blunt and provocative I am sure my old friend Dr. Clements and my younger friend Professor Phillips will forgive me. Bluntness makes for conciseness and has other advantages, always provided that it is not malicious and does not overstep the line which separates it from rudeness. And at the outset let me express my conviction that Dr. Clements has given us a theory of vegetation which has formed an indispensable foundation for the most fruitful modern work. With some parts of that theory and of its expression, however, I have never agreed, and when it is pushed to its logical limit and perhaps beyond, as by Professor Phillips, the revolt becomes irrepressible. But I am sure nevertheless that Clements is by far the greatest indi-

From *Ecology* 16 (1935): 284–309. Notes omitted.

vidual creator of the modern science of vegetation and that history will say so. For Phillips's work too, and particularly for his intellectual energy and single-mindedness, I have a great admiration.

Phillips's articles remind one irresistibly of the exposition of a creed—of a closed system of religious or philosophical dogma. Clements appears as the major prophet and Phillips as the chief apostle, with the true apostolic fervor in abundant measure. Happily the *odium theologicum* is entirely absent: indeed the views of opponents are set out most fully and fairly, and the heresiarchs, and even the infidels, are treated with perfect courtesy. But while the survey is very complete and almost every conceivable shade of opinion which is or might be held is considered, there is a remarkable lack of any sustained criticism of opponents' arguments. Only here and there, as for instance in dealing with Gillman's and Michelmore's specific contentions, and in a few other places, does the author present scientific *arguments*. He is occupied for the most part in giving us the pure milk of the Clementsian word, in expounding and elaborating the organismal theory of vegetation. This exposition, with its very full citations and references, is a useful piece of work, but it invites attack at almost every point.

The three articles are respectively devoted to "Succession," "Development and the Climax," and "the Complex Organism." The greater part of the third article is mainly concerned with the relation of this last concept to the theory of "holism" as expounded by General Smuts [1926] and others, and is really a confession of the holistic faith. As to the repercussions of this faith on biology I shall have something to say in the sequel. But first let me deal with "Succession" and "Development and the Climax."

Succession

My own views on succession are given fairly fully in my two papers already mentioned. In the first place I consider that the concept of succession can be given useful scientific significance only if we can trace in the sequences of vegetation "certain uniformities which we can make the subject of investigation, comparison, and the formulation of laws" (Tansley 1929). . . .

In 1926 (680) I proposed to distinguish between *autogenic succession*, in which the successive changes are brought about by the action of the plants themselves on the habitat, and *allogenic succession[,]* in which the changes are brought about by external factors. "It is true of course (I wrote) and must never be forgotten, that actual successions commonly show a mixture of these two classes of factors—the external and the internal" (678). I think now that I should have gone farther than this and applied my suggested new terms in the first place to the factors rather than to the successions. It is the fact, I think, that autogenic and allogenic factors are present in all successions; but there is often a clear preponderance of one or the other, and where this is so we may fairly apply the terms, with any necessary qualifications, to the successions themselves. I went on to contend, as indeed I had already done in 1920 (136 – 39) though without using the terms, that only to autogenic succession can we apply the concept of development of what I called a "quasi-organism" (= climax vegetation), but that this developmental (or autogenic) succession is the normal typical process in the gradual production of climax vegetation.

Phillips, following Clements, contends, on the other hand, that "succession is due to biotic reactions only, and is always progressive . . . succession being developmental in nature, the process must and can be progressive only" (1934: 562); and again, "succession is the expression of development" (1935a: 214).

Now here we are concerned first of all with the use of words. If we choose to confine the use of the term *succession* to the series of phases of vegetation which lead up to a climatic climax, for example the various "priseres" from bare rock or water to forest, then it naturally follows that the process is "progressive only." If in addition we conceive of vegetation as an organism, of which the climax is the adult and the earlier phases of the prisere are successive larval forms, then also succession is clearly "developmental in nature," is "the expression of development." But if, on the other hand, we apply the term, as I do, and as I think most ecologists naturally do, to *any* series of vegetational phases following one another in one area, repeating themselves everywhere under similar conditions, and clearly due in each case to the same or a similar set of causes, then to say that "succession must and can be progressive only," or that it is always and everywhere developmental, is clearly contrary to the fact.

Most of the controversy about the possibility of "retrogressive succession" depends simply on this difference in the use of the word. It is true that Clements (1916: 146 – 63 [reproduced in this volume]) successfully showed that the phenomena represented by some of the looser uses of "retrogression" were more properly described as destruction of (for example) the climax phase, or of the dominants of the climax phase, a destruction which would normally initiate a subsere leading again to the climax if the vegetation were then let alone. But if on the other hand there is what Phillips would call a "continuative cause" at work which gradually leads to the degradation of vegetation to a lower type[,] it seems to me that the phenomenon is properly called retrogressive succession. Here I should include the continuous effect of grazing animals which may gradually reduce forest to grassland, the gradual leaching and concomitant raw humus formation which may ultimately reduce forest to heath, gradual increase of drainage leading to the replacement of a more luxuriant and mesophytic by a poorer and more xerophytic vegetation, or a gradual waterlogging which also leads to a change of type and usually the replacement of a "higher" by a "lower" one. All these are perfectly well-established vegetational processes. To me they are clear examples of allogenic retrogressive successions, and I cannot see how their title can be denied except by an arbitrary and unnatural limitation of the meaning of the word *succession.* All the processes mentioned certainly involve destruction, but they also involve the invasion, ecesis, and growth of new species. "Destruction" by itself is not a criterion: does not all *progressive* succession, as Cooper (1926: 402) has pointed out, involve constant destruction of the plants of the earlier phases? . . .

Catastrophic destruction, whether by "natural" agencies or by man, does, I think, remove the phenomena from the field of the proper connotation of succession, because catastrophes are unrelated to the causes of the vegetational changes involved in the actual process of succession. They are only initiating causes, as Clements rightly insists: they clear the field, so to speak, for a new succession. That is why I have insisted on gradualness as a character of succession. Gradualness in effect is the mark of the action of "continuative" causes.

Development and the Quasi-Organism

The word *development* may be used in a very wide sense: thus we speak of the development of a theme or of the development of a situation, though always, I think, with the implication of becoming more complex or more explicit. Always, too, it is some kind of *entity* which develops, and in biology it is particularly to the growth and differentiation of that peculiarly well defined entity the individual organism that we apply the term. Hence we can perfectly well speak in a general way of the development of any piece of vegetation that has the character of an entity, such as marsh or forest, and in common language we actually do so; but we should use the term as part of the theory of vegetation, of a body of well-established and generally acceptable concepts and laws, only if we can recognize in vegetation a number of sufficiently well-defined entities whose development we can trace, and the laws of whose development we can formulate.

In 1920 I inquired whether we could recognize such entities in vegetation, and I analyzed the whole topic in considerable detail and with considerable care. To the best of my knowledge that analysis has not been seriously criticized or impugned, and I may be permitted to think it holds the field, though various divergent opinions unsupported by arguments have since been expressed. Briefly my conclusion was that mature well-integrated plant communities (which I identified with plant associations) had enough of the characters of organisms to be considered as *quasi-organisms,* in the same way that human societies are habitually so considered. Though plant communities are not and cannot be so highly integrated as human societies and still less than certain animal communities such as those of termites, ants, and social bees, the comparison with an organism is not merely a loose analogy but is firmly based, at least in the case of the more complex and highly integrated communities, on the close interrelations of the parts of their structure, on their behavior as wholes, and on a whole series of other characters which Clements (1916 [reproduced in this volume]) was the first to point out. In 1926 (679) I called attention to another important similarity which, it seems to me, greatly strengthens the comparison between plant community and organism—the remarkable correspondence between the species of a plant community and the genes of an organism, both aggregates owing their "phenotypic" expression to development in the presence of all the other members of the aggregate and within a certain range of environmental conditions. But this position is far from satisfying Clements and Phillips. For them the plant community (or nowadays the "biotic community") *is* an organism, and he who does not believe it departs from the true faith. . . .

There is no need to weary the reader with a list of the points in which the biotic community does *not* resemble the single animal or plant. They are so obvious and so numerous that the dissent expressed and even the ridicule poured on the proposition that vegetation *is* an organism are easily understood. Of course Clements and Phillips reply that no one asserts that the plant community is an *individual* organism. In the more recent phrase it is a "complex organism"—a thoroughly bad term, as it seems to me, for it is firmly associated in the minds of biologists with the "higher" animals and plants— the mammals and spermaphytes [*sic*]. In any case it is, in my judgment, impossible to get the proposition generally accepted. Whether it is true or untrue depends entirely on the connotation of *organism*, and as to that the present generation of biologists have a firmly established use from which they will not depart—and I think they are right. We need a word for the peculiarly definite, sharply limited and unique type of organization embodied in the individual animal or plant, and *organism* is the accepted term. . . .

I can only conclude that the term *quasi-organism* is justified in its application to vegetation, but that the terms *organism* or *complex organism* are not.

Climaxes

Professor Phillips's treatment of the concept of climax is open to nearly the same criticism as his treatment of succession. Just as he will only have one kind of succession, which is always progressive, and entirely caused by the "biotic reactions" of the community, so he will have only one kind of climax, the climatic climax, of which there is only one in each climatic region.

He rather ingenuously suggests that the adjective "climatic" had better be dropped: it is misleading to the uninitiated. Since there is only one kind of climax why qualify the word? The suggestion would be unanswerable if we all agreed with him! . . .

Clements realized from the first (1916 [reproduced in this volume]) that vegetation existed which was neither climatic climax nor part of a sere actually moving toward it, but might be in a permanent or quasi-permanent condition in some sense "short of" the climax, and all such vegetation he called *subclimax*. He used this term in two senses, for an actual seral stage which would normally lead to the climatic climax, and for a type of climax "subordinate to" the climatic climax. It was pointed out that this double use was undesirable, and that if we confined the term *subclimax* to the former case, terms were wanted for permanent or quasi-permanent vegetation which did not closely represent a particular phase of a sere leading to the climatic climax, but were dominated by species that did not enter into any of the "normal" seres. For such climaxes Clements has now (1934: 45) proposed the word *proclimax*, i.e., vegetation which appears *instead of* the climatic climax, or as he would say, instead of *the* climax. This I think is an unobjectionable term, but it does not specify the factors which have differentiated the different types of this sort of climax. . . .

As expounded by Phillips the "monoclimax theory" explains away the existence of what some of us are accustomed to call edaphic and physiographic climaxes within a climatic region in two ways. Either these supposed climaxes are not climaxes at all but stages in a sere leading to *the* climax, whose movement has been *retarded*, perhaps for a long time, by the edaphic or physiographic factors, or they are mere variations of "the formation" (the climatic climax). It is not to be supposed and is not in fact the case, it is argued, that either climate or soil will be absolutely uniform within a great climatic region, which often extends for many hundreds of miles. The climatic formation (*the* formation according to the "monoclimax theory") is often "a veritable mosaic" of vegetation (Clements). This of course is quite true: the only question is, *how great differences* are we to admit as mere variations within the formation? The difficulty disappears of course if we *define* a formation—a climatic climax—as *all* permanent vegetation within the climatic region and are therefore willing to swallow such differences, however great. But is this sound empirical method? [Is it] not rather a case of making the facts fit the theory? Is it not sounder scientific method *first* to recognize, describe, and study all the relationships of actually existing vegetation, and *then* to see how far they fit or do not fit any general hypothesis we may have provisionally adopted? . . .

I plead for empirical method and terminology in all work on vegetation, and avoidance of generalized interpretation based on a theory of what *must* happen because "vegetation is an organism."

"The Complex Organism"

Professor Phillips's third article (1935b) is devoted to a discussion of the "complex organism," otherwise known as "the biotic community" (or "biome" of Clements) in the light of the doctrines of emergent evolution and of holism. On the biotic community he had already written (1931) and so also have Shelford (1931) and others.

I have already expressed a certain amount of skepticism of the soundness of the conception of the biotic community (1929: 680), without giving my reasons at all fully. It seems necessary now to state the grounds of my skepticism, and at the same time to make clear that I am not by any means wholly opposed to the ideas involved, though I think that these are more naturally expressed in another way.

On linguistic grounds I dislike the term biotic *community*. A "community," I think it will be generally agreed, implies *members,* and it seems to me that to lump animals and plants together as *members* of a community is to put on an equal footing things which in their whole nature and behavior are too different. Animals and plants are not common members of anything except the organic world (in the biological, not the "organicist" sense). One would not speak of the potato plants and ornamental trees and flowers in the gardens of a human community as *members* of that community, although they certainly enter into its constitution—it would be different without them. There must be some sort of *similarity,* though not of course *identity,* of nature and status between the members of a community if the term is not to be divorced too completely from its common meaning. . . .

Animal ecologists in their field work constantly find it necessary to speak of *different* animal communities living in or on a given plant community, and this is a much more natural conception, formed in the proper empirical manner as a direct description of experience, than the "biotic community." Some of the animals belonging to these various animal communities have very restricted habitats, others much wider ones, while others again such as the larger and more active predaceous birds and mammals range freely not only through an entire plant community but far outside its limits. For these reasons also, the practical necessity in field work of separating and independently studying the animal communities of a "biome," and for some purposes the necessity of regarding them as external factors acting on the plant community—I cannot accept the concept of the *biotic* community.

This refusal is[,] however[,] far from meaning that I do not realize that various "biomes," the whole webs of life adjusted to particular complexes of environmental factors, are real "wholes," often highly integrated wholes, which are the living nuclei of *systems* in the sense of the physicist. Only I do not think they are properly described as "organisms" (except in the "organicist" sense). I prefer to regard them, together with the whole of the effective physical factors involved, simply as "*systems.*"

I have already criticized the term *organism* as applied to communities of plants or animals, or to "communities" of plants *and* animals, on the ground that while these aggregations have *some* of the qualities of organisms (in the biological sense) they are too different from these to receive the same unqualified appellation. And I have criticized the term *complex organism* on the ground that it is already commonly applied to the species or individuals of the higher animals and plants. Professor Phillips's third article (1935b) is largely devoted to an exposition and defense of the concept of "the complex organism." According to the organicist philosophy, which he seems to espouse, though he does not specifically say so, he is perfectly justified in calling the whole formed by an integrated aggregate of animals and plants (the "biocenosis," to use the continental term) an "organism," provided that he includes the physical factors of the habitat in his conception. But then he must also call the universe an organism, and the solar system, and the sugar molecule, and the ion or free atom. They are all organized "wholes." The nature of what biologists call living organisms is wholly irrelevant to this concept. They are merely a special kind of "organism."

With the philosophical aspects of Phillips's discussion I cannot possibly deal adequately here. They involve, as indeed he recognizes, some of the most difficult and elusive problems of philosophy. The doctrine of "emergent evolution," stated in a particular way, I hold to be perfectly sound, and some, though not all, of the ideas contained in Smuts's [1926] holism I think are acceptable and useful. But on the scientific, as distinct from the philosophical plane, I do think a good deal of fuss is being made about very little. For example—"newness springing from the interaction, interrelation, integration and organisation of qualities . . . could not be predicted from the sum of the particular qualities or kinds of qualities concerned: integration of the qualities thus results in the development of a whole different from, unpredictable from, their mere summation." Can one in fact form any clear conception of what "mere summation" can mean, as contrasted with the actual relations and interactions observed between the components of an integrated system? Has "mere summation" any meaning at all in this connection? What we *observe* is juxtaposition and interaction, with the resulting emergence of what we call (and I agree *must* call) a "new" entity. And who will be so bold as to say that this new entity, for example the molecule of water and its qualities, would be unpredictable, if we really understood *all* the properties of hydrogen and oxygen atoms and the forces brought into play by their union? Unpredictable by us with our present knowledge, yes; but *theoretically* unpredictable, surely not. When an inventor makes a new machine, he is just as certainly making a new entity, but he can predict with accuracy what it will be and what it will do, because within the limits of his purpose he *does* understand the whole of the relevant properties of his materials and knows what their interactions will be, given a particular set of spatial relations which he arranges.

In discussing General Smuts's doctrine of "holism" Phillips lays stress on the whole as a *cause*, "'holism' is called the fundamental factor operative towards the creation of wholes in the universe." It is an "operative cause" and an "inherent, dynamic characteristic" in communities. All but those who take "a static view of the structure, composition and life of communities—cannot fail to be impressed with the fundamental nature of the *factor of holism* innate in the very being of community, a factor of *cause*" (italics in the original)....

Is the community then the "cause" of its own activities? Here we touch the very difficult philosophical question of the meaning of causation, which I cannot possibly attempt to discuss here. In a certain sense however, the community as a whole may be said to be the "cause" of its own activities, because it represents the aggregation of components the sum (or more properly the synthesis) of whose actions we call the activities of the community—actions which would not be what they are unless the components were associated in the way in which they are associated. So far we may concede Phillips's contention. But it is important to remember that these activities of the community are *in analysis* nothing but the synthesized actions of the components in association. We have simply shifted our point of view and are contemplating a new entity, so that we now, quite properly, regard the totality of actions as the activity of a higher unit.

It is difficult to resist the impression that Professor Phillips's enthusiastic advocacy of holism is not wholly derived from an objective contemplation of the facts of nature, but is at least partly [motivated] by an imagined future "whole" to be realized in an ideal human society whose reflected glamour falls on less exalted wholes, illuminating with a false light the image of the "complex organism."

The Ecosystem

I have already given my reasons for rejecting the terms *complex organism* and *biotic community*. Clements's earlier term *biome* for the whole complex of organisms inhabiting a given region is unobjectionable, and for some purposes convenient. But the more fundamental conception is, as it seems to me, the whole *system* (in the sense of physics), including not only the organism-complex, but also the whole complex of physical factors forming what we call the environment of the biome—the habitat factors in the widest sense. Though the organisms may claim our primary interest, when we are trying to think fundamentally we cannot separate them from their special environment, with which they form one physical system.

It is the systems so formed which, from the point of view of the ecologist, are the basic units of nature on the face of the earth. Our natural human prejudices force us to consider the organisms (in the sense of the biologist) as the most important parts of these systems, but certainly the inorganic "factors" are also parts—there could be no systems without them, and there is constant interchange of the most various kinds within each system, not only between the organisms but between the organic and the inorganic. These *ecosystems*, as we may call them, are of the most various kinds and sizes. They form one category of the multitudinous physical systems of the universe, which range from the universe as a whole down to the atom. The whole method of science, as H. Levy (1932) has most convincingly pointed out, is to isolate systems mentally for the purposes of study, so that the series of *isolates* we make become the actual objects of our study, whether the isolate be a solar system, a planet, a climatic region, a plant or animal community, an individual organism, an organic molecule, or an atom. Actually the systems we isolate mentally are not only included as parts of larger ones, but they also overlap, interlock, and interact with one another. The isolation is partly artificial, but is the only possible way in which we can proceed.

Some of the systems are more isolated in nature, more autonomous, than others. They all show organization, which is the inevitable result of the interactions and consequent mutual adjustment of their components. If organization of the possible elements of a system does not result, no system forms or an incipient system breaks up. There is in fact a kind of natural selection of incipient systems, and those which can attain the most stable equilibrium survive the longest. It is in this way that the dynamic equilibrium, of which Professor Phillips writes, is attained. The universal tendency to the evolution of dynamic equilibria has long been recognized. A corresponding idea was fully worked out by Hume and even stated by Lucretius. The more relatively separate and autonomous the system, the more highly integrated it is, and the greater the stability of its dynamic equilibrium.

Some systems develop gradually, steadily becoming more highly integrated and more delicately adjusted in equilibrium. The ecosystems are of this kind, and the normal autogenic succession is a progress toward greater integration and stability. The "climax" represents the highest stage of integration and the nearest approach to perfect dynamic equilibrium that can be attained in a system developed under the given conditions and with the available components.

The great regional climatic complexes of the world are important determinants of the primary terrestrial ecosystems, and they contribute *parts* (components) to the systems, just as do the soils and the organisms. In any fundamental consideration of the ecosystem it is arbitrary and misleading to abstract the climatic factors, though for purposes of separation and classification of systems it is a legitimate procedure. In fact the climatic complex has more effect on the organisms and on the soil of an ecosystem than these have on the climatic complex, but the reciprocal action is not wholly absent. Climate acts on the ecosystem rather like an acid or an alkaline "buffer" on a chemical soil complex.

Next comes the soil complex which is created and developed partly by the subjacent rock, partly by climate, and partly by the biome. Relative maturity of the soil complex, conditioned alike by climate, by subsoil, by physiography, and by the vegetation, may be reached at a different time from that at which the vegetation attains its climax. Owing to the much greater local variation of subsoil and physiography than of climate, and to the fact that some of the existing variants prevent the climatic factors from playing the full part of which they are capable, the developing soil complex, jointly with climate, may determine variants of the biome. Phillips's contention that soil never does this is too flatly contrary to the experience of too many ecologists to be admitted. Hence we must recognize ecosystems differentiated by soil complexes, subordinate to those primarily determined by climate, but none the less real.

Finally comes the organism-complex or biome, in which the vegetation is of primary importance, except in certain cases, for example many marine ecosystems. The primary importance of vegetation is what we should expect when we consider the complete dependence, direct or indirect, of animals upon plants. This fact cannot be altered or gainsaid, however loud the trumpets of the "biotic community" are blown. This is not to say that animals may not have important effects on the vegetation and thus on the whole organism-complex. They may even alter the primary structure of the climax vegetation, but usually they certainly do not. By all means let animal and plant ecologists study the composition, structure, and behavior of the biome together. Until they have done so we shall not be in possession of the facts which alone will enable us to get a true and complete picture of the life of the biome, for both animals and plants are components. But is it really necessary to formulate the unnatural conception of biotic *community* to get such cooperative work carried out? I think not. What we have to deal with is a *system,* of which plants and animals are components, though not the only components. The biome is determined by climate and soil and in its turn reacts, sometimes and to some extent on climate, always on soil.

Clements's "prisere" (1916 [reproduced in this volume]) is the gradual development of an ecosystem as we may see it taking place before us today. The gradual attainment of more complete dynamic equilibrium (which Phillips quite rightly stresses) is the fundamental characteristic of this development. It is a particular case of the universal process of the evolution of systems in dynamic equilibrium. The equilibrium attained is however never quite perfect: its degree of perfection is measured by its stability. The atoms of the chemical elements of low atomic number are examples of exceptionally stable systems — they have existed for many millions of millennia: those of the radioactive elements are decidedly less stable. But the order of stability of all the chemical elements is of course immensely higher than that of an ecosystem, which consists of components that are themselves more or less unstable — climate, soil, and organisms. Relatively to the more stable systems the ecosystems are extremely vulnerable, both on account of their own unstable components and because they are very liable to invasion by the components of other systems. Nevertheless some of the fully developed systems — the "climaxes" — have actually maintained themselves for thousands of years. In others there are elements whose slow change will ultimately bring about the disintegration of the system.

This relative instability of the ecosystem, due to the imperfections of its equilibrium, is of all degrees of magnitude, and our means of appreciating and measuring it are still very rudimentary. Many systems (represented by vegetation climaxes) which appear to be stable during the period for which they have been under accurate observation may in reality have been slowly changing all the time, because the changes effected have been too slight to be noted by observers. Many ecologists hold that *all* vegetation is *always* changing. It may be so: we do not know enough either to affirm or to deny so sweeping a statement. But there may clearly be minor changes within a system which do not bring about the destruction of the system as such.

Owing to the position of the climate-complexes as primary determinants of the major ecosystems, a marked change of climate must bring about destruction of the ecosystem of any given geographical region, and its replacement by another. This is the *clisere* of Clements (1916 [reproduced in this volume]). If a continental ice sheet slowly and continuously advances or recedes over a considerable period of time all the zoned climaxes which are subjected to the decreasing or increasing temperature will, according to Clements's conception, move across the continent "as if they were strung on a string," much as the plant communities zoned round a lake will move toward its center as the lake fills up. If on the other hand a whole continent desiccates or freezes[,] many of the ecosystems which formerly occupied it will be destroyed altogether. Thus whereas the prisere is the development of a single ecosystem *in situ*, the clisere involves their destruction or bodily shifting.

When we consider long periods of geological time we must naturally also take into account the progressive evolution and rise to dominance of new types of organism and the decline and disappearance of older types. From the earlier Paleozoic, where we get the first glimpses of the constitution of the organic world, through the later Paleozoic where we can form some fairly comprehensive picture of what it was like, through the Mesozoic where we witness the decline and dying out of the dominant Paleozoic groups and the rise to prominence of others, the Tertiary with its overwhelming dominance of Angiosperms, and finally the Pleistocene ice age with its disastrous results for much of the life of the northern hemisphere, the shifting panorama of the organic world presents us with an infinitely complex history of the formation and destruction of ecosystems, conditioned not only by radical changes of land surface and climate but by the supply of constantly fresh organic components. We can never hope to achieve more than a fragmentary view of this history, though doubtless our knowledge will be very greatly extended in the future, as it has been already notably extended during the last thirty years. In detail the initiation and development of the ecosystems in past times must have been governed by the same principles that we can recognize today. But we gain nothing by trying to envisage in the same concepts such very different processes as are involved in the shifting or destruction of ecosystems on the one hand and the development of individual systems on the other. It is true, as Cooper insists (1926), that the changes of vegetation on the earth's surface form a continuous story: they form in fact only a part of the story of the changes of the surface of this planet. But to analyze them effectively we must split up the story and try to focus its phases according to the various kinds of process involved.

Biotic Factors

Professor Phillips makes a point of separating the effect of grazing herbivorous animals *naturally* belonging to the "biotic community," e.g., the bison of the North American prairie or the antelopes, etc., of the South African veld, from the effect of grazing animals introduced by man. The former are said to have cooperated in the production of the short-grass vegetation of the Great Plains, which has even been called the *Bison-Bouteloa* climax, and to have kept back the forest from invading the edges of the grassland formation. The latter are supposed to be merely destructive in their effects, and to play no part in any successional or developmental process. This is perhaps legitimate as a description of the ecosystems of the world before the advent of man, or rather with the activities of man deliberately ignored. It is obvious that modern civilized man upsets the "natural" ecosystems or "biotic communities" on a very large scale. But it would be difficult, not to say impossible, to draw a natural line between the activities of the human tribes which presumably fitted into and formed parts of "biotic communities" and the destructive human activities of the modern world. Is man part of "nature" or not? Can his existence be harmonized with the conception of the "complex organism"? Regarded as an exceptionally powerful biotic factor which increasingly upsets the equilibrium of preexisting ecosystems and eventually destroys them, at the same time forming new ones of very different nature, human activity finds its proper place in ecology.

As an ecological factor acting on vegetation the effect of grazing heavy enough to prevent the development of woody plants is essentially the same effect wherever it occurs. If such grazing exists the grazing animals are an important factor in the biome . . . whether they came by themselves or were introduced by man. The dynamic equilibrium maintained is primarily an equilibrium between the grazing animals and the grasses and other hemicryptophytes which can exist and flourish although they are continually eaten back.

Forest may be converted into grassland by grazing animals. The substitution of the one type of vegetation for the other involves destruction of course, but not merely destruction: it also involves the appearance and gradual establishment of new vegetation. It is a successional process culminating in a climax under the influence of the actual combination of factors present and since this climax is a well-defined entity it is also the development of that entity. It is true of course that when man introduces sheep and cattle he protects them by destroying carnivores and thus artificially maintains the ecosystem whose essential feature is the equilibrium between the grassland and the grazing animals. He may also alter the position of equilibrium by feeding his animals not only on the pasture but also partly away from it, so that their dung represents food for the grassland brought from outside, and the floristic composition of the grassland is thereby altered. In such ways *anthropogenic ecosystems* differ from those developed independently of man. But the essential formative processes of the vegetation are the same, however the factors initiating them are [different].

We must have a system of ecological concepts which will allow of the inclusion of *all* forms of vegetational expression and activity. We cannot confine ourselves to the so-called "natural" entities and ignore the processes and expressions of vegetation now so abundantly provided us by the activities of man. Such a course is not scientifically sound, because scientific analysis must penetrate beneath the forms of the "natural" entities, and it is not practically useful because ecology must be applied to conditions brought about by human activity. The "natural" entities and the anthropogenic derivates alike must be analyzed in terms of the most appropriate concepts we can find. Plant community, succession, development, climax, used in their wider and not in specialized senses, represent such concepts. They certainly involve an abstraction of the vegetation as such from the whole complex of components of the ecosystem, the remaining components being regarded as factors. This abstraction is a convenient isolate which has served and is continuing to serve us well. It has in fact many, though by no means all, of the qualities of an organism. The biome is a less convenient isolate for most purposes, though it has some uses, and it is not in the least improved by being called a "biotic community" or a "complex organism," terms which are illegitimately derived and which introduce misleading implications.

Methodological Value of the Concepts Relating to Successional Change

There can be no doubt that the firm establishment of the concept of succession has led directly to the creation of what is now often called dynamic ecology and that this in its turn has greatly increased our insight into the nature and behavior of vegetation. The simplest possible scheme involves a succession of vegetational stages (the prisere of Clements) on an initially "bare" area, culminating in a stage (the climax) beyond which no further advance is possible under the given conditions of habitat (in the widest sense) and in the presence of the available colonizing species. If we recognize that the climax with its whole environment represents a system in relatively stable dynamic equilibrium while the preceding stages are not, we have already the *essential framework* into which we can fit our detailed investigations of particular successions. Unless we use this framework, unless we recognize the universal tendency of the system in which vegetation is the most conspicuous component to attain dynamic equilibrium by the most complete adjustment possible of all the complexes involved[,] we have no key to correct interpretation of the observed phenomena, which are open to every kind of misinterpretation. From the results of detailed investigations of successions, which incidentally throw a great deal of new light on existing vegetation whose nature and status were previously obscure, we may deduce certain general laws and formulate a number of useful subsidiary concepts. So far the concept of succession has proved itself of prime methodological value.

The same can scarcely be said of the concept of the climax as an organism and all that flows from its strict interpretation. On the contrary this leads to the dogmatic theses that development of the "complex organism" can *never* be retrogressive, because retrogression in development is supposed to be contrary to the nature of an organism, and that edaphic or biotic factors can *never* determine a climax, because this would cut across the conception of the climatic climax as *the* "complex organism."

Phillips says (1935a: 242) that "the utility of the climax in Clements' sense would be greatly impaired were we to attempt to isolate from it the concept of the community as a complex organism. Its natural dynamic utility for orientation of research in succession, development and classification would be distinctly diminished." And again (1935b: 503), "The biotic community is an organism, a highly complex one: this concept is fundamental to a natural setting and classification of the profoundly important processes of succession, development and attaining of dynamic equilibrium."

What is the justification for such statements? What researches have been stimulated or assisted by the concept of "the complex organism" *as such*? Professor Phillips seems to have in mind cooperative work in which plant and animal ecologists take part. But nobody denies the necessity for investigation of *all* the components of the ecosystem and of the ways in which they interact to bring about approximation to dynamic equilibrium. That is the prime task of the ecology of the future.

We cannot escape the conclusion that the supposed methodological value of the concept of the "complex organism," contrasted with the value of succession, development, climax, and ecosystem, is a false value, and can only mislead. And it is false because it is based either on illegitimate extension of the biological concept of organism (Clements) or on a confusion between the biological and "organicist" uses of the word (Phillips).

CHAPTER 4

A Succession of Paradigms in Ecology: Essentialism to Materialism and Probabilism

Daniel Simberloff

The Rise of Probabilism and Materialism in Ecology

Ecology has undergone, about half a century later than genetics and evolution, a transformation so strikingly similar in both outline and detail that one can scarcely doubt its debt to the same materialistic and probabilistic revolutions. Many major events in this transformation have been described by Ponyatovskaya (1961) and McIntosh (1975, 1976), but the relationship to developments both inside and outside biology seems not to have been noticed. An initial emphasis on similarity of isolated communities, replaced by concern about their differences; examination of groups of populations, largely superseded by study of individual populations; belief in deterministic succession shifting, with the widespread introduction of statistics into ecology, to realization that temporal community development is probabilistic; and a continuing struggle to focus on material, observable entities rather than ideal constructs; all parallel trends [in] genetics and evolution.

Ecology's first paradigm was the idea of the plant community as a superorganism, propounded by Clements in the first American ecology book (1905) and elaborated by him in numerous subsequent publications. The crux of this concept was that single-species populations in nature are integrated into well-defined, organic entities, and key subsidiary aspects were that temporal succession in a sere is utterly deterministic, analogous to development of an individual, and leads inevitably to one of a few climax communities. The relationship between the stylized, integrated superorganism and the deterministic successional development producing it is organic and fundamental, as pointed out by Tansley (1920): "When we have admitted the necessity of first determining empirically our natural units, we have to find ways of grouping them. This

From *Synthese* 43 (1980): 3–39.

way we can only find in the concept of development. Development of vegetation is a concrete fact equal with its structure." McIntosh (1975, 1976) illustrates the extent to which this paradigm dominated ecology until recently with a series of well-chosen quotes and an analysis of ecology texts.[1] Suffice it here to cite Shelford (1913), one of the leading animal ecologists of the first half of this century: "Ecology is the science of communities. A study of the relations of a single species to the environment conceived without reference to communities and, in the end, unrelated to the natural phenomena of its habitat and community . . . is not properly included in the field of ecology." Although the superorganismic community concept with its deterministic succession arose in plant ecology almost as a logical consequence to de Candolle's pioneering descriptions of plant formations, it quickly won acolytes among zoologists (e.g., Shelford) and limnologists (e.g., Naumann). Allee as early as 1931 had aligned his work on animal symbioses with the superorganismic paradigm, observing "a more or less characteristic set of animals which are not mere accidental assemblages but are integrated communities." Emerson's similar focus also led him to the notion of an "integrated ecological community" (1939). Thienemann described a lake community as "a unity so closed in itself that it must be called an organism of the highest order" (vide McIntosh 1975). Probably even more important than this zoological and limnological support in solidifying the superorganism concept as a paradigm was Elton's description of the food chain as a conduit for community energy flow. Although Elton himself was not an adherent of the superorganism view, his discovery provided such a diagrammatic analogy to the physiology of an individual organism that it was readily incorporated as an integral part of the superorganism, in fact, one of the forces giving it organismic cohesion. Citing Elton (1927), Clements and Shelford (1939) stress that trophic structure studies "can be utilized to reveal the significance of each process in the working of the community as a whole." Further, "the universal role of coaction [including trophic interaction] is to be seen in the integration of plant and animal relations to constitute an organic complex, which is characterized by a certain degree of dynamic balance in number and effects." . . .

Surely the ultimate philosophical basis for the superorganism paradigm is Greek metaphysics, and this explains its strong appeal in the face of data-based objections by Gleason and others. For the superorganism, one of a small number of distinct climax communities, is an explicitly typological construct which allows immediate classification of an observed community into an already described category. Differences among individuals within that category are viewed as less important than the similarities which cause them to be classified together, and are ontogenetically different from differences between categories. The latter are viewed as a reflection of different organizing relationships (such as the "multiple stable equilibria" (Sutherland 1974) in a currently popular incarnation of the superorganism concept (cf. Holling 1973)). The former, as in pre-Mendelian genetics, are rather viewed as "noise," probably the result of minor differences in physical environment, like soil chemistry, during development. And the deterministic path of succession in the strictest Clementsian monoclimax formulation is as much an ideal abstraction as is a Newtonian particle trajectory. There is a tidiness, an ease of conceptualization, to well-defined ideals moving on perfect paths that is as appealing, both aesthetically and functionally, in ecology as it was in genetics and evolution. Unfortunately, it is as poor a description of ecological as of evolutionary reality. That the superorganism paradigm did not lead to mechanistic understanding of the operation and structure of nature is not surprising. . . .

The watershed year for the materialistic and probabilistic revolution in ecology was 1947, in which three respected plant ecologists (Egler, Cain, and Mason) all published papers in *Ecological Monographs* forcefully attacking the Clementsian paradigm and citing Gleason's "individualistic concept of plant association" as the first articulation of their view (McIntosh 1975). The formal analogy to Mendel's resurrection is patent, but even more enlightening is an examination of the specific reasons given for this dramatic change. Egler cites Raup (1942) and Cain (1944) to the effect that the Clementsian assumption of cause-and-effect in community development is an *a priori* explanation, rather than an empirically derived mechanism, and he claims that his own extended observations on a series of Hawaiian communities are completely in accord with the individualistic concept of Gleason's "all but forgotten paper." Cain avers his interest in "actual, concrete, specific communities on the ground," and scorns the "hypothetical" Clementsian community. Later he suggests that the monoclimax theory as originally stated was wrong, had subsequently become a panchreston, and that focus on local studies will demonstrate the correctness of Gleason's individualistic hypothesis. In a "heretical" section entitled, "Does the Association Have Objective Reality?" he bases his negative answer on materialism ("Species are facts. . . . Environments are facts") while lambasting the "preconceptions of the reality of the association in the abstract." He is at pains to stress that a specific stand, having material existence, is real while the association, only a hypothetical ideal, is not. In the final analysis, for Cain, the superorganism is a fictitious construct because (1) unlike a species, it has no continuity by descent, and (2) there is no objective criterion for determining when two stands are similar enough to belong to the same association. Mason also stresses that genetic continuity renders the species population a real entity, and lack of it renders the community a fortuitous abstraction, limited only by the "coincidence of tolerance" of environmental factors by its component species. He, too, credits Gleason as the first proponent of this notion.

The spate of ecology texts early in this decade (Colinvaux 1973; Collier et al.

1973; Poole 1974; Krebs 1978) all agree that, twenty-five years after its rehabilitation, the Gleasonian paradigm had overthrown the Clementsian one (McIntosh 1975). If one asks why the revolution occurred when it did, two convergent lines of research appear to have necessitated it. First, the facts that real stands generally lack well-defined boundaries, and when such boundaries do exist they are frequently associated with abrupt changes in the physical environment, were often noted by Gleason and Ramensky, but by 1947 constituted an intolerable contradiction of dogma by observed fact. Egler, Cain, and Mason all attack the superorganism paradigm on this basis using data from field studies. This type of observation was greatly extended in the next decade by two independent groups; Curtis's "vegetational continuum" (Curtis and McIntosh 1951; Curtis 1955, 1959) and Whittaker's "gradient analysis" (1956, 1967) both describe the spatial distribution of plants as a consequence of the individual, relatively uncoordinated responses of individual species to gradients in the physical environment, without need to invoke groups of species' persisting or dying as a unit. It may be observed that this work also demonstrates that discrete populations acting individually produce a community continuum (whatever statistic is used to characterize communities) much as the Morgan group's understanding of polygenicity, position effects, etc.[,] resolved the apparent Mendelian paradox that phenotypes are, for the most part, continuous while the alleles of a gene are discrete.

The second line of research which, I would argue, contributed to the demise of the superorganism was a shift in emphasis within the study of animal and microbial populations which led, inevitably, to focusing on individual populations as proper objects of study, or at most two or three of them together, rather than the entire community in which they are embedded. The failure of Haskell in 1940 to attract interest in a hypervolume concept of the species' niche (McIntosh 1976) is a reflection of the lack of interest in studying populations per se. That Hutchinson's identical suggestion in 1957 has generated an enormous literature on niche parameters and relationships is an indication of the extent to which the plant population research by Egler, Cain, Mason, Curtis, and Whittaker and animal population studies by Nicholson and Park had legitimized the population or few-species interaction as an object of study, independent of the community. Experiments on real populations were key to this shift of focus from community to population. . . .

The Current Struggle

One ought not to be left with the feeling that the materialistic, probabilistic revolution in ecology is a *fait accompli*. Essentialism, idealism, and determinism are, if not dominant, still rampant (Slobodkin 1975). Their persistence is partly a reflection of ecologists' diffidence because of the apparent sloppiness of their field compared to the physical sciences; "physics-envy is the curse of biology" (Cohen 1971). The large scatters of points and jagged trajectories which typify ecology (e.g., the colonization curves for new island communities (Simberloff and Wilson 1970) and the dispersion of the number of herbivores on different plant species (Strong and Levin 1979)) seem to foster the view that ecology is not quite so scientific as chemistry and physics, and militate for a search for more ideal models, often from the physical sciences themselves. For example, a topical endeavor during this decade has been the erection of deterministic rules for packing species into communities, as if some clearly and permanently bounded physical entity could be denoted "community" and studied in isolation. In MacArthur's summary work (1972), he presents an analogy of species packing to crystal packing, suggested by Gordon Lark, a biochemist. Such an ideal, deterministic approach appears to be symptomatic of a wide variety of proposals on species packing, and May (1974; vide McIntosh 1976), a physicist-turned-ecologist who has quickly become the leading figure in analytic ecological modeling (e.g., May 1973a), raises the metaphor to an even higher level, envisioning the eventual establishment of many "perfect crystal" models in ecology and the consequent emergence of ecology as coequal to "the more conventional (and more mature) branches of science and engineering."

The crystal-packing and related models for species packing, however, do not appear to have been strikingly successful; certainly they constitute a retreat to idealism. The community matrix of interaction coefficients from the deterministic logistic equations (Levins 1968a, 1968b) is still used to characterize communities and to explain community properties (e.g., Culver 1975). May's treatment of the matrix, and the community in general (1973a), appears to be revolutionary since it incorporates statistical noise to produce a stochastic neighborhood outcome of community dynamics, rather than a single deterministic point. But the underlying equations are distressingly ideal, and the noise distribution is an ad hoc suggestion. MacArthur's quadratic form Q, which he claims competition minimizes (1969, 1970), is not only a deterministic consequence of logistic equations, but as metaphysical an entity as Darwin's gemmules, Maxwell's Demon in statistical mechanics, and Adam Smith's hidden hand in economics. Surprisingly, it is condoned by Lewontin (1969). Limiting similarity L of coexisting species was calculated from logistic equations by MacArthur and Levins (1967); despite a data-based, mechanistic demonstration of its incorrectness by Dayton (1973) and a cogent, damning theoretical treatment by Heck (1976) it is still cited as a possible characteristic of nature (e.g., Fraser 1976). Neill's experiments showing that the multispecies form of the logistic model rests on untenable assumptions (1974) seem not to have been heeded. MacFadyen (1975) deplores the autonomy and independence from the biological world which the essentialist strain in ecology appears to have achieved, but his explanation—frequent failure to propose truly testable hypotheses—is only the proximate cause. The ultimate difficulty is the tenacity of the Greek metaphysical worldview.

Another manifestation of this tenacity is the relative independence of the nascent stochastic school of population and community ecology, discussed above, from the mathematical school epitomized by logistic-based differential equations. A striking text of ecology from the probabilistic viewpoint was produced by E. C. Pielou in 1969 and though it attracted generally favorable reviews and a few new adherents to the stochastic cause (e.g., Wangersky 1970), its ideas and methods are given short shrift by the logistic-oriented texts which dominate American ecology in this decade; earlier stochastic treatments by Chiang (1954, 1968), Skellam (1955), Bartlett (1957, 1960), Bartlett et al. (1960), Leslie and Gower (1958), and Leslie (1958, 1962) were even more summarily consigned to oblivion in these quarters. Systems-analysis ecologists were not more enthusiastic than mathematical ones. For example, Watt (1968) predicts that stochasticizing systems models of communities and large populations will not appreciably improve their performance. Cohen's review (1970) of Pielou's text states, "This book should liberate those who assess work in mathematical ecology according to its projection along an axis from Princeton to Davis by informing them in the very substantial efforts, accomplishments, and opportunities in orthogonal directions."[2] The liberation is still in the future, and will come only with the completion of a materialist revolution in ecology. Recent papers by Tiwari and Hobbie (1976a, 1976b) stochasticizing differential equations which describe a simple aquatic ecosystem may be in its vanguard.

The unease of ecologists vis-à-vis physics and the zeal with which they seek deterministic physical science models are misplaced. What physicists view as noise is music to the ecologist; the individuality of populations and communities is their most striking, intrinsic, and inspiring characteristic, and the apparent indeterminacy of ecological systems does not make their study a less valid pursuit. Mayr (1961) suggests that the uniqueness of biological entities and phenomena constitutes one of the major differences between biology and the physical sciences, and that this difference makes it particularly difficult for physical scientists to understand biological concerns. There are three types of indeterminacy at issue here. One is at least as fundamental to ecology as Heisenberg's uncertainty is to physics, and is in fact grounded in the latter. . . . A second form of apparent indeterminacy has only recently been addressed (May 1974; Oster 1975; Poole 1977) and takes the form of "chaotic" behavior of populations governed by certain nonlinear, apparently quite realistic, systems of difference equations. Despite the fact that the underlying equations are completely deterministic, the resulting trajectories may be so complex as to appear random, and Oster (1975) suggests that it may be impossible with biological data to distinguish among true stochasticity, experimental error, and complex flows of a deterministic model. However, similar situations arise in meteorology (May 1974) and even in that quintessential physical ideal, the Newtonian billiard table (Oster 1975), so ecology can hardly be relegated to the status of second-rate science on these grounds.

The third type of ecological indeterminacy is probably most foreign to the physical sciences and is the primary cause of ecologists' defensiveness. This is apparent indeterminacy engendered by the enormous number of entities even in simple ecological systems themselves rather than by the form of the equations describing the systems. Further, these entities may be interacting, and the interactions are often subtle. Whether we believe with Mayr (1961) and Wangersky (1970) that this complexity will forever preclude completely deterministic ecological description or feel, more optimistically, that better and better instrumentation and effort will bring us to within Heisenberg's limits of a perfect description of ecological systems (Holling 1966), we must agree that we will not, in the near future, have sufficient information or insight to produce equations as predictive as those of most physicists and engineers. On the other hand, neither are meteorologists able to predict weather patterns with remarkable precision, and the amount of money and manpower committed to the study of single ecological systems pales compared to the effort involved in, say, scooping up a few moon rocks. With sufficient resources ecologists have been notably more successful and precise in their predictions; many examples are given by DeBach (1974) for biological control. In any event, the nature of genetic systems of living organisms and the fact that evolution constantly occurs ensures a certain amount of variability in the outcome of ecological events (Pimentel 1966), and this variability is among the most interesting aspects of ecological phenomena. Further, the amount of variation itself can be predicted, as in Schaffer and Elson's study of salmon life history phenomena (1975). . . .

I end this section on the status of the materialistic revolution in ecology with the observation that the first ecological ideal, Clements's superorganism, is not dead, but rather transmogrified into a belief that holistic study of ecosystems is the proper course for ecology (Watt 1966b; Levins 1968a, 1968b; Lane et al. 1975; Johnson 1977; [E. P.] Odum 1977 [reproduced in this volume]; cf. McIntosh 1976). [Eugene] Odum (1964) views the ecosystem as bearing the same relation to ecology as the cell does to molecular biology, a clearly superorganismic conception. Patten (1975) sees the ecosystem as a "holistic unit of coevolution," and argues that ecosystems evolve toward linear good behavior: "Nonlinearity is a mathematical property, not an ecological one, and no ecosystem process is nonlinear until someone writes a relation that describes it so." Aside from the fact that linearity is as much a mathematical property as is nonlinearity, the well-behavedness is an artifact of Patten's convention of defining an ecosystem out of existence when it is egregiously ill-behaved; Holling (1973) views the same behavior as evidence for multiple domains of attraction in the same system. The true measure of the validity of the holistic, well-behaved ecosystem concept is whether it provides insight into community mechanisms, and its record here is equivocal (Mitchell et al. 1976; Auerbach et al. 1977). Well known instances of abrupt fluctuations in single populations are ignored, yet Preston (1969) gives sufficiently many examples that one might reasonably claim that poor behavior characterizes nature. The widely cited, dramatic increases of *Acanthaster planci* in the Pacific (Branham 1973; Glynn 1974) and *Sphaeroma terebrans* on Florida mangroves (Rehm and Humm 1973) are two more recent examples of ill-mannered ecosystems.

One may ask why focus on ecosystems has seduced so many ecologists in the face of its failure to add substantially to our understanding of the workings of nature. Indeed, even when this failure is noted, it is ignored or explained away, as Kuhn (1970) suggests is typical for a paradigm before it is finally overthrown by a scientific revolution. Admitting the predominant failure of the U.S. International Biological Program (IBP), the most massively supported American ecological effort and one wholly conceived in a holistic, ecosystem vein, Odum (1977 [reproduced in this volume]), a leading ecosystem adherent, suggests that it is not the paradigm which is faulty, but rather the consistency with which the paradigm was used in organizing research. In short, the IBP effort was not holistic enough!

One suggestion for the apparent paradigmatic status of the ecosystem concept in the face of conflicting data is that it provides support for the notion of self-regulatory powers inherent in unfettered capitalism (Leigh 1971). For if a community of organisms, naturally selected each to maximize the representation of its own genes, can be shown to be analogous to a single organism whose parts all work to a common purpose, so ought a competitive capitalism to produce a unified whole which benefits all. This is an old notion; Adam Smith's metaphor was that of a hidden hand converting the profit-maximizing activities of individuals into the good of the whole. That this should be true for ecological systems is questionable on both evolutionary grounds (Levins 1975) and the grounds of observed ecological irregularity described above. Even were it true for ecology, I suspect that it is not the primary attraction of the ecosystem paradigm; but one ought always to recognize the strength with which a basic philosophy, even an economic one, structures our perception of apparently unrelated phenomena. Perhaps the most convincing argument that the main attraction of holism is not as a subtle justification of capitalism is that it has adherents with long-standing, impeccable Marxist credentials (e.g., Levins 1968a, 1968b; Lewontin and Levins 1976).

McIntosh (1976) documents the transformation in the 1950s and 1960s of American ecology into a big-money operation, the era of "Grant Swinger," without observing that the big money is primarily in the area of ecosystems. For example, the IBP was followed by the creation of a new program, Ecosystem Studies, in the National Science Foundation. . . . To the extent that grant funding is an important determinant of academic advancement, and economic well-being a general goal (Merton 1973; Storer 1973), one might reasonably argue that the ecosystem paradigm is seductive on economic grounds alone, independent of either philosophical or biological considerations.

Yet another attraction of the ecosystem is that it lends itself to cybernetic interpretation via systems analysis, a vogue vocation in the United States for about two decades (McIntosh 1976). Indeed, the primary thrust of ecosystematists has been systems analysis (Patten 1959, 1971; Watt 1966a) and the glamour of turning ecology into a space-age science, replete with the terminology of engineering and physics, must itself have been a powerful inducement of the ecosystem approach, fitting hand-in-glove with the economic appeal. The concurrent rise of computer technology further augmented the appeal of a systems analytic study of ecosystems, and though McIntosh (1976) observes that the tide appears to have crested, nevertheless this aspect of ecosystem research remains a powerful force in ecology today. Odum's (1964) analogy, cell:molecular biology = ecosystem:ecology, may be relevant not only for the light it throws on the relationship of the ecosystem concept to metaphysical thought and its intellectual debt to the superorganism concept, but also as an expression of the desire of ecologists to achieve the respectability, even glamour, of molecular biology in its heyday, when Odum wrote.

But I suggest that the chief reason for the persistence of the ecosystem paradigm is that it accords with Greek metaphysics. The attractiveness of holism, the notion that "everything affects everything else" (Watt 1966a), includes not only its tidiness but its determinism, for if all components are included in the system and linked to all others by deterministic equations, then no exogenous, random input is possible. The myth of the balance-of-nature persists in the popular consciousness, and takes systems ecological form in Barry Commoner's condensation of all ecology into "You can't change just one thing" (1971); Colwell (1970) also notes the identity of the ecosystem paradigm and the balance-of-nature. That an idea so readily accepted by the lay public attracts professional adherents as well is not surprising, particularly when the idea has two-thousand-year-old roots. Even Albert Einstein, a founder of stochastic quantum mechanics, viewed it only as an instrument for dealing with atomic systems, not as a true representation of the universe: "I am absolutely convinced that one will eventually arrive at a theory in which the objects connected by laws are not probabilities, but conceived facts" (Born 1949). That Einstein's objection to a "dice-playing God" was irrational, however, he readily admitted: "I cannot provide logical arguments for my conviction, but can only call on my little finger as a witness, which cannot claim any authority to be respected outside my own skin" (ibid.). Small wonder that Greek metaphysics continues to influence ecologists! Certainly there is something profoundly disturbing about a nature in which random elements play a large role. Just as much of the opposition to Darwinian evolution powered by natural selection was engendered by the large role assigned to chance, so the idea of an unbalanced, stochastically driven natural community inspires distrust.

The ecosystem paradigm purports to have corrected the superorganism's shortcomings, primarily by explicitly noting that succession need not always lead to the same climax and by focusing on certain individual characteristics of ecosystems. But its most fundamental features are determinism and interest in a high-level ideal entity; in these it is squarely in the camp of the essentialists. The success of the materialist revolution in other disciplines, particularly evolution and genetics, augurs well for ecology, but Greek metaphysics will not vanish easily.

Notes

1. The power of the paradigm was such as to preclude gathering of data oriented toward individual species. Margaret B. Davis, in her long-term studies of long-term vegetation changes, has been frustrated by this lacuna: "We do not know what the virgin vegetation of the pioneer days was like because all the ecologists were so busy looking for a nonexistent climax that they forgot to record what was actually growing there" (Colinvaux 1973).

2. Princeton, the institution of Robert MacArthur, was widely viewed as a center of mathematical ecology, while the University of California at Davis housed K. E. F. Watt's leading ecosystem systems analysis group.

CHAPTER 5

Life after Newton:
An Ecological Metaphysic

Robert E. Ulanowicz

Ecology never has nestled comfortably among the traditional sciences. The uniqueness of ecology was characterized by Arne Naess (1988) when he wrote about "deep ecology" as something that affects one's life and perception of the natural world in a profound way. Numerous others sense that ecology is useful for addressing phenomena in fields that are well-removed from the meadow or savannah. Thus, one encounters books on "the ecology of computational systems" (Huberman 1988) or discovers whole institutes devoted to the "ecological study of perception and action" (Gibson 1979).

It is well and good to laud ecology through poetry or metaphor, but most scientists demand a more rational and systematic comparison of ideas emerging from ecology with those comprising the established modern synthesis. In what follows I will attempt to compare the historical lines of ecosystems thinking against the axioms that undergird the Newtonian treatment of natural phenomena. I will argue that these elements of Newtonianism fail as a framework upon which to hang the manifold facets of the process called life (i.e., life post-Newton.) It appears that a full and coherent apprehension of living systems can be achieved if one reconsiders, in addition to a certain Newtonian element, some ancient Aristotelian concepts in conjunction with the more contemporary notion of a world that is ontologically open, as proposed by Peirce (1877) and Popper (1982). The ensuing unitary description of natural dynamics is cast in the mold of postmodern constructivism (Ferré 1996; Griffen 1996) but nonetheless retains marked similarity to the fundamental laws outlined by Newton in his *Principia* (life in the image of Newton). The axioms that later thinkers appended to Newton's work, however, do not appear consonant with ecological thinking and perhaps should be replaced by a new set of assumptions.

The Ecological Triptych

Conventional wisdom holds that there exists no single theoretical core to ecological thinking as, say, Maxwell's laws provide for the study of elec-

From *BioSystems* 50 (1999): 127–42. Edited by the author.

tricity and magnetism. The reason is that, historically, ecosystems theorists have divided roughly into three camps: (1) organicists, (2) mechanists, and (3) nominalists, or stochastics (Hagen 1992). The rise of systems ecology early in the twentieth century tracks closely the writings of Frederic Clements (1916 [reproduced in this volume]) and his manifold analogies between ecosystems and organisms. The metaphor of ecological community as "superorganism" is frequently used to characterize Clements's ecology (although a perusal of his writings will reveal he used this neologism only rarely). Clements credited Smuts (1926) for his interest in holism, and Clements, in turn, motivated two of ecology's most renowned figures, G. Evelyn Hutchinson and Eugene P. Odum. We note in passing that organic imagery hints broadly of Aristotle, and that Francis Bacon devoted his life to purging Aristotelian thought from natural philosophy.

Organicism appears heterodox to most scientists, because scientific orthodoxy demands a purely mechanical view of the world. Although Darwin did introduce history into biology, he otherwise hewed closely to Newtonian strictures, which he inherited via Thomas Malthus and Adam Smith (Depew and Weber 1994). In the United States, this mechanical view of the living world was bolstered by the technocratic movement of the 1920s and 1930s, which strongly influenced the ecology of Howard Odum (1960). George Clarke (1954), for example, in his textbook on ecology, went so far as to depict ecosystem populations and processes as the figurative gears and wheels of a machine. Connell and Slatyer (1977) provide a more recent example of the attempt to portray ecosystem succession as a constellation of basic mechanisms.

Inherent in the mechanical picture of ecosystems is a rigidity and determinism not always found in organisms and other living systems. At the other extreme stand the stochastics, or nominalists, who regard any notion of organization in ecosystems as pure illusion. For some, even ecosystems themselves appear as pure artifice (Engleberg and Boyarsky 1979; Cousins 1987). The stochastic vein in ecology runs deep and is at least as old as the holist. Gleason (1917), for example, published his contention that plant communities are stochastic assemblies at about the same time that Clements (1916 [reproduced in this volume]) was advocating that ecosystems approach their climax configurations in virtually deterministic fashion. Stochasticism was slow to catch on in American ecology until the 1950s, when changing political fashions may have herded many into this camp (Barbour 1996). The nominalist scenario seems to require fewer mathematical applications, so it is perhaps understandable that some of the more gifted ecological writers are attracted to this movement (e.g., Simberloff 1980 [reproduced in this volume]; Shrader-Frechette and McCoy 1993; Sagoff 1997).

Metaphors by definition are not precise, so it is not surprising that ecological descriptions often borrow from more than one of these analogies. Most renditions of ecological succession, for example, include both organismal and stochastic elements. Other fields of inquiry also provide hybrid metaphors. Thus, neo-Darwinian evolution, ecosystem modeling, and thermodynamics all include both mechanisms and chance events, whereas developmental biology, cybernetics, and teleonomy (*sensu* Mayr 1992) combine proportions of organicism with mechanism. An example has yet to appear in ecology of any scientific model that amalgamates all three metaphors (but see Salthe 1993), yet such an admixture would be required of any theory that would coalesce ecology into a unified discipline. Such an overarching paradigm would have seemed unthinkable several decades ago, but recent insights by both philosophers and ecologists have made the goal of a unified theory of ecosystems seem much closer.

The Newtonian Metaphysic

To begin a systematic comparison of the assumptions behind the three prevailing ecological metaphors, we focus initially upon the most prevalent—the Newtonian or mechanical worldview. It is widely accepted that the foundations for the modern view of nature, which had its beginnings with Bacon, Galileo, Hobbes, Gassendi, and Descartes, were made fast by Newton's publication of *Principia*. (Less well-known is how the actual format of *Principia*, which was largely accidental, set the stage for the materialist/mechanical revolution that was to follow (Westfall 1993; Ulanowicz 1995a)).

To review, the three laws that Newton himself formulated can be summarized as:

1. *A body once set in motion continues in straight-line motion until acted upon by another force.* This is a statement about what happens to a body when it is isolated from any external influences. It was the revolutionary idea of Descartes to accord primacy to straight-line motion over curvilinear pathways, such as the circle. The Greeks, for example, had regarded the circle as the most natural trajectory for heavenly bodies, because it was considered to be the most perfect geometrical form.

2. *The rate of change of the momentum of a body is proportional to the applied force.* This law tells what happens when external agencies intervene. The law is far more phenomenological than most seem willing to admit, and it is central to the definitions of force and mass.

3. *Every action is opposed by an equal and opposite reaction.* Like most of his contemporaries, Newton seemed to value conservation quite highly, and one of the ways to impose conservation is to require *symmetry*.

The mathematical forms of these three laws were sufficient to prescribe all of classical mechanics. It was their rigorous minimalism, more than Newton's three principles, that came to characterize the ensuing "Newtonianism." Somewhat surprisingly (given the legion of print that is devoted to the scientific method), one rarely finds the minimalist assumptions behind Newtonianism spelled out in any detail. One exception is by Kampis (1991) and another is the attempt by Depew and Weber (1994) to formulate the Newtonian canon in terms of four fundamental postulates. According to these latter authors:

1. Newtonian systems are *deterministic*. Given the initial position of any entity in the system, a set of forces operating on it, and stable closure conditions, every subsequent position of each particle or entity in the system is in principle specifiable and predictable. This is another way of saying that mechanical causes are everywhere ascendant.

2. Newtonian systems are *closed*. They admit of no outside influences other than those prescribed as forces by Newton's theory.

3. Newtonian systems are *reversible*. The laws specifying motion can be calculated in both temporal directions. There is no inherent arrow of time in a Newtonian system.

4. Newtonian systems are strongly decomposable or *atomistic*. Reversibility presupposes that larger units must be regarded as decomposable aggregates of stable least units—that what can be built up can be taken apart again. Increments of the variables of the theory can be measured by addition and subtraction.

Closer analysis reveals that implicit in the four items cited by Depew and Weber lies a fifth assumption that should be made explicit, namely,

5. Newtonian laws are *universal*. They are applicable everywhere, at all times and over all scales (Ulanowicz 1997). Not only is time considered to be uniform throughout the universe (Matsuno and Salthe 1995); but, in principle, no complications should arise if laws pertaining to very small dimensions, such as those governing strong intranuclear forces, were to be applied at galactic scales, where gravitation operates.

It should be noted in passing that there are considerable overlaps among the five attributes of the Newtonian metaphysic. That is, the Newtonian ideal itself is not strongly decomposable.

Broader Horizons

The advantage in spelling out the framework of the Newtonian worldview item-by-item is that we may now proceed to compare these five assumptions with any counterparts that might pertain to the other two ecosystem metaphors:

1. As for *determinism*, we recognize immediately that the other viewpoints

do not share this Newtonian assumption. Stochasticism is tautologous with chance. Of course, the Newtonian is likely to counter immediately with the belief that such chaos is only apparent and can always be resolved by analyzing the system with greater precision and in finer detail. Ultimately, however, such reductionistic regression leads to a dead end as one approaches the scale of molecular or subatomic particles, where indeterminism reigns.

Ontogenists might question whether the organic analogy is anything but deterministic. Within certain bounds (see Griffiths and Knight 1998), one can predict how a specific instance of organic development will play out. But one should note that the type of organic development ascribed to ecosystems is decidedly weaker than what is normally observed in ontogeny. As Depew and Weber (1994) put it, "Clements had it backwards. Ecosystems are not super-organisms; organisms are superecosystems." (Clements actually referred to ecological communities, as the term *ecosystem* was not coined until 1935 by Tansley.) Certainly, the limits on how accurately one can predict the outcome of ecological succession are far broader than what is possible with ontogeny—too broad, in fact, to allow the claim that such succession is a deterministic process.

2. Concerning causal *closure*, the point is almost moot in nominalism. Certainly, nominalism allows no cause other than the material or the mechanical. But the nominalist would go further and question whether it is useful, or even possible, to trace every event to its mechanical/material origins. Chance is, after all, the very crux of nominalism. All of which points directly at the Achilles heel of Newtonianism, namely, that chance is an unwelcome interloper in a Newtonian world. For, if there are only material and mechanical causes at work, a chance event that cannot be subsumed by the law of averages would disrupt the reductionists' scenario, making it impossible to predict higher-level phenomena. What, then, keeps the world from coming apart? It took some major backtracking for science to reconcile the idea of chance at the microscopic level with predictable Newtonian behaviors at macroscopic levels. Ronald Fisher (1930) pointed out, however, that such reconciliations are predicated on the assumption that "the reliability of physical material flows not necessarily from the reliability of its ultimate components, but simply from the *fact* that these components are very numerous and largely independent." Fact? The real fact is there is no guarantee that components at very small scales act independently of each other—only the desire on the part of some to cling to the Newtonian metaphysic.

It is worth noting that the assumption of independence is favored by individuals who, like most biologists, profess a falsificationist stance regarding scientific propositions. That is, one is expected to pull no punches whatsoever in a continuing attempt to falsify hypotheses. Yet cherished Newtonian beliefs appear to be exempt from such scrutiny. Far better to focus upon a narrow range of conditions under which Newtonian postulates can be retained and ignore the rest.

If the nominalists preserve the logical coherence of their beliefs by boldly proclaiming all order (except that imposed by observation) to be illusory, what options are available to the organicist, for whom form and function are essential attributes of the systems they study? One possibility would be to borrow from antiquity and explicitly acknowledge form and function as agents behind development. That is, they could return to the ideas of Aristotle, so fervently eschewed by Bacon, and reconsider opening the window of causality to admit once again the existence of natural formal and final causalities (Rosen 1985, 1991; Ulanowicz 1990). For these two types of cause are capable of restraining the disordering effects of chance events at lower levels from unbridled propagation up the hierarchy of scales. (See Salthe 1985, 1993; and Wimsatt 1994 for related formulations.)

That formal and final causes occur over longer times and larger scales can be seen from the (unsavory) example of a military battle. The material causes of a battle include the weapons and munitions that the armies employ against each other. The soldiers who use these implements of destruction comprise the efficient causes behind the conflict. The juxtaposition of the armies in the context of the physical landscape constitutes two formal agencies that strongly influence the course and outcome of the fray. Lastly, the final causes for the battle appear among the set of social, political, and economic events that initiated the war. That the temporal and spatial scales associated with each cause are hierarchically ordered should be obvious and becomes literal when one notes that efficient, formal, and final causes are the domains of the private, general, and prime minister, respectively. The legitimate interests of these agents map onto progressively broader geographic areas over longer times.

The existence of higher-level causes means that a chance event at any level need no longer set organization at larger scales collapsing like some house of cards. As will be argued below, formal and final agencies are capable of exerting top-down selection upon stochastic events below (Salthe 1985), both mitigating the detrimental consequences of microevents and nurturing those random happenings that enhance functioning. Inevitably, the rehabilitation of formal, and especially final, causalities will elicit strong, but misdirected criticism from those who abhor teleology in biology.

3. It should be apparent that *reversibility* has no place in the chaotic world painted by the nominalist. Although, in the organic worldview, formal and final agencies can contain the effects of disturbance and restore functioning, it

does not necessarily follow that the system always can be returned to its original state. Organic systems are not fully reversible; most chance events leave behind finite alterations in systems structure. Organic systems are historical in nature (Brooks and Wiley 1986).

4. Nominalists regard ensembles as *atomistic* by definition. The antithesis to this view is the organic system, which it is assumed can function (exhibit an intrinsic *telos*) only by acting as a whole. The Newtonian ideal, as espoused in elementary systems theory, is to break the system into parts, study the behaviors of the parts in isolation, and to reconstitute the behavior of the whole from the combined descriptions of such atomistic behavior. This stratagem does not seem to apply to biological systems for several reasons. First, the component processes may simply cease to function once an element is separated from the whole. Even if they should continue, the repertoire of component responses that can be elicited in isolation most likely will not include those most relevant when the unit is imbedded in its organic matrix. Finally, any adaptation that a component might undergo when under the selective influence of the ensemble is simply unknowable whenever the element is observed in isolation (Abrams 1996).

5. *Universality* is antithetical to nominalism (stochasticism), and its relevance to organic systems seems dubious as well. The idea that a law or phenomenon formulated within a particular window of time and space should be applicable all the time and everywhere seems limited to circumstances where system parts are rarified and interact weakly. Only a theoretical physicist would dare to imagine the coupling of quantum phenomena, relevant at atomic scales, with the gravitational forces exerted over light-years (Hawking 1988). Ecology teaches its practitioners somewhat more humility. When objects and processes crowd upon one another, it becomes difficult to project the influence of one event at a particular scale over remote domains of time and space. Too many other elements and processes interfere along the way (and it has just been argued how recourse to atomism is unlikely to salvage matters).

Under these considerations the most realistic stance for the organicist appears to be the hierarchical approach (Allen and Starr 1982 [reproduced in this volume]; Salthe 1985). While the organic approach is more likely to encompass rules and laws than is nominalism, organic principles seem to pertain to limited ranges of space and time. One must remain wary of any attempt to stretch explanations across scales, as, for example, when the reasons behind all social behavior are sought in genetics, or when the autonomy of human thought processes is purported to arise in the quantum phenomena occurring in molecules of the brain. Although through the hierarchical perspective one perceives an organic world that is more loosely coupled than the Newtonian clockwork universe, the picture is certainly not one without a modicum of order. Whenever a rule or principle wanes in explanatory power as the scale of observation shifts, another is sure to emerge that maintains structure at the new level. Although such a "granular" view of reality is at odds with the Newtonian vision, it is wholly consistent with the picture of organic causality portrayed above.

Enlarging the Newtonian Edifice

Ecology is not the sole heterodox discipline in science. Dissatisfaction with the rigidity that Newtonianism forces upon the neo-Darwinian synthesis itself has been expressed by Chomsky (1996) and also has been voiced by notable developmental biologists, such as Sidney Brenner and Guenther Stent (Lewin 1984). The late Sir Karl Popper, regarded by many as a conservative philosopher of science, likewise has suggested that practitioners of science need to reconsider their views on basic causality if they are ever to achieve a truly "evolutionary theory of knowledge" (Popper 1990). His own opinion was that the universe is causally open: that chance operates, not only in the netherworld of quantum phenomena, but at all levels (Popper 1982). Popper was no iconoclast, however, and he urged not that we abandon Newtonian forms, but rather that we expand upon them. In his view Newtonian forces are but a small subset of a more general universe of agents he called "propensities." Briefly, a propensity is the tendency for a certain event to occur in a particular context. Propensities are related to, but not identical with, conditional probabilities. Suppose, for example, that one is considering a Newtonian force that relates antecedent A with consequence B. Then every time that A occurs, it is followed by B, without exception. In the language of probabilities one may say that the conditional probability that B will occur, given that A has happened, is unity (1) or certainty. In the larger non-Newtonian world we might observe that whenever A happens, B usually ensues—but not always. Whenever the coupling between A and B is not isolated, interferences can intervene to affect the outcome. Whence, the conditional probability that B will occur, given that A has happened, is usually less than unity. This means that the probabilities for other outcomes, say $C, D, E, \ldots$ are not zero.

As an example, we consider an events frequency table (table 1) that reports the number of times each "cause" $a_1, a_2, a_3,$ or a_4 is followed by any of five possible outcomes $b_1, b_2, b_3, b_4,$ or b_5. From among the one thousand events recorded in table 1, one sees that the "joint probability" that, say, a_1 and b_3 occur together is sixteen out of one thousand events, or 0.016. The joint probability is not the same as the conditional probability, however. To calculate the latter, one must, according to Bayes (1908), normalize the joint probability by the probability, $p(a_1)$, that a_1 occurs under any circumstances. Since a_1 occurs

TABLE 1.
Frequency table of the hypothetical number of joint occurrences that
four "causes" ($a_1 \ldots a_4$) were followed by five "effects" ($b_1 \ldots b_5$).

	b_1	b_2	b_3	b_4	b_5	Sum
a_1	40	193	16	11	9	269
a_2	18	7	0	27	175	227
a_3	104	0	38	118	3	263
a_4	4	6	161	20	50	241
Sum	166	206	215	176	237	1,000

a total of 269 times, the conditional probability that a_1 occurs, given that b_3 has already happened, is the quotient 16/269, or 0.059.

From table 1 it is apparent that whenever a_1 happens, there is a good likelihood that b_2 will follow. Similarly, b_5 is likely to result from a_2 and b_3 from a_4. The situation is less clear as to what ensues from a_3, but b_1 and b_4 are more likely to occur than the rest. The events not mentioned (e.g., (a_1, b_3), (a_1, b_4), etc.) result from what Popper terms "interference." Presumably, if it were possible to isolate phenomena, then each of the 269 times that a_1 happens, it would be followed by b_2. Similarly, if all phenomena could be strictly isolated, the results might look something like those shown in table 2. Under isolation, propensities degenerate into mechanical-like forces, and the associated conditional probabilities approach unity.

Interestingly, b_4 never occurs under isolated (rarified or laboratory) conditions. It arises purely as the result of interferences among propensities. That the propensity associated with this interference phenomenon depends entirely upon its proximity to (ability to interact with) other system propensities is an illustration of Popper's assertion that propensities, unlike forces, never occur in isolation, nor are they inherent in an object. They always arise out of a context, which almost invariably includes other propensities. Thus, Popper concludes that the fall of an apple is a decidedly non-Newtonian event. The

TABLE 2.
Frequency table as in Table 1, except that care was taken to isolate causes from each other.

	b_1	b_2	b_3	b_4	b_5	Sum
a_1	0	269	0	0	0	269
a_2	0	0	0	0	227	227
a_3	263	0	0	0	0	263
a_4	0	0	241	0	0	241
Sum	263	269	241	0	227	1,000

tendency for an apple to fall and where it might land depend, not just upon the weight of the apple and the gravitational constant, but also upon biochemical conditions in the stem, the speed of the blowing wind, etc.

One concludes that whenever propensities occur in propinquity, interferences and new propensities are likely to arise. Conversely, one must add constraints in order to "organize" the more indeterminate configuration represented in table 1 into the more mechanical-like system depicted in table 2. That is, the transition from the loose configuration into its rigid counterpart is an example of what is meant by "organization" (cf. Skyrms 1980; Matsuno 1986). But conditional probabilities by themselves do not quantify propensities in any way that bears analogy with Newtonian laws. Popper was quite aware of this lack of connection, and so he noted simply that "we need to develop a calculus of conditional probabilities." More about the attempt to develop such a calculus presently, but we turn our attention first to elaborating what sort of natural agency conceivably might be behind the transition from the behavior inherent in table 1 to that shown in table 2. What causes systems to grow and develop?

Autocatalysis: A Unitary Agency

A clue to one agency behind growth and development comes from considering what happens when propensities act in close proximity to each other. Any one process will either abet ($+$), diminish ($-$), or not affect (o) another. The second process in turn can have any of the same three effects upon the first. Out of the nine possible combinations for reciprocal actions, one is very different from all the rest—mutualism ($+$, $+$). There is a growing consensus that some form of positive feedback is responsible for much of the order and structure we perceive in living systems (e.g., Eigen 1971; DeAngelis et al. 1986; Haken 1988; Kauffman 1995). Sometimes the positive feedback is considered to take a particular form, such as autopoeisis (Maturana and Varela 1980) or autocatalysis (Ulanowicz 1986, 1997). It is this latter form of mutualism that we wish to consider here.

By autocatalysis is meant any cyclical concatenation of processes wherein each member has the *propensity* to accelerate the activity of the succeeding link. Suppose, for example, that the action of some process A has a propensity to augment a second process B. B, in turn, tends to accelerate a third process C, which then promotes the initial action A. The consequence of any process, when followed around the cycle, is self-stimulation. An ecological example of autocatalysis exists among the biotic associations formed around aquatic plants belonging to the genus *Utricularia* (Ulanowicz 1995b). The surface of the *Utricularia* plant supports the growth of a fast-growing film of diatomaceous algae, generically called "periphyton." This periphyton is consumed by any number of waterborne heterotrophs referred to as "zooplankton." The cycle is completed when the *Utricularia* captures and absorbs many of the zooplankton in small bladders or utricules that are scattered along its featherlike leaves and stems.

It is important to note two things about autocatalytic loops: (1) The members are not always linked in rigid fashion. That is, the action of A does not have to augment that of B at every instant—just most of the time. There is simply a *propensity* for A to augment B. (2) The members of the cycles, being biotic elements and processes, are capable of variation. Whereas autocatalysis in simple chemical systems can justifiably be regarded as a mechanism (discrete stoichiometry of the reactions and simple, unchangeable reactants keep the process mostly mechanical in nature), as soon as chance and variation enter the scene, autocatalysis begins to exhibit some behaviors that are decidedly nonmechanical in nature.

Autocatalysis among indeterminate processes gives rise to a form of selection pressure that the ensemble exerts upon its components. If, for example, some characteristic of process B should change in some chance way, and if that change should either increase the catalytic effect of B upon C or make B more sensitive to catalysis by A, then the change in B will be rewarded and retained. If, however, the change should decrement B's effect upon C or make B less sensitive to A, then B subsequently will receive less support from A and the change is most likely to atrophy. Formally, such selection is unlike what normally is considered under the rubric of natural selection.

In particular, such selection in autocatalytic systems engenders a centripetal movement of material and energy toward the loop itself. For any change in a constituent process that happens to bring in more material or energy to abet that process will be rewarded. This selection applies to any and all elements of the loop, so that the cycle itself becomes the focus of an inward migration of material and energy that is *actively* induced by the kinetic configuration.

But the centripetal flow of resources represents a siphoning of vital materials away from system members that do not engage as effectively in autocatalysis. There will also be competition between autocatalytic loops. The net result is that the topology of the exchanges connecting system elements is gradually pruned of those members that least effectively participate in autocatalysis. At the same time, flows over the links that remain will increase appreciably, due to the acceleration that is inherent in autocatalysis. An example of the growth and development of a network engendered by autocatalysis is depicted schematically in figure 3. In figure 3a the system begins with many largely equiponderant connections. As autocatalysis prunes the system, however, some of the links shrink or disappear, and an overall greater level of activity is channeled

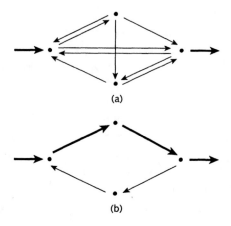

(a)

(b)

Figure 3. Schematic representation of the major effects that autocatalysis exerts upon a system: (a) Original system configuration with numerous equiponderant interactions. (b) Same system after autocatalysis has pruned some interactions, strengthened others, and increased the overall level of system activity (indicated by the thickening of the arrows).

(constrained) along these pathways that most effectively engage in autocatalysis (figure 3b).

Quantifying Growth and Development

Paramount among the advantages of Newtonian science has been that it is strictly quantitative. If Newtonian dynamics are to be extended successfully to an open universe, the effects of autocatalysis just described must be given concrete mathematical expression. Toward this end, we note how the terms *growth* and *development*, although their meanings overlap considerably, nonetheless emphasize different aspects of a unitary process. *Growth* highlights the increase in system size or activity, whereas *development* lays more stress upon the increase in system organization.

The extensive nature of growth is rather easy to quantify. To do so we denote the magnitude of any transfer of material or energy from any donor (prey) i to its receptor (predator) j by T_{ij}. Then one measure of total system activity is the sum of all such exchanges—a quantity referred to in economic theory as the "total system throughput" and denoted simply as T. If reckoning the "size" of a system by its level of activity seems at first a bit strange, one should recall that such is common practice in economic theory, where the size of a country's economy is gauged by its "gross domestic product."

Quantifying the intensive process of development is a good deal more complicated, and involves concepts derived under information theory. The goal is to quantify the transition from a very loosely coupled, highly indeterminate collection of exchanges to one in which exchanges are more constrained along the most efficient pathways. We begin, as did Boltzmann (1872), not by quantifying the constraint directly, but rather by quantifying the opposite notion, that of system *indeterminacy.* Roughly speaking, the indeterminacy is correlated with how surprised the observer will be when B_j occurs. For example, if event B_j is almost certain to happen, its associated probability will be a fraction near 1, and the indeterminacy will be quite small. Conversely, if B_j happens only rarely, the probability of B_j will be a fraction very near zero, and the indeterminacy will become a large positive number. In the latter instance the observer is very surprised to encounter B_j.

By common understanding, constraint removes indeterminacy. Therefore, the indeterminacy of a system with constraints should be less than what it would be in unconstrained circumstances. Suppose, for example, that an *a priori* event A_i exerts some constraint upon whether or not B_j subsequently occurs. Then the conditional probability that B_j will happen in the wake of A_i should give rise to a presumably smaller indeterminacy than if B_j had remained unconstrained. It follows, therefore, that one may use the *decrease in indeterminacy* as one measure of the intensity of the *constraint* that A_i exerts upon B_j. (We note here—without demonstration—that the constraint that A_i exerts upon B_j is formally equal to the constraint that B_j exerts on A_i, that is, one may speak of the *mutual* constraint that A_i and B_j exert on each other.)

One may use this measure of constraint between any arbitrary pair of events A_i and B_j to calculate the amount of constraint inherent in the system as a whole: one simply weights the mutual constraint of each pair of events by the associated joint probability that the two will co-occur and then sums over all possible pairs. This procedure yields a measure of the average mutual constraint, A. In order to calculate the value of A, it is necessary to know (or estimate) all of the exchanges T_{ij} occurring in the system.

That the ensuing A indeed captures the extent of organization created by autocatalysis can be see from the example in figure 4. In figure 4a there is equiprobability that a quantum will find itself in the next time step in any of the four compartments. Little is constraining where medium may flow. The average mutual constraint in this kinetic configuration is appropriately zero. One infers that some constraints are operating in figure 4b, because medium that leaves any compartment can flow to only two other locations. These constraints register some amount k units of A. Finally, figure 4c is maximally constrained. Medium leaving a compartment can flow to one and only one other node, and the average mutual constraint rises to $2k$, which happens to be the maximum possible for a four-compartment system.

The reader may be puzzled why we should have chosen to measure con-

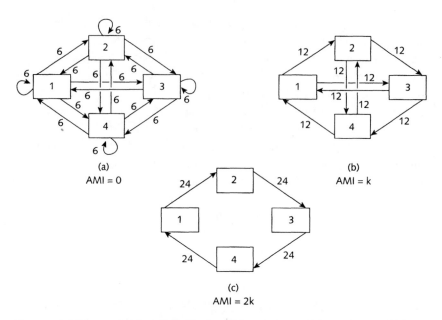

Figure 4. (a) The most equivocal distribution of ninety-six units of transfer among four system components. (b) A more constrained distribution of the same total flow. (c) The maximally constrained pattern of ninety-six units of transfer involving all four components. AMI = Average mutual information.

straint in units of *k*. That is because the formula for the measure (not given here) derives from information theory. The conventional practice in information theory, which heavily involves the use of logarithms, is to designate the base to be used in calculating those logarithms (usually 2, e, or 10) and set the value of $k = 1$. The units of *A* would then appear as "bits," "napiers," or "hartleys," respectively. The problem with this convention is that the calculated value conveys no indication as to the physical size of the system. Conceivably, one could apply the same measure to an ecosystem of microbes in a petri dish and to the ecosystem of the Serengeti Plain and arrive at nearly the same number. The goal here, however, is to capture both the extensive and intensive consequences of autocatalysis in a single measure. One convenient way of introducing size into the measure of constraint is to give physical dimensions to *k* (Tribus and McIrvine 1971). That is, we set $k = T$, and the ensuing dimensions of *A* will contain the units used to measure the exchanges. Now if we apply the scaled measure to the petri dish and the plain, the latter will be orders of magnitude larger than the former.

To signify that the scaled measure is now qualitatively different, we choose to rename *A* as the system "ascendancy" (Ulanowicz 1980). It measures both the *size* and the *organizational status* of the network of exchanges that occur in an ecosystem. In an attempt to characterize what it means for an ecosystem to develop, Eugene Odum (1969) catalogued ecosystem attributes that were observed to change during the course of ecological succession. His list of twenty-four properties can be subgrouped as pertaining to speciation, specialization, internalization, or cycling—all of which tend to increase during system development. But increases in these same four features of network configurations lead, *ceteris paribus*, to increases in ascendancy. Whence, Odum's phenomenology can be quantified and condensed into the following principle: *In the absence of major perturbations, ecosystems exhibit a propensity toward configurations of ever-greater network ascendancy.*

The reader may correctly object that real ecosystems are never free of perturbations, and under many natural conditions a rise in systems ascendancy hardly even seems probable. That is because ascendancy tells only half of the story. A complementary narrative that quantifies the freedom and indeterminacy of a given network can be formulated with the help of other variables from information theory in terms of what has been called the systems "overhead." Components of the system overhead play very important roles in the evolution and sustainability of ecosystems, but elaboration on this topic would only distract from the search here for generalized Newtonian images. Those interested in the persistence of ecosystems are encouraged to read about the significance of adaptability in ecosystems (Conrad 1983) and about the importance of overhead (Ulanowicz 1986, 1997; Ulanowicz and Norden 1990).

Propensities as Generalized Forces

Returning to the notion of propensity, the reader may recall how propensity was explicitly folded into the definition of autocatalysis. It is reasonable to expect, then, that one can identify the propensities within the definition of the ascendancy. As it turns out, the ascendancy can be shown to be the flow-weighted sum of the propensities of all the processes that occur within the system (Ulanowicz 1996). In other words, if one multiplies each flow by its associated propensity and sums all such products, the result is the system ascendancy.

To anyone with a passing familiarity [with] irreversible thermodynamics, the procedure just described should be very familiar. It is exactly how one calculates the total dissipation in an ensemble of processes. For, very near to thermodynamic equilibrium, the theory of irreversible thermodynamics postulates that, for each observable process or flow, one may identify a conjugate "thermodynamic force." For example, mass diffusion is assumed to flow "downhill" along any spatial gradient in the chemical potential of the medium in question. Similarly, thermal conduction flows down a gradient in temperature of the conducting medium, electrical current follows a gradient in voltage, chemical reaction occurs in response to a difference in Gibbs free energy, etc. It is assumed that these "forces" all can be cast in appropriate dimensions such that the product of each flow times its conjugate "force" yields a product with the dimensions of power (Onsager 1931). The sum of all such products pertaining to each process in a system yields what is known as the power (or dissipation) function, which is believed to characterize the overall dynamics of the system. Prigogine (1945), for example, has hypothesized that, very near to equilibrium, any ensemble of processes takes on the particular configuration that results in the smallest value of the dissipation function. In the generalized formulation the role of the thermodynamic forces comes to be occupied by the propensities.

Popper (1990) appealed for the development of "a calculus of conditional probabilities." We note here that it is conditional probabilities, not marginal or unconditional probabilities, that are most germane to the meaning of information (Tribus and McIrvine 1971). Hence, I wish to suggest that information theory already satisfies Popper's desiderata for a calculus of conditional probabilities (Ulanowicz 1996).

It is worth noting that by deriving an explicit formula for Popper's propensities, it renders his concept fully operational whenever all the T_{ij}'s in a system are known. For example, Ulanowicz and Baird (1999) estimated the transfers of carbon, nitrogen, and phosphorus among the major taxa of the ecosystem inhabiting the mesohaline reach of Chesapeake Bay and utilized the resulting propensities to identify those exchanges that most influence the nutrient dynamics of that community.

In the Image of Newton

This discovery of a convenient thermodynamical analogy opens one's eyes to still further analogs. We recall that the measure of constraint which a donor (A_i) exerts on its receptor (B_j) is equal in value to that which the receptor imposes upon the donor. That is, Newton's third law (symmetry) has its counterpart in our probabilistic reformulation. (N.b., the propensities, p_{ij} themselves are *not* symmetric with respect to donor and recipient, any more than the force with which one boxer striking a second is obliged by Newton's third law to equal the force with which the second might simultaneously strike the first.)

Still further analogies appear. Reconsideration of the formula for A in light of the new identification of propensities reveals that the community ascendancy takes the form of a flow-weighted average propensity for the system as a whole. Since A is itself a propensity, Odum's phenomenological principle can be restated as: *In the absence of major perturbations, there is a propensity for the flow-weighted ecosystem propensity to increase in value.* The "propensity of a propensity" relationship appearing in the restated principle is reminiscent of Newton's second law, which deals with the time rate of change of momentum. Momentum, in its turn, is a time rate of change of position, i.e., the second law treats the rate of change of a rate of change (Ulanowicz 1998). If at first this resemblance should seem a bit far-fetched, then one should pause to consider the contexts of Newton's first and second laws. As mentioned earlier, the first law tells what happens in the absence of external influence—the rate of change of the rate of change is identically zero. With mechanical systems, no input means no change. This differs from ecological phenomenology, which indicates that, in the absence of external disturbance, the propensity of the system propensity assumes a positive value. An organic system can exhibit change from within. This is a radical departure from Darwin, who, as intellectual grandson of Newton, took great pains to locate selection pressure *outside* the developing system (Depew and Weber 1994).

Newton's second law states that, in response to an external interference, the system will follow the disturbance, i.e., the response is positive. In the biological realm, after any immediate negative impact that a disturbance might have on an organic system, the system response looks at first glance qualitatively indistinguishable from how it behaves when the ensemble is unperturbed, i.e., there is again a propensity for the system to increase in ascendancy. But closer scrutiny of the usual behaviors in perturbed and unperturbed situations reveals that the dynamics in each case differ significantly.

For one, the scales of the responses with and without intervention usually differ markedly. What happens after intervention (the analog to the second law) is likely to occur rapidly and in such a way that the interference is absorbed locally and with minimal dissipation (Lubashevskii and Gafiychuk 1995). Quite often, the operative control is via localized negative feedbacks. This response has been labeled "self-regulation" (Gafiychuk and Ulanowicz 1996) and is depicted as a Venn diagram in figure 5. The system A had adapted to some extent to its environment, B, as represented by the intersection between A and B. Suddenly, the environment changes from B to B', and some structural constraint is irretrievably lost, as indicated by the dotted region on the diagram. Adaptation ensues, and ascendancy increases, however, by progressively greater overlap with the new environment, as indicated by the striped area on figure 5.

A system in relative isolation (all living systems must remain open to some degree) will undergo "development" in a way that is qualitatively different from recovery from intervention. It will develop slowly over time and usually

AFTER INTERVENTION

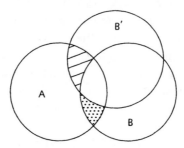

Figure 5. Venn diagram depicting the process of self-regulation. System A had accommodated to environment B to an extent indicated by the overlap of the two circles. Intervention brings about a changed environment, represented by B'. Some modes of accommodation are irretrievably lost (dotted area), whereas regulation proceeds by expanding the striped area (and the ascendency) as much as feasible.

involve positive feedbacks that span most of the system. Usually, development is accompanied by progressively more overall dissipation (Ulanowicz and Hannon 1987). In terms of Venn diagrams this slower transition resembles that shown in figure 6.

An Ecological Metaphysic

Just as Schroedinger used the form of Newton's second law as a point from which to embark upon an entirely unmechanical view of the submicroscopic world, we have just discussed formal connections between the laws that Newton himself exposited and at least one version of contemporary

RELATIVE ISOLATION

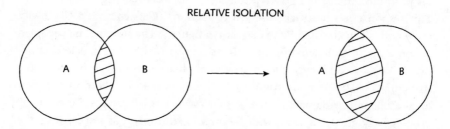

Figure 6. Venn diagram depicting the process of semiautonomous development when the system is in relative isolation from major perturbations. (a) System A and environment B at a given time. (b) System at later time has developed to harmonize more with its environment, as indicated by increased overlap between A and B.

ecosystems science. It does not follow, however, that the assumptions that were added by later practitioners of Newtonian science translate as well into the biological realm, Darwin and his successors notwithstanding. If one wishes to understand the development of biological systems in full hierarchical detail and is not content with the abrupt juxtaposition of pure stochasticity and determinism found in neo-Darwinism (Ulanowicz 1997), then one must abandon the assumptions of closure, determinism, universality, reversibility, and atomism and replace them by the ideas of openness, contingency, granularity, historicity, and organicism, respectively. That is, one must formulate a new metaphysic for how to view living phenomena. By way of summarizing the foregoing exposition, I propose below a set of rough counterparts to the five elements of the Newtonian metaphysic elaborated earlier (albeit in a different order):

1. Ecosystems are ontologically *open:* Indeterminacies, or "genetic events," can arise anytime, at any scale. Mechanical, or efficient, causes usually originate at scales inferior to that of observation and propagate upward; formal agencies appear at the focal level; and final causes arise at higher levels and propagate downward (Salthe 1985; Ulanowicz 1997).

2. Ecosystems are *contingent* in nature: Biotic actions resemble propensities more than mechanical forces.

3. The realm of ecological phenomena is *granular* (Allen and Starr 1982 [reproduced in this volume]) in the hierarchical sense of the word: An event at any one scale can affect matters at other scales only with a magnitude that diminishes as the scale of the effect becomes farther removed from that of the eliciting event. It follows that genetic events at lower levels do not propagate unimpeded up the hierarchical levels, because they become subject to constraint and selection by formal and final agencies extant at higher levels.

4. Ecosystems are *historical* entities: Genetic events often constitute discontinuities in the behaviors of systems in which they occur. As such, they engender irreversibility and degrade predictability. The effects of genetic events are retained in the material and kinetic forms that result from adaptation. The interactions of propensities in organic systems create a more likely direction or telos in which the system develops.

5. Ecosystems are *organic:* Genetic events often appear simultaneously at several levels. Propensities never exist in isolation from their context, which includes other propensities. Propensities in communication grow progressively more interdependent, so that the observation of any part in isolation (if possible) reveals ever less about its behavior when acting within the ensemble.

None of the foregoing statements is entirely new to ecology. Most have appeared in the literature, either singly or in combination with several others. The intention here has been to portray each element as part of a complete and coherent framework for viewing the ecological world. It is possible the meta-physic could pertain as well to the broader biological and social sciences.

It is appropriate to note in closing that the indefinite article appears in the title of this essay modifying the word *metaphysic*. No one is pretending to have developed "the" metaphysic for ecology, much less for all higher-level phe-nomena. Other combinations may be possible; however, the encouraging fea-ture of the perspective just formulated is that it appears to reconcile disparate schools of ecological thought into one overarching, coherent structure. As a unified vision, it offers the promise for a fecund, new outlook that, it is hoped, will elicit more penetrating insights into ecosystem behaviors.

PART 2

Community, Niche, Diversity, and Stability

There is a popular saying that ecology has few, if any, principles, but many concepts. Frequently ecologists find hunting for exceptions to generalizations to be good sport, but equally frequently the quarry escapes the hunter because the argument often hinges on a definition of terms. Since ecology has a complex history of being formed from several sciences, its language can be ambiguous and confusing. Concepts are loose enough to be the subject of this type of sniping, but, of course, they are exceedingly useful because they bridge many observations and link interpretations.

The relationship between a concept and a principle is not a matter of either/or; rather, these two forms of generalization represent a continuum. A principle has withstood tests and is widely recognized as certain in an empirical sense. The areas of application of the principle usually are well understood. In contrast, a concept is a generalization based on fewer cases than a principle, and seldom are concepts tested through experimentation or formal observation. A concept represents the usual circumstances and may be weak in certain applications. We are less sure of a concept, although it represents a useful generalization that can guide research and application. Well-established concepts may even be recognized and used outside the field, although with qualifications.

For part 2, we have chosen four concepts that are central to ecological science, being generalizations of long standing or answers to fundamental definitional problems of ecology. These concepts are community, niche, diversity, and stability. They represent only a few of the many generalizations used in the science.

The Community Concept

The word *community* means "entities having interests or characteristics in common." In human terms *community* often refers to those who live together in a village or city and share a common space. The study of human communities has always been a central theme in human ecology, just as the study of plant and animal communities has been a recurring feature of the ecology of nature.

Early ecologists automatically adopted the term *community* for assemblages of plants and animals. For example, Anton Kerner von Marilaun, professor of botany at the University of Vienna, reported on his travels through the Danube Basin in 1863: "Wherever the reign of nature is not disturbed by human interference the different plant-species join together in communities, each of which has a characteristic form, and constitutes a feature in the landscape of which it is part" (1951). In the late nineteenth century, Karl Möbius, professor of zoology at Kiel, invented the word *biocœnosis* (meaning "biota living together") for the living organisms on an oyster reef in the Baltic Sea. In "An Oyster Bank Is a Biocönose, or a Social Community" (1881), reprinted here as chapter 6, Möbius describes plants and animals in terms of a single biotic community, recognizing the interconnections between taxa representing different trophic groups.

These two examples illustrate the two main options in community analysis. One might focus on the taxa, as Kerner did, or on the area shared by the taxa, as Möbius did. These different foci yield different approaches to community ecology because they lead toward problems of identification and quantification of taxa, or toward questions of the ways species and individuals share space, their trophic relationships, and so on.

Initially ecologists in their description of diversity merely listed the plants or animals observed within a habitat. But the invention of numerical methods of sampling meant that quantitative representation of species and the abundance of individuals could be in statistical form. Standardized sampling led toward theories of abundance.

Robert McIntosh, in *The Background of Ecology* (1985), traces the development of the numerical theme in community studies. As far back as 1789, Gilbert White, the naturalist parson of Selborne, England, recognized that the larger the area examined, the more species would be found (White 1981). The relationship describes a curvilinear pattern in which new species are encountered rapidly early in the sampling, but as the sample becomes larger, fewer new species appear (figure 7). Eventually all the species present in the community appear in the sample and the curve levels off. There are exceptions. Tropical rain forests' species-area curves sometimes continue to increase. Highly polluted habitats may exhibit declining diversity since fewer species can live in the inhospitable conditions.

August Thienemann (1939), director of the German Limnological Station at Plön, observed that the longer conditions at a site remained the same, the more species were present and the more stable the community. Thus there is a connection between stability and biological diversity.

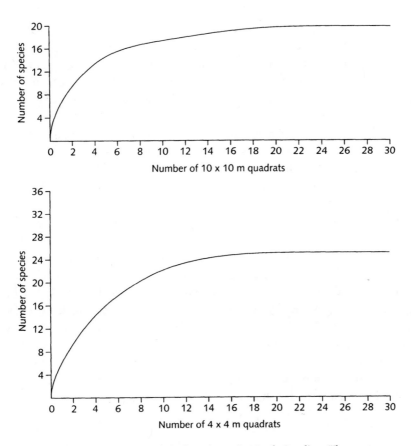

Figure 7. Species-area curves for oak-hickory forests in North Carolina. The upper curve represents trees sampled on 10 × 10–meter plots. The lower curve represents shrubs sampled on 4 × 4–meter plots. (From Oosting 1950.)

The resonance between the concepts of the human community and the natural community that must have influenced the language of ecology is clearer when one reflects on the history of human settlements. At the smallest scale, human adaptation is to microscale environmental conditions. The dwellings of a small village seem to be placed higgledy-piggledy with little relationship to other dwellings. At this level of scale the human has so little control over the environment that it is only through careful observation of the habitat that permanent habitation of a site becomes possible. Tiny raised areas are chosen and tiny depressions avoided to ensure a dry house site. As the settlement grows in scale, space becomes more organized until the stage of the large town or city is reached. At this stage, abstract principles of architecture and design can be used to organize activities and habitations. The classical design of Washington, D.C., by Pierre L'Enfant and the imperial Chinese cities designed to represent the physical body of the Buddha are illustrations. At the time ecological science was born, ecologists had all scales of settlement visible before them. Now that is no longer true, especially in developed countries. For example, in the United States even the smallest space is designed on geometrical principles, and the natural environment has relatively little influence on the design of the site or the buildings.

Community ecology shares with human ecology some of these issues. First, the focus in community ecology is on the biota (organized by species), not on the interaction of the biota and its environment, just as the focus in human ecology is on humans. Second, the community is often treated as a bounded entity. Third, the community is sometimes treated as having a single scale. Fourth, the presence of species or individuals is assumed to have causes that come from the life history of the organism or the environment. All of these assumptions raise questions that lead to research. How does selection operate on the individuals and species within a community? Can we recognize community boundaries? How is the microenvironmental scale treated? How do species respond to the diversity of environments within the habitat occupied by a community? Is the presence of species due to chance (à la Gleason) or are there design rules that lead to the most efficient or most powerful combinations of species (à la Clements)? Since biotic communities may contain thousands of species, unpacking communities and understanding the ways they are organized is a formidable challenge.

Niche

The ecologist's concept of niche is an invention of the science itself. It has, as far as we know, no deeper origin. It is concerned with a fundamental problem in ecology: how to integrate the two parts of the ecological system: the biotic element and the environmental element. The puzzle has three parts, as we have described earlier: the biotic, the environment, and the links between biota and environment. Each element in the triad is highly dynamic and changes over space/time. For example, the biota, through sexual reproduction and genetic processes, produce phenotypes that vary in fitness. The environment, through variation in the energy dynamic of the planet, changes over space/time. As a consequence, the links between the biota and the environment vary as well. There are few satisfactory models that allow the ecologist to relate the individual parts of the triad to the whole ecological system.

In "Niche, Habitat and Ecotope" (1973), Robert Whittaker, Simon Levin, and Richard Root review the history of the concept of niche; and in "On the Reasons for Distinguishing *Niche, Habitat,* and *Ecotope*" (1975), reprinted here as chapter 7, they provide a rationale for distinguishing the three concepts.

Initially, the variable in the life history of the biota that reflected its dynamic character was the capacity of organisms to move and occupy space. The distribution of organisms was related to environmental conditions. Together, these patterns represented the niche of the organism. Some ecologists equated the niche to the habitat and its properties. Other ecologists, especially those concerned with food as a resource, associated the niche with the organism.

These conversations led to a variety of insights. For example, the movement of a species into a new habitat, as occurs when an exotic animal appears for the first time on a continent, sometimes results in a successful invasion, leading eventually to its incorporation into the fauna or flora. In this case the ecologist might claim that the species moved into a vacant niche that existed in the habitat. *Niche* used in this sense seems to be a property of the place.

The Russian scientist G. F. Gause (1964) investigated the competition between species of yeast in closed chambers. In these experiments both species competed for the same resource, and one species eventually won out over the other. Gause's experiments were interpreted to mean that two species cannot occupy the same niche. In this sense niche is associated with a place but also reflects the species' ability to utilize the resources in that place. This usage comes closer to representing an integrated concept.

Finally, the English animal ecologist Charles Elton (1927, 1930), who was concerned with the feeding relationships of species, suggested that *niche* should be used in the same way that *profession* is used in everyday English, as when we observe someone walking in the street and say, "There goes the vicar." The word *vicar* conveys to the listener a particular picture of a person (especially, in this case, if one is British). Niche is the "profession" of a species. *Niche* used in the Eltonian sense is attached mainly to the species.

The modern ecologist knows of many more processes that describe the dynamic properties of the biota and the physical environment than the distribution of organisms or their feeding rates. The genotypic differences that underlie differences in the phenotype; the life history strategies that make use of cooperative, competitive, parasitic, predatory, and other relations; and our concepts of climatic and biogeochemical patterns may all contribute to a modern concept of niche. In modern ecological research, the challenge is to integrate the separate elements into a system in order to predict the success or failure of a species in an environment.

Biological Diversity

Awareness of the diversity of life is ancient. Clarence Glacken (1967: 5) attributes the term to Arthur Lovejoy (1964), who traced the term to Plato's *Timaeus* (1982). The modern concept of diversity is related to the ancient idea of plenitude. Glacken comments: "The principle of plenitude thus presupposes a richness, an expansiveness of life, a tendency to fill up, so to speak, the empty niches of nature: implicit in it is the recognition of the great variety of life and perhaps its tendency to multiply. When the principle of plenitude was fused with the Aristotelian idea of continuity, the richness and fecundity of all life was seen as manifesting itself in a scale of being from the lowest to the highest forms, and revealing itself in a visible order of nature" (6). Glacken continues by noting the importance of the concept in the Christian interpretation of nature, in natural history, and in Thomas Malthus's theory of population (1926), especially in his emphasis on fertility. One of the characteristics of Western thought about nature has been the emphasis on nature's fecundity, its capacity to multiply and overcome drought, deluge, plague, and destructive events. Many writers have observed that if a single species were allowed to breed freely, it could, given sufficient time, cover the entire earth. Darwin used this idea in his observation that even slow-breeding elephants could "stock the world in a few thousand years."

Thus, the ancient idea of plenitude has two parts. First, it recognizes the enormous variety of life that we observe in nature. And, second, it explains this variety by the tendency to propagate. If unchecked, the proliferation of life would have no end. These themes are directly relevant to ecology because the ecologist is interested in how the plenitude of life is checked and controlled by environmental factors.

When ecologists use the word *diversity,* they refer to taxonomic variety and the number of species in a community, area, or sample. As Ruth Patrick observes in "Biological Diversity in Ecology" (1983), reprinted here as chapter 8, ecologists also have applied the term to the variety of functions in an ecosystem or to the variety of ecosystems in a landscape. *Diversity* also has become a widely used word in conservation ecology as it relates to concerns about the loss of species through extirpation and extinction due to human activity. Biological diversity is one of the major interests of the world conservation movement.

While ecologists consider diversity an important parameter and an indicator of the health and well-being of a biotic community, diversity is difficult to measure. Very few complete censuses have been made in communities (though in unusual situations, such as hot springs, the diversity is often greatly reduced and as a consequence may have been completely sampled). Overall, we lack a good understanding of the size and variety of the biota in our forests, grasslands, and rivers. The problem with inventories usually lies with the microorganisms, such as bacteria and protozoa. In some instances we are not even certain if the species concept can be applied to these organisms. Edward O. Wilson, citing a Norwegian study of a single gram of beech forest soil in which between four and five thousand species were found, says: "The bacteria await biologists as the black hole of taxonomy. Few scientists have even tried to dream of how all that diversity can be assayed and used" (1992: 148). As a consequence, diversity studies often focus on groups of organisms that can be sampled by conventional techniques, such as plants in quadrats or insects in Malaise traps. Ecologists assume that the patterns demonstrated for these easily recognized partial samples represent the whole community—that, for example, the diversity of herbivores is related to the diversity of the plants that serve as their food. But it is difficult to test such assumptions.

Stability

Stability is a fundamental concept in ecology that touches upon essentially every research project and crops up in every textbook. The idea that the earth is elegantly teleological comes from the earliest recorded statements of the ancient world. In Glacken's words:

> Geographically, it was a most important idea: if there were harmonious relationships in nature . . . the spatial distribution of plants, animals and man conformed to and gave evidence of this plan; there was a place for everything and everything was in its place . . . the idea of a design with all its parts well in place and adapted to one another in an all-embracing harmony implied stability and permanence; nature and human activity within it were a great mosaic, full of life and vigor, conflict and beauty, its harmony persisting among the myriads of individual permutations, an underlying stability. (1967: 147–48)

An orderly and harmonious cosmos is a stable cosmos.

Frank Egerton, who has made the most complete study of the balance of nature, comments that in ecology the concept has been a "background assumption rather than a hypothesis or theory" (1973: 324). And, as a consequence, the concept has been poorly articulated and defined and therefore is seldom examined explicitly. Like Daniel Simberloff (1980 [reproduced in this volume]), Egerton pins the origin of the idea of teleology to Greek philosophy and science which assumed that nature was "constant and harmonious." In the *History*, Herodotus (1862), for example, notes the correlation between reproductive capacity and the habits of species, which were evidence for the design of nature by divine Providence. In the *Timaeus*, Plato (1982) accounts for the creation of the universe by an intelligent being, meaning that the universe itself has an intelligible design. These classical ideas, supported by evidence from natural history, provided the basis for the Christian concept of an orderly nature operating by divine plan. Much of theoretical ecology's emphasis on stability and equilibrium, in the words of philosopher Mark Sagoff, "blurs the line between science and religion" (1997: 888).

As Egerton shows, the balance of nature is at the core of Carolus Linnaeus's famous concept of an "economy of nature." In a 1744 essay, Linnaeus attempts to explain how the world was stocked with plants and animals. Linnaeus imagines the Garden of Eden as a tropical mountain with Arctic species at the top, temperate species in the middle, and tropical species at the bottom, in a pattern familiar to ecologists as life zones. Species in these zones increased in number and spread out over the earth. In a later (1749) essay, he speculates that the "economy of nature" is maintained by the propagation, preservation, and destruction of plants and animals. The balance between these regulatory functions became an organizing theory of the new science. Frederic Clements's superorganism ontology (1916 [reproduced in this volume]) is founded on the *a priori* assumption of design and balance in nature.

In contemporary ecology the balance-of-nature concept has been largely rejected. Charles Elton is perfectly clear: "'The balance of nature' does not exist and perhaps never has existed" (1930). The problem is that species vary continuously in time and space and their regulation comes from factors both within and without the organism. The notion of specific variation with each case study fits the nature of observation in a postmodern world in which relativity is a major explanatory element. Yet the concept of balance or stability continues to reemerge. The field naturalist and ecologist continue to find in nature order and continuity at one scale and variety and change at other scales. For example, Eugene Odum states, "Questions of stability versus aging of mature systems may be academic in many situations where disease, storms, fires and so on hasten the death of the community at or before climax and start a new cycle of several stages. . . . But acute perturbations can also be stabilizing if they occur in the form of regular pulses that can be utilized by adapted species as an extra energy subsidy" (1993: 203, 202).

So, even though the teleological view of nature has been severely criticized, the connection between biotic diversity and system stability remains an important theme in ecology. David Tilman and John Downing (1994) have demonstrated that ecosystems with high species diversity are more likely to contain species that can do well under a perturbed environment and will compensate for species that are negatively affected. In their studies of grasslands affected by an unusual drought, they found that drought resistance of the system was significantly related to the predrought species richness of the community. Both resistance to drought and resilience, or the capacity to recover from drought, were related nonlinearly to species richness.

Thus, the expression of a modern concept of the balance of nature or system stability is still alive in its general sense. But it is changed from the original definition because it recognizes change and variation; and as Odum and Tillman and Downing suggest, change maintains balance. In "Stability in Ecological Communities" (1987), reprinted here as chapter 9, Andrew Redfearn and

Stuart Pimm identify five different meanings of *stability* and demonstrate how the early stability-diversity of Elton (1958) and Robert MacArthur (1955)—i.e., that more complex communities are more stable than simpler ones—has been supplanted by the view "that there is nothing inherently unstable about simple systems."

Where some ecologists see a degree of balance or stability in nature, others see chaos. Chaos theory derives from the mathematics of meteorology. It proposes that complex systems, such as ecosystems, behave in unpredictable, nonlinear, nondeterministic ways. Chaos theory replaces conventional stability or equilibrium theory with nonlinear, nonequilibrium behavior. The behavioral sequences of such systems partly depend upon the initial conditions. Very small differences, which were ignored as unimportant to system behavior under earlier theory, in chaos theory are recognized as causing widely different system patterns. Entropicists also believe chaos is creative. James Gleick, for example, states: "Unpredictability was only the attention grabber. Those studying chaotic dynamics discovered that the disorderly behavior of simple systems acted as a creative process. It generated complexity; richly organized patterns, sometimes stable and sometimes unstable, sometimes finite and sometimes infinite, but always with the fascination of living things" (1987: 43). Stuart Kauffman (1995) suggests that a new kind of order emerges at "the edge of chaos"—to use Christopher Langton's (1989) phrase—between fixed deterministic order, which is a static or deathlike condition, and chaos, which is unpredictable behavior. Between these two conditions, patches of order in a chaotic landscape create conditions for new forms of dynamic order to emerge, exist, and eventually disappear.

Chaos theory has not yet been applied to the problems of community theory except in a negative and critical way. If we are to find a new model that fits our explanation more effectively, then it must also be expressed in ways that fit both our scientific experience and our personal observations. Until that happens we have a job of translation before us.

Conclusion

Consideration of the background and history of several concepts in wide use in ecology indicates that there is a rich area of philosophic exploration within ecological science. Some of the concepts have deep roots that take us back to antiquity. The ideas of plenitude and order continue to inform our interpretations of nature. In addition, however, ecologists have invented new concepts to describe or make clear their insights. The concept of niche is an example of this theme. Other concepts are created by analogy from other sciences or from human experience generally. A case in point is the community concept in ecology. Möbius was among the first to write about the community, but he gave it a unique name, the *biocönose*, to distinguish the ecological community from the human community. Thus, ecological concepts are grounded in many ways other than the evolutionary, systems, and natural history themes discussed in the earlier chapters. Each concept deserves analysis, which will lead to greater precision in thought and method.

CHAPTER 6

An Oyster Bank Is a Biocönose, or a Social Community

Karl Möbius

The history of the impoverishment of the French oyster-beds is very instructive. When the beds of Cancale had been nearly deprived of all their oysters, by reason of excessive fishing, with no protection, the cockle (*Cardium edule*) came in and occupied them in place of the oyster; and vast hordes of edible mussels (*Mytilus edulis*) under similar circumstances appeared upon the exhausted beds near Rochefort, Marennes, and the island of Oléron. The territory of an oyster-bed is not inhabited by oysters alone but also by other animals. Over the Schleswig-Holstein sea-flats, and also along the mouths of English rivers, I have observed that the oyster-beds are richer in all kinds of animal life than any other portion of the sea-bottom. As soon as the oyster-men have emptied out a full dredge upon the deck of their vessel, one can see nimble pocket-crabs (*Carcinus moenas*) and slow horn-crabs (*Hyas aranea*) begin to work their way out of the heap of shells and living oysters, and try to get to the water once more. Old abandoned snail-shells begin to move about, caused by the hermit-crabs (*Pagarus bernhardus*), which have taken up their residence in them, trying to creep out of the heap with their dwelling. Spiral-shelled snails (*Buccinum undatum*) stretch their bodies as far out of the shell as they can, and twist from side to side, trying, with all their power, to roll themselves once more into the water. Red starfish (*Asteracanthion rubens*), with five broad arms, lie flat upon the deck, not moving from the place, although their hundreds of bottle-shaped feet are in constant motion. Sea-urchins (*Echinus miliaris*), of the size of a small apple, bristling with greenish spines, lie motionless in the heap. Here and there a ring-worm (*Nereis pelagica*), of a changeable bluish color, slips out of the mass of partially dead, partially living, animals. Black edible mussels (*Mytilus edulis*) and white cockles (*Cardium edule*) lie there with shells as firmly closed as are those of the oysters.

From *The Oyster and Oyster-Culture*, trans. H. J. Rice. In *Documents of the Senate of the United States for the Third Session of the Forty-sixth Congress and the Special Session of the Forty-seventh Congress (1880–1881)*, pp. 721–24. Notes omitted.

Even the shells of the living oysters are inhabited. Barnacles (*Balanus crenatus*), with tent-shaped, calcareous shells and tendril-shaped feet, often cover the entire surface of one of the valves. Frequently the shells are bedecked with yellowish tassels a span or more in length, each of which is a community of thousands of small gelatinous bryozoa (*Alcyonidium gelatinosum*), or they are overgrown by a yellowish sponge (*Halichondria panicea*), whose soft tissue contains fine silicious spicules. Upon many beds the oysters are covered with thick clumps of sand which are composed of the tubes of small worms (*Sabellaria anglica*). These tubes, called "sand-rolls," resemble organ-pipes, and are formed from grains of sand cemented into shape by means of slime from the skin of the worm. The shell forms a firm support upon which the worms can thus live close together in a social community. Upon certain beds near the south point of the island of Sylt, where the finest-flavored oysters of our sea-flats are to be found, there lives upon the oyster-shells a species of tube-worm (*Pomatoceros triqueter*) whose white, calcareous, three-sided tube is very often twisted about like a great italic s. The shells of many oysters upon these beds also carry what are called "sea-hands" (*Alcyonium digitatum*), which are white or yellow communities of polyps of the size and shape of a clumsy glove. Often the oyster-shells are also covered over with a brownish, clod-like mass, which consists of branched polyps (*Eudendrium rameum* and *Sertularia pumila*), or they may be covered with tassels of yellow stems which are nearly a finger long and have at their distal ends reddish polyp-heads (*Tubularia indivisa*). Among these polyps, and extending out beyond them, are longer stems, which bear light yellow or brown polyp-cups (*Sertularia argentea*). Within the substance of the shell itself animals are also found. Very often the shells are penetrated from the outside to the innermost layer, upon which the mantle of the living oyster lies, by a boring sponge (*Clione cleata*), and in the spaces between the layers of the shell in old oysters is found a greenish-brown worm (*Dodecaceraea concharum*), armed with bristles, and bearing twelve large tentacles upon its neck. I once took off and counted, one by one, all the animals living upon two oysters. Upon one I found 104 and upon the other 221 animals of three different species. The dredge also at times brings up fish, although it is not very well adapted for catching them. Soles (*Platessa vulgaris*), which seek by jumping to get out of the vessel and once more into the water, stone-picks (*Aspidophorus cataphractus*), and sting-rays (*Raja clavata*), which strike about with their tails, are abundant upon the oyster-banks. Besides those already mentioned, there are many other larger animals which are taken less frequently in the dredge. There are also a host of smaller animals covered up by the larger ones, and which can be seen only with a magnifying glass. Very few plants grow upon the banks. Upon only a single one of the oyster-beds of the sea-flats has eel-grass (*Zostera marina*) taken root. Upon other beds reddish-brown algae (*Floridiae*) are found, and, floating in the water which flows over the beds, occur micro-scopic algae (*Desmidiae* and *Diatomaceae*), which serve as nourishment to the oysters. If the dredge is thrown out and dragged over the sea-flats between the oyster-beds, fewer and also different animals will be found upon this muddy bottom than upon the sand. Every oyster-bed is thus, to a certain degree, a community of living beings, a collection of species, and a massing of individ-uals, which find here everything necessary for their growth and continuance, such as suitable soil, sufficient food, the requisite percentage of salt, and a temperature favorable to their development. Each species which lives here is represented by the greatest number of individuals which can grow to maturity subject to the conditions which surround them, for among all species the num-ber of individuals which arrive at maturity at each breeding period is much smaller than the number of germs produced at that time. The total number of mature individuals of all the species living together in any region is the sum of the survivors of all the germs which have been produced at all past breeding or brood periods; and this sum of matured germs represents a certain quantum of life which enters into a certain number of individuals, and which, as does all life, gains permanence by means of transmission. Science possesses, as yet, no word by which such a community of living beings may be designated; no word for a community where the sum of species and individuals, being mu-tually limited and selected under the average external conditions of life, have, by means of transmission, continued in possession of a certain definite terri-tory. I propose the word *biocœnosis* for such a community. Any change in any of the relative factors of a biocönose produces changes in other factors of the same. If, at any time, one of the external conditions of life should deviate for a long time from its ordinary mean, the entire biocönose, or community, would be transformed. It would also be transformed, if the number of individuals of a particular species increased or diminished through the instrumentality of man, or if one species entirely disappeared from, or a new species entered into, the community. When the rich beds of Cancale, Rochefort, Marennes, and Oléron were deprived of great masses of oysters, the young broods of the cockles and edible mussels which lived there had more space upon which to settle, and there was more food at their disposal than before, hence a greater number were enabled to arrive at maturity than in former times. The bio-cönose of those French oyster-banks was thus entirely changed by means of over fishing, and oysters cannot again cover the ground of these beds with such vast numbers as formerly until the cockles and edible mussels are again reduced in number to their former restricted limits, because the ground is already occupied and the food all appropriated. The biocönose allows itself to be transformed in favor of the oyster, by taking away the mussels mentioned above, and at the same time protecting the oysters so that the young may become securely established in the place thus made free for them. Space and food are necessary as the first requisites of every social community, even in the great seas. Oyster-beds are formed only upon firm ground which is free from mud, and if upon such ground the young swarming oysters become attached in great numbers close together, as happened upon the artificial receptacles in the Bay of Saint Brieux, their growth is very much impeded, since the shell of one soon comes in contact with that of another, and they are thus unable to grow with perfect freedom. Not only are they impeded in growth in this manner, but each oyster can obtain less nourishment when placed close together than when lying far apart.

CHAPTER 7

On the Reasons for Distinguishing
Niche, Habitat, and *Ecotope*

Robert H. Whittaker, Simon A. Levin,
and Richard B. Root

In response to George Kulesza's (1975) comment [on Whittaker et al. 1973], we first restate the concepts and their relations to one another. The ecotope "describes the species' response to the full range of environmental variables to which it is exposed" and "is the ultimate evolutionary context of a species. . . . Species' distributions over ranges of habitats and migrations between communities . . . are to be understood in terms of the ecotope. The niche may moreover be regarded as the restriction of the ecotope to a particular community, however that community is defined" (Whittaker et al. 1973: 334). Niche refers to the functional relationships of a species within a community (ibid.: 332), and habitat to its distributional response to environmental factors at different points in the landscape (ibid.: 328).

Many investigations focus on the species within a community (it is in this connection that the term *niche* is most often used). Therefore, it is useful to identify the factors within a community to which species respond as "niche variables." Similarly, another tradition emphasizes the distributions of species over the landscape. When the points in the landscape are arranged along gradients of environmental factors, the distribution of a species can be analyzed in terms of "habitat variables." Clearly niche and habitat variables intergrade; distinction between them depends largely on the investigator's scale of consideration. We agree with Kulesza's point that temperature, for example, is not simply a niche or habitat factor. If one takes a forest stand as the unit of study, the temperature differences in the different strata of the community, and the daily and seasonal temperature changes to which species respond, are niche variables. Conversely, if the scale of study is larger than a community (e.g., an elevation gradient within mountains), temperature changes that characterize different environments are habitat variables.

There is indeed no discontinuity between the two groups of variables, as we

From *American Naturalist* 109 (1975): 479–82.

have sought to indicate (Whittaker et al. 1973: 327, 334). However, the fact that niche and habitat factors sometimes overlap does not in itself argue against distinguishing the concepts. A great many of the concepts of science and other human discourse are based on useful distinctions recognized within (or at the poles of) continua. Clarity in both field research and ecological and evolutionary theory is served by distinguishing niche and habitat as concepts. Kulesza observes, for example, that competitive displacement can occur in different ways. When two species that are closely related, and that are closely similar in ecological relationships when allopatric, also occur sympatrically, their sizes or other characteristics may diverge in the area where they occur together. This divergence, which implies difference in niche by which the species can coexist in the sympatric area, is character displacement in the usual sense (Brown and Wilson 1956). Two closely similar species can also occur sympatrically, however, if their habitats diverge in the area of overlap, so that in this area they occur primarily in different communities. This habitat divergence can occur without morphological divergence. It is worthwhile to recognize the latter as a "habitat shift" and as an adaptive tactic permitting sympatry that is rather different from morphological divergence expressing niche difference (Schoener 1975). More broadly, it is difference in ecotope that makes sympatry possible.

We agree with Kulesza that microhabitat factors should be treated as part of the niche hypervolume. Microhabitat factors vary, in connection with the internal structure and patterning of what we reasonably interpret as a given community; these are niche variables. Species can maintain themselves within individual communities by utilizing resource patches, such as those formed by local disturbance (Levin 1974; Levin and Paine 1974; Root and Chaplin 1976). This exploitation of patchiness, which is a natural feature of most environments, is a fundamental aspect of *niche*. (Macro)habitat factors, in contrast, vary in space and relate to one another different communities in different biotopes. The distinction is not pointless because it is a matter of scale, and it is part of Hutchinson's (1958b, 1967) formulation.

We agree also that competition and other species interactions are important in determining habitat distribution. Evolutionary responses to niche and habitat factors are not separate, but there are often differences in the kinds of adaptations from which result differences in niche, in contrast to differences in habitat (Whittaker et al. 1973: 324). Niche differentiation can permit coexistence of species within a biotope. Species with nonoverlapping habitats occur in different biotopes; they do not coexist and need not evolve adaptation to one another, but they have evolved differences in response to some spatially extensive environmental factor(s). As Kulesza observes, there is a connecting case—species populations that are centered in different biotopes but meet with possible competition or other interaction where they are in contact. Furthermore, the relationships between species occupying different habitats may be subject to continued testing as propagules or immigrants of one species are wafted by passive dispersal or driven by intraspecific contest into the habitat of the other species, where they attempt to establish themselves. These types of encounter, especially where competition is involved, strengthen the interest of distinction between niche and habitat. Competition between species can lead to evolution toward: different niches that are contiguous in the niche hyperspace, with competition reduced by resource use or other difference where the species occur together; different habitats that are contiguous along extensive environmental gradients, with competition reduced by occurrence in different biotopes; or different ecotopes, with competition reduced by simultaneous divergence in niche and habitat.

The ultimate arena for consideration of species relations to environment is, of course, the ecotope. The ecotope concept is independent of the notion of community and is thus a useful concept even when niche and habitat factors intergrade. Therefore it may have most utility to ecologists dealing with motile animal populations. Thus Root (1967), while studying gnatcatcher niches within a California woodland, showed that the ability of the gnatcatcher *Polioptila caerulea* to utilize different plant formations was a critical adaptation permitting the population to ride out local fluctuation in food supply. Kulesza's citations of Grinnell (1904) and Caccamise (1974) provide other examples. The ecotope concept also offers a framework for dealing with the fact that, although species persistence in a community is often discussed in terms of niche difference, species can maintain themselves in suboptimal habitats if they are superior elsewhere and have sufficient dispersal ability (Levin 1974).

On the other hand, even for higher animals, many interesting questions concern niche relationships within particular communities, such as the birds of an oak woodland or a spruce-fir forest, the rodents or lizards of a desert, or the insects and their trophic structure on a mangrove island. The study of niches in communities is not made less important by the fact that a "community" is in many cases an arbitrarily bounded segment of a continuum. The ecotope is the true evolutionary context, but the niche is the appropriate focus of many investigations. We feel that understanding will be much advanced if biologists use *ecotope* in discussing the broader evolutionary context and restrict *niche* to the role of species within a community. In restricting *niche* to one meaning, we seek to emphasize the importance of all three concepts (niche, habitat, ecotope) by providing each with its own name.

Our article was written because of our feeling that confused usage threatened the useful life of the term *niche*. We do not seek the enshrinement of particular definitions, but we feel that science would be poorly served if individual users simply chose whatever definitions suited them; such a practice would only compound confusion. Scientific concepts develop by interaction between (a) applications that indicate directions of greatest usefulness and (b) periodic efforts at formal statement and systematic interrelation of concepts. Our effort had as its purpose strengthening the usefulness of the concepts discussed by clarifying their relations to one another and stating them in ways appropriate for research using measurements of population variables. The best test of what we have written will be its usefulness in application. We hope that application may show that our formulation of three valuable concepts—niche, habitat, and ecotope—will in practice serve well both the field researcher and the ecological and evolutionary theorist.

CHAPTER 8

Biological Diversity in Ecology

Ruth Patrick

In the field of ecology the term *diversity* is commonly used to describe the assemblage of species that interact . . . and form . . . a community. Species in the complex do not merely respond to a particular environment but create new conditions through their interactions with each other. For example, one species may modify an environmental factor such as light so that another species, or group of species, can live more successfully, or one species may be the food source for another or produce oxygen by photosynthesis, which is necessary for respiration of both. Through such interactions the community develops its identity and carries out its characteristic functions.

Diversity is a generalized term that refers to the structure of the community. In a sense, it expresses the genetic variability existing in the taxa that occur together and, therefore, the adaptive capacity of the assemblage. Thus, the measure of diversity is not merely a count of presence but rather it is a measure of the structural and functional interactions of the community.

When one considers the structure of communities of organisms, the first question that arises after one has the list of species in hand is why are there so many species with such different characteristics? Reasoning from our human experience, we might think that a single-species community structure might be more efficient. However, this is not the case. For example, at the herbivore level of an aquatic community we might find insects and fish of a variety of species, genera, and families feeding upon the plants. In another comparable community we might find protozoa, as well as insects and fish, serving as herbivores. Intuitively one would say that the gene pool present in these herbivore taxa was greater in the second case than in the first. Furthermore, the second set might consume a greater variety in size and taste preference. Its tolerance to natural products produced by various plants also might be broader; therefore, nutrient transfer from the primary production level might be more, not less, efficient (Freeland and Jansen 1974).

In general we find that species seem to prefer a variety of food rather than a

From *Diversity: Benchmark Papers in Ecology.* Volume 13. Stroudsburg, Pa.: Hutchinson Ross Publishing Company, 1983, pp. 8–13.

single species. Of course, a notable exception is parasitism. In the aquatic world, organisms from protozoa to fish generally prefer many species of diatoms as a food source. In contrast, blue-green algae and green algae such as *Cladophora glomerata* are the least preferred food sources. This may be due to lower food value and the presence of toxic chemicals in *Cladophora*. These data indicate that the characteristics of the prey as well as the preferences of the predator determine the efficiency of nutrient transfer in a diverse community.

One might consider diversity of organisms from a functional rather than a taxonomic viewpoint, although usually they go hand in hand. That is, in an environment that is favorable for many species, one will find an association composed of a large number of species each of which can utilize the environment in a somewhat different way and thus they can cosurvive. MacArthur (1969) and others have pointed out that in such a case there may be a large invasion rate of species into the favorable habitat, with the result that more species will become established and will ultimately pack the area with the greatest number of species that can coexist by utilizing the resources in different ways. From this standpoint the number of species gives us the most information about the diversity of the environment. The most diverse environments exposed an adequate time to invasion would be characterized by many species with relatively small populations. The presence of disproportionately large populations might mean that the environment or the multidimensional habitats were not as diverse as they might be if there were more species with smaller populations, and therefore less redundancy. For example, in a riffle of a stream there is a range of size of rocks that will support the greatest number of species if they are distributed so as to produce a large variety of current patterns and protection against predators. Rocks that are too small roll or are shifted too much and are poor habitats. Rocks that are too large will produce a redundancy of habitats and will not support any more species, but may support more individuals of the same species. Thus, the diversity of organisms tells us about the diversity of the environment and vice versa.

In considering diversity of a community it is also relevant to know whether the species are mainly species with high population growth rates and high productivity ("r" selected species) or "K" species, which are more efficient at utilizing resources and have lower rates of population growth and production (MacArthur and Wilson 1967). The "r" species often have short life cycles and may use resources less efficiently than "K" selected species. Most communities have a variety of "r" and "K" species. Areas in which most of the ecological factors are highly variable tend to have mostly "r" selected species, and species replacement is correlated with the shifting environment. Typically the existing species are not eliminated by unfavorable environmental conditions, but are greatly reduced in population size or occur in life-cycle stages that are dormant, cryptic, or are not collectable.

One may also think of diversity of species in the terms of their ability to disperse and invade communities (Diamond 1975). Very stable "S" species are found only in species-rich communities and represent the extreme of "K" selection in MacArthur and Wilson's (1967) terminology. Then follows a series—A-B-C-D tramps—and finally supertramps that are widely distributed, do well in harsh environments, but can maintain themselves in species-rich communities.

Any discussion of diversity in ecosystems will find itself intertwined with the concept of stability (see Levin 1970; Botkin and Sobel 1975). In the last thirty years the terms *diversity* and *stability* have been the subjects of many scientific papers in response to questions such as: Why are there so many species? (Hutchinson 1958a); to findings by Patrick (1949) that in similar ecological habitats the numbers of species were similar and remained similar over time if severe perturbation did not occur; and to the Odums' (E. Odum 1962; H. Odum 1957) concepts of community homeostasis derived from study of the energy flow through communities and the relationship between community structure and function. It has often been proposed that a direct relationship exists between diversity and stability, and research on this relationship reached a peak with the 1969 Brookhaven Symposium on *Diversity and Stability in Ecological Systems* (Woodwell and Smith, eds.).

Stability has many definitions but the two that pertain to our use are, first, a quality of endurance without alteration and, second, the ability to return to the original form after alteration. It is this double meaning that has contributed to the different interpretations of the meaning of stability in considering community structure.

Stable communities composed of a few species are found in harsh environments such as the *Spartina* marshes of the east coast of the United States. Niering (personal communication) has found by boring these marshes that the same species have been dominant over hundreds of years. There are many reasons why this is true. For example, the concentration of salt in the water and its high variability and the very high evapotranspiration rates in summer limit the number of species that can establish themselves in this environment. A second consideration is the characteristics of the *Spartina* plants. The species are perennials with well-branched rhizomes. In *Spartina alterniflora*, the most common species, the rhizomes are 4–7 mm thick. These rhizomes and roots form a tough mat and thus reduce the invasion of other species. These various factors would tend to mitigate invasion and, coupled with the harsh environment, produce stability of the species over time. This is an example of a highly productive community, in terms of carbon fixed, that is simple and stable.

A similar type of a relatively stable simple community has been observed in polluted streams. In this condition the nutrients may be adequate to support a diversified species community, but the pollutant contains a toxic substance in concentrations that only a few species can tolerate. These grow well and dominate the stream over time. An example is the dominant growth of *Stigeoclonium lubricum* in Lititz Run (Lancaster County, Pennsylvania). The stream receives a variety of toxic materials, particularly heavy metals. The absence of predators, which are killed by the toxic substances, allows the species to develop large standing crops. As long as the pollutant is released this simplified and productive (in terms of ^{14}C fixation) community is stable. Under natural conditions in this drainage basin other species are common and *Stigeoclonium lubricum* does not develop large populations.

These are examples of what might be classed as communities that live in harsh environments where stability is not correlated with diversity. There are, of course, other examples where stability and diversity in the community are highly correlated. Tropical rain forests and coral reefs are often cited as examples.

The other definition of stability—that is, the ability to return to the original state after displacement—is exemplified by communities found in natural streams. These communities are composed of many species forming interlocking chains or nets of nutrient transfer. Typically the species are "r" selection species, although a few may be "K" selection types. The communities are in a quasi-equilibrium state with significant inputs and outputs. The stream is a multidimensional resource area, with the resources fluctuating in an unpredictable manner. The species with short turnover time rapidly adjust to the variable environment. Usually the number of species performing various functions remains fairly similar, although the kinds of species vary greatly in similar ecological habitats at the same time or in the same habitat during the same season over time (Patrick et al., 1969). Furthermore, these stream communities—because of the large available species pool, rapid invasion rates, and inputs and outputs—can recover rather quickly from severe perturbation; whereas smaller perturbations, such as increased nutrients, produce a readjustment in population sizes of the existing species rather than a shift in species.

McNaughton (1977) similarly found in his studies of the grassland of the Serengeti-Mara in Tanzania and Kenya that areas with large numbers of species adjust to the perturbation of chemical fertilization of the vegetation by shifts in the population sizes rather than shifts in species. Such communities were more stable than those with fewer species. It is the natural oscillations of the multidimensional environment that help to maintain the large number of species, which in turn promotes functional stability.

From these comments it is evident that when one talks about species diversity one may be concerned with taxonomic diversity, functional diversity in the community, diversity in autecology of the species, or in reproductive strategy.

In each case the primary datum is a count of species and individuals, and depending upon our knowledge of the taxa these counts may represent any relevant biological or ecological feature of the community.

It is interesting to note that the formation of diverse communities, whether they are bird or diatom communities, involves the same factors—that is, size of species pool, invasion rate, size and diversity of the area to be invaded (MacArthur and Wilson 1963). Furthermore, the maintenance of a diverse community is dependent on density-independent factors, density-dependent factors, and predator pressure. The relative importance of these forces depends on the type of community being studied. A community in or near equilibrium is usually more dependent on density-dependent factors and predator pressure for its maintenance, whereas a quasi-equilibrium community (with substantial inputs and outputs as compared with storage) often is more dependent on the type of oscillation of density-independent factors and on predator pressure for its maintenance.

CHAPTER 9

Stability in Ecological Communities

Andrew Redfearn and Stuart L. Pimm

Meanings of "Instability"

Both Elton (1958) and MacArthur (1955) argued forcefully that there might be a relationship between a population's dynamics and the intrinsic properties of the species, as well as those of the community to which it belongs. Both attributed instability to system simplicity. Elton argued that pest outbreaks were just one of many manifestations of instability and that they were more likely to occur in simple, agricultural systems than in complex, natural systems. He relayed a conversation with some tropical foresters that (perhaps mistakenly) led him to believe that pest outbreaks are a feature of simple, temperate forests but not of complex, tropical ones.

MacArthur developed his ideas somewhat more formally. He defined instability in these terms: "Suppose, for some reason, that one species has an abnormal abundance, then we shall say that the community is unstable if the other species change markedly in abundance as a result of the first. The less effect this abnormal abundance has on the other species, the more stable the community." MacArthur defined the correlate of stability, complexity, as "the amount of choice of the energy in going through the [food] web."

It is tempting to be highly critical of these early studies. Elton's arguments are heterogeneous and often based on scant evidence. Agricultural systems differ in many ways from natural ones, and there are some remarkably simple natural systems that are not devastated annually by insect herbivores. (Examples include the large stands of bracken fern, *Pteridium,* studied extensively by J. H. Lawton. Strong et al. 1984b, provide a review.) MacArthur's argument is fine as it stands, but is incomplete; it considers changes in abundance of species at the base of food chains. Changing abundances of top predators might (and indeed do) have the opposite effect: the more complex the web, the more widely disturbances may propagate. Yet we consider these early studies to be particularly important. They argue that examining the characteristics of a single species is not enough. We must look at the system to which it belongs.

From "Insect Outbreaks and Community Structure." In *Insect Outbreaks,* ed. Pedro Barbosa and Jack C. Schultz. New York: Academic Press, 1987, pp. 100–108.

They also point to a wide variety of possible meanings, not just of instability, but of community features that may correlate with instability.

Clearly, what we must do first is to look at the definitions of population stability. Then we must ask these questions: To what extent do the various kinds of instability correspond to pest outbreaks? How do these kinds of instability vary with the properties of the systems to which the species belong? Is there any evidence that outbreaks are more likely to occur in systems with certain trophic structures—simple ones, for example?

In reviewing the meanings of *stability,* we have recognized five major ideas: stability (in the strict, mathematical sense), resilience, persistence, resistance, and variability (Pimm 1984a).

Stability exists if and only if the species densities in a system tend to return to their equilibrium values following disturbances to the densities. In a variable, uncertain world, equilibrium levels may not be the population levels at which species remain; in such cases, equilibrium is better defined as the level below which the population tends to increase and above which the population tends to decrease (Tanner 1966; Pimm 1984b). Resilience is a measure of how fast a population returns to equilibrium. *Resilience* is measured in models by the characteristic return time—the time taken for the perturbation (equilibrium density minus the population density) to fall to $1/e$ ($\sim$37%) of its initial value. A resilient system has a short return time. *Persistence* measures the time a system lasts before it is changed to a different one—for example, how long a system may last before one equilibrium is replaced by another. *Resistance* is the tendency for a system to remain unchanged by a disturbance. *Variability* includes such measures as the variance, standard deviation, or coefficient of variation of population densities over time. . . .

The Stability-Complexity Question

Stability is well defined mathematically, and most theoretical studies examine it alone. Early studies (such as Gardner and Ashby 1970; May 1972) found that a smaller proportion of models of multispecies systems were stable when there were more species, when a greater proportion of those species interacted (high connectance), and when the species interacted more strongly. This seemed so contradictory to the notions of Elton and MacArthur that considerable efforts were made to evaluate the many unrealistic assumptions these early models required. The patterns of the interactions were made more realistic, as were the parameters and even the form of the equations.

Some reviews of this literature are given by May (1973a, 1979) and Pimm (1982, 1984a); to report them in detail here would be repetitious. The initial results, however, seem fairly robust. They can be reversed most easily by using models in which the predators have no effect on their prey's population growth rate (so-called donor-controlled models) (DeAngelis 1975). This can happen if predators take those prey that are most likely to die from other causes—starvation, for example. There is a large body of literature on removing predatory species from communities or, as in the case of biological control, introducing them. The vast majority of these studies show that predators do have an impact on the densities of their prey (Pimm 1980). For insects, this impact can be very large, with predators depressing prey populations to a fraction of 1% of the levels in the predator's absence (Beddington et al. 1978). In short, the donor-controlled assumption does not seem to be a good one, and so we are faced with the conclusion that more complex systems are less likely to have a stable equilibrium than simple ones.

On closer inspection, this conclusion does not contradict the ideas of MacArthur and Elton as much as it might superficially appear (Pimm 1982, 1984a). There is no difficulty with the idea that systems with stable equilibria are likely to persist. Those with unstable equilibria can have one of two initial fates: They can lose species and settle to a new stable equilibrium, or the populations may persist, oscillating probably in some complex manner. For many natural systems these two initial fates are really the same. Large-amplitude oscillations will eventually mean that populations will be driven to such low levels that they will not be able to recover. Thus, stable systems will persist, whereas unstable systems will tend to lose species and simplify to the point where they will contain a stable species assembly.

Although the quantitative results of stability analyses depend on the various assumptions made, there is a qualitative prediction that seems relatively robust: to retain stability, the product of two measures of community complexity, species number n and connectance C, should be smaller than a critical value (which depends on the strength of the species interactions). If the systems we observe in the real world are those that are stable, observed values of C and n should fall in a region *below* a hyperbolic function, as suggested by figure 8a. Data from a variety of communities, including the aphid-dominated systems shown in figure 8b, show this to be the case (Rejmanek and Stary 1979). What this reveals is straightforward: the systems we observe in nature are relatively simple ones. There is now considerable evidence that the patterns of trophic interactions we observe in nature are simpler, in a variety of ways, than we would expect by chance (Pimm 1982). There seems to be no reason to expect that simple crop ecosystems should not be stable.

Species Deletion Stability

Introduction and Model Results

What happens when we simplify a system, say by reducing the number of species present? How often are we likely to retain a stable system, and

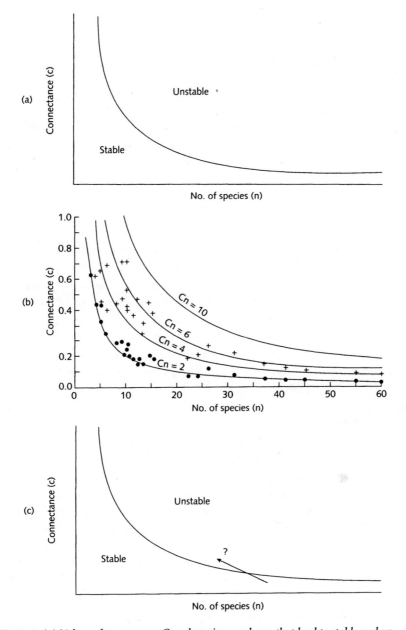

Figure 8. (a) Values of connectance C and species number n that lead to stable and unstable systems in food web models. This result suggests that natural systems should have values of C and n bounded by a hyperbola. (b) Four hyperbolic approximations and the observed values of C and n for aphids, their plant hosts, and their parasitoids. Discovered trophic connectance (●) was calculated on the basis of only the discovered interactions, whereas potential total C (+) also includes all potential competitive interactions. (c) When species number is reduced, it is possible that the resultant system is unstable, even though, other things being equal, simple systems are more likely to be stable. (From Rejmanek and Stary 1979.)

how often will the system become unstable (figure 8c)? Simple systems may often be stable, but there is no guarantee that we will produce stable systems from simplifying existing stable, complex ones. Answering these questions requires an examination of what we have called "species deletion stability" (Pimm 1979).

A system is deemed "species deletion stable" if, after the removal of a species, the remaining $n - 1$ species can coexist at a new, stable equilibrium (Pimm 1979, 1980). We can determine a system's species deletion stability repeatedly for the same species and the same model structure, but over ranges of interaction parameters designed to mimic those found in nature. This gives a probability of species deletion stability for that species' removal and for that particular web. These probabilities, averaged over all the species in a food web, vary in much the same way with connectance, species number, and interaction strength as does simple stability. The more complex the model community, the more likely it is that the loss of a species will cause further species losses. Most model systems with complexities anything near those observed in nature are not species deletion stable, nor are natural systems. The vast majority of natural systems cannot withstand species removals without changes in species composition (a review of this literature can be found in Pimm 1980).

From these studies it might seem that we have an explanation of pest outbreaks in accord with the view of Elton and MacArthur: instability is caused by simplification (rather than just by simplicity). This may be so, but on close inspection it is not anywhere near as clear-cut as it might seem. Species deletion stability varies markedly depending on which species are removed from a community (Pimm 1980). The reviews of species removals mentioned earlier tend to focus on the removals of predators or top predators. For these we expect, and find, further species losses. For plant removals, however, particularly those of plants fed on by generalized herbivores, models predict fewer losses. Moreover, these losses should become increasingly less likely with more complex systems. There is less evidence to support this result, but one cannot help notice the lack of an effect of removing chestnut, *Castanea dentata*, . . . from eastern North America. Chestnuts occupied more than 40% of the canopy in some areas in the early twentieth century and have now almost totally disappeared (Krebs 1978). Although the disappearance may have caused the loss of seven insect species that fed only on chestnuts, most insects that fed on chestnuts also fed on other tree species (Opler 1978). There do not seem to have been any losses of vertebrate species.

The simplification practiced in agriculture leads us to ask, How often does the removal of an unwanted plant species cause the crop species' load of insect herbivores to increase? Thus, in reducing the competitors of the plants we wish to harvest, do we make the crop species more vulnerable to attack by insects?

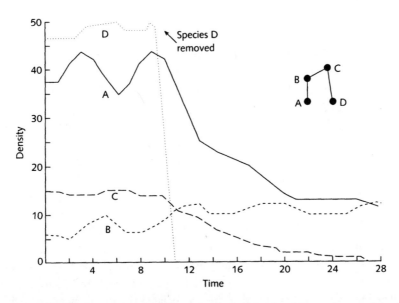

Figure 9. Effect of removing a plant species (D) on the densities of the species remaining in the system. Note that the other plant species (A) may end with a lower density because of the increased attention of its herbivore (B).

This can certainly happen theoretically. A simulation is shown in figure 9, where the removal of a plant causes the loss of a generalist predator on a specialist herbivore; the remaining plant species goes to a lower equilibrium than before. However, the models do not tell us the frequency of this occurrence in practice; it is certainly not inevitable, and it may be unlikely. Common sense dictates the conditions under which it will be a likely event—when we remove plant species essential to the survival of generalist predators that have a controlling effect on the herbivores feeding on the crop (just as in figure 9). Moreover, the loss due to these herbivores must be greater than the gain obtained from competitive release.

In short, taking an existing system and simplifying it by removing species will usually cause further species losses. It is far from certain, however, whether removing one plant species will cause a decrease in the other remaining plant species by increasing their vulnerability to insect herbivores. How do these results match our observations and intuition about the real world?

Some Field Studies

The kinds of studies that have tested the ideas about simplification have asked, How do a crop's insect numbers differ if that crop is grown singly or in a multispecies planting? Root's (1973) work is an early example of such a study. *Brassica* were grown in a single-species planting and also among many other plant species. In multispecies plantings, more species of insects were present throughout the planting, and on the *Brassica* itself, insect herbivores did not reach such high levels. From this, we might conclude that simplification caused a pest outbreak. But how general is this result, and exactly what is being simplified?

Answering these questions requires many other studies. More than 150 such studies have been compiled in a timely and important review by Risch et al. (1983). In a highly significant proportion of cases, insect herbivores were more likely to reach high densities in single-species plantings, but there were some important patterns of variation. Risch et al. argued that increased density in single-species plantings might occur for one of two reasons. First, reduced predator diversity and impact might make herbivore outbreaks more likely. Second, on the basis of the phenomenon described in the "resource concentration" hypothesis (Root 1973), the plants associated with the crop in a multispecies planting might have a direct effect on the ability of insects to find and utilize the crop. They argued that these associated plants might mask the herbivore's host-finding stimuli, generally reduce movement between individual plants, or in various other ways lower herbivore colonization rates.

The two hypotheses make different predictions about the effects of plant diversification on monophagous and polyphagous herbivores. Both groups might be expected to suffer from the increased attention of predators in multispecies plantings, if this is the cause of the reduced densities. Monophagous species, however, should decline far more than polyphagous species, if the "resource concentration" hypothesis is correct, because for polyphages the multispecies plantings will not represent such a dilution of resources.

The data support the "resource concentration" hypothesis. For monophagous species, 61% of studies showed a decrease in density with multispecies plantings, 10% showed an increase, and the rest were equivocal. For polyphagous species, 27% of the studies showed a decrease and 44% an increase. The differences were highly significant.

Risch et al. (1983) went on to consider the differences between herbivores on annual and perennial plant species. They argued that annual species might rely more heavily on escape in time from their herbivores, whereas perennials might rely on chemical defenses to slow herbivore growth. In the latter case, herbivores would be subject to longer periods of exposure to predation. If the reduced numbers in multispecies plantings were due to the effects of predators, we might expect differences between annuals and perennials. Risch et al. were unable to detect such an effect. Thus, monophagous herbivores were less abundant in diversified plantings of annuals in 58% of the cases and in 67% of the cases for perennials. For polyphagous species, the corresponding figures were 27 and 28%, respectively.

In short, Risch et al. (1983) make a persuasive case that complexity reduces herbivore densities, but the complexity is that of the plant species and the physical effects spacing plants might have. It seems to have little to do with the trophic structure of the insect communities.

Summary

Early studies suggested that pest outbreaks in agricultural systems might be due to simplification of the system. Later theoretical studies suggested that there is nothing inherently unstable about simple systems. Indeed, it is the sufficiently complex systems that should be unstable. We might expect such systems to become simplified through species losses. The result should be that the systems we observe in nature are relatively simple compared with what chance dictates (figure 8a). This seems to be the case.

Models show that the actual process of simplifying a system by removal of species from it can be expected to cause further species losses and changes in the densities of the remaining species. Removal of plant species can lead to increased herbivore levels on some of the remaining plant species, but this is not inevitable. There is now a large collection of studies that show the effects on insect herbivores of simplifying a system by removing plant species. The insect herbivores are generally more abundant on plant species grown in monoculture, but this seems to have far more to do with the difficulty of getting from host plant to host plant in the multispecies planting than to any trophic interactions.

PART 3

Rationalism and Empiricism

How do we attain knowledge of the world? Romantics, fideists, and mystics maintain that what we know about reality is learned through *irrational* means, such as intuition and religious insight. Scientists and others maintain that we know about reality through *rational* ways, such as logic and mathematical methods.

During the Renaissance and Enlightenment, a historic disagreement arose amongst philosophers regarding exactly *how* the rational type of knowing occurs. Empiricists (e.g., Francis Bacon, John Locke, George Berkeley, John Stuart Mill, and David Hume) contended that truths are known only after (*a posteriori*)—and never independently of—experience. Rationalists (e.g., René Descartes, Baruch Spinoza, and Gottfried Leibniz) contended that truths can be known prior to (*a priori*)—and independently of—experience. Admittedly, these labels are somewhat confusing, because both rationalists and empiricists advocate rationalism in the sense of opposing irrationalism (figure 10). The important point is that the distinction between empirical and rationalist methodologies persists to this day.

The difference in methodology is nicely illustrated in scientific ecology by two stereotypes: the field-worker and the theoretical modeler. What is the best way to learn about the fundamental relationships between biota and the abiotic environment? Does one get out in the field, forest, or fen and indefatigably compile data from observation, trying to make sense of it later? Or is it better to work on computers and develop theoretical models? Robert MacArthur (1962) elaborates: "We can divide ecologists into two camps. One, so aware of the complexities of nature that it is critical of simplifying theory, is content to document observations at endless length. The second, primarily interested in making a science of ecology, arranges ecological data as examples testing the proposed theories and spends most of its time patching up the theories to account for as many of the data as possible." Clearly, both methodologies are well represented in ecology. Part 3 takes a close look at rationalism, empiricism, and alternative methodologies in ecology.

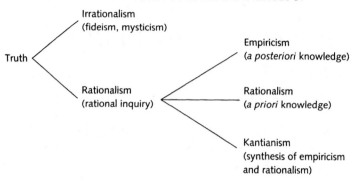

Irrationalism
(fideism, mysticism)

Truth

Empiricism
(*a posteriori* knowledge)

Rationalism
(rational inquiry)

Rationalism
(*a priori* knowledge)

Kantianism
(synthesis of empiricism
and rationalism)

Figure 10. Modern Occidental epistemology

Hypothetico-Deductive Empiricism in Ecology

As we saw in part 1, ecology grew out of the tradition of natural history. For this reason, empiricist proclivities run strong in ecology.

"The Bucket and the Searchlight: Two Theories of Knowledge" (1977), reprinted here as chapter 10, introduces the work of Austrian-born philosopher Karl Popper, whose ideas on empirical methodology have influenced many ecologists. Due to the complexity of his thinking, a brief sketch of his philosophy of science will be useful.

Popper rejects any "naïve empiricism" that strives to amass as much data as possible (the "bucket theory" of knowledge) because there is no such thing as pure, passive, "objective" observation. Rather, the scientist always *actively directs* his or her attention according to subjective impulses and preexperiential expectations. The effective scientist is able to efficiently select and organize the data in order to discover something new (the "searchlight theory" of knowledge). This emphasis on the importance of subjectivity in directing observation and organizing knowledge reveals the influence of Immanuel Kant (1965) on Popper's thought.

Although Popper's Kantianism distances him from earlier empiricists, Popper underscores the central importance of empirical testing in the scientific method. Popper, like the logical positivist philosophers of the Vienna Circle, seeks to identify the necessary requirement or condition of scientific theory but diverges from positivism in practice. In line with party dogma, positivists assert that *empirical verifiability* is the necessary condition for meaningful discourse.[1] Unfortunately, scientific theories cannot be logically substantiated in this manner; the only way to verify unrestrictedly general theories—such as scientific theories—by observed instances is through induction.

Unfortunately, as David Hume showed (1992: 98–106), induction is logically invalid (Popper 1968: 27–30). An inference is valid if and only if it is impossible to accept the premise(s) and at the same time deny the conclusion. In the case of inductive arguments, it is always possible to deny the conclusion while at the same time accepting the premises. No possible amassment of data is ever sufficient to guarantee unequivocally the certainty of a conclusion arrived at by inductive generalization: "It is far from obvious, from a logical point of view, that we are justified in inferring universal statements from singular ones, no matter how numerous," Popper says (1968: 27). To make things worse, a theory might be verified in some instances and falsified in others, and theories that are known to be false (e.g., persons born when the stars and planets are configured a specific way will have tragic lives) have positively confirming instances. For these reasons, Popper asserts, empirical verifiability is not the necessary condition of a scientific theory, because science is not based on induction.

To overcome the problems of induction, Popper asserts that empirical *falsifiability*—not verifiability—is the essential element of scientific theories. The criterion of falsifiability measures the empirical (and hence scientific) content of a theory. Simple statements (i.e., those without stipulations) are more sweeping in scope and thus say more about the world than limited statements. The greater the scope of a statement, the greater potential there is for disproving instances. For example, the statement "All Z have Y" is greater in scope and more likely to be falsified by specimens of Z that do not have Y than the statement "Some Z have Y." This means that an ecological hypothesis such as "the energy lost by moving up one trophic level is 10 percent" has more empirical content (and hence more scientific value) than the hypothesis "the energy lost by moving from the insect trophic level to the bird trophic level is 10 percent." Therefore the proper method of science is to come up with falsifiable hypotheses, Popper claims, and the more falsifiable, the better.

The scientific enterprise begins with generating an idea or hypothesis. The process of hypothesis generation is radically subjective: the flash of insight or foggy apprehension that spawns new ideas depends on individual idiosyncrasy and uniqueness. For example, the chemist Friedrich Kekulé hit upon the idea of representing the molecular structure of benzene as a hexagonal ring after dozing off by the fire and dreaming of a snake dancing in the flames, grabbing its tail, and spinning around wildly (Findlay 1965: 39–40). Oddly enough, Popper claims, the germ of the scientific enterprise is personal creativity and cannot be reduced to a rule. Inherent to the scientific method is something not explainable by the scientific method: science has a nonscientific foundation! Robert Pirsig makes the same point in *Zen and the Art of Motorcycle Maintenance* (1974: 108): "If the purpose of scientific method is to select from among

I.   Hypothesis generation by subjective imagination:
     E.g., (a) "All Z have Y."
II. Deduction of single empirically testable proposition from hypothesis (a):
     E.g., (b) "This particular Z (viz. E) has Y."
III. Test: is (b) falsifiable?
     Yes: Reject (b)—generate new hypothesis (I).
     No: Proceed with further testing (IV).
IV. Further testing: if hypothesis continues to withstand continued serious attempts to falsify
     it, then the hypothesis can be provisionally accepted first as scientific theory and then
     scientific law.

Figure 11. The hypothetico-deductive method

a multitude of hypotheses, and if the number of hypotheses grows faster than experimental method can handle, then it is clear that all hypotheses can never be tested. If all hypotheses cannot be tested, then the results of any experiment are inconclusive and the entire scientific method falls short of its goal of establishing proven knowledge." To avoid drowning in a sea of hypotheses, the scientist must use imagination, intuition, and gut reaction in order to select one hypothesis that might turn out to be especially interesting.

Once a hypothesis is generated, it needs to be tested. To test a hypothesis, it is necessary to derive a singular descriptive statement from the hypothesis by ordinary deductive logic. For example, from the hypothesis (i) "All ecosystems require solar radiation to function" we can deductively derive the singular statement (ii) "This particular ecosystem E requires solar radiation to function." Does the singular statement have any observable disproving instances? If not, then (ii) is evidence for the truthfulness of (i). We can continue to deduce singular statements from hypothesis (i):

  (iii) "Ecosystem F requires solar radiation to function."
  (iv) "Ecosystem G requires solar radiation to function."
  (v) "Ecosystem H requires solar radiation to function."
  (vi) "Ecosystem I requires solar radiation to function."
  etc.

After numerous singular statements are found not to be counterinstances of (i), then (i) achieves the status of theory and eventually law. (Figure 11 summarizes Popper's hypothetico-deductive method.)

Popper conveys Hume's skepticism about the ultimate certainty of scientific law: that some law is unrefuted does not entail the conclusion that it is *irrefutable*. Scientific laws are "happy guesses" that end up not having any disproving instances. We can never know whether scientific laws are absolute truth. As Popper says in *The Logic of Scientific Discovery:* "Science is not a system of certain, or well-established, statements. . . . Our science is not knowledge (episteme): it can never claim to have attained truth, or even a substitute for it. . . .

We do not know: we can only guess" (1968: 278; emphasis in original). In the last analysis, the hypothetico-deductive method paradoxically has the flavor of the inductive method Popper ostensibly dismissed: science turns out to be a matter of probabilities; nothing is ever certain, including the prediction that the sun will rise tomorrow.

Plant ecologists (e.g., Cain 1947; Elger 1947; Mason 1947) began to assimilate Popperian methodology into their science in the 1940s in the form of support for Henry Gleason's (1917, 1926, 1939 [reproduced in this volume]) probabilistic model of vegetation communities against Frederic Clements's (1916 [reproduced in this volume], 1936) deterministic model. Frank Elger (1947: 389) goes so far as to state that he "adopts wholeheartedly and without exception the 'individualistic concept' of the plant community as developed by Gleason." The unifying theme of these plant ecology papers is that Clementsian ecology is an *a priori* assumption rather than an *a posteriori* fact, and that plant ecology needs to become more empirically rooted in the spirit of the hypothetico-deductive method.

Despite Popper's influence, however, not all ecologists have been convinced of the utility of the hypothetico-deductive method. Amyan MacFadyen, for example, says: "Those of us who work in [ecology] often find it hard to see quite how our kind of science fits into the Popper . . . hypothetico-deductive image" (1975: 352).

Rationalism in Ecology

Rationalism is well represented in scientific ecology. Ecologists have claimed over and over again that the science needs unifying principles—a well-developed and coherent body of theory.

As early as 1880, Stephen Forbes noted the importance of theoretical models in ecological investigation: "To determine the primitive order of nature by induction alone requires such a vast number of observations in all parts of the world, for so long a period of time, that more positive and satisfactory conclusions may perhaps be reached if we call in the aid of first principles, traveling to our end by the *a priori* road" (1880: 5). Rationalist methodologies are a necessity for the ecologist for two obvious reasons. In practice, the development of ecological theory simply outpaces observation, as Robert May points out in "The Role of Theory in Ecology" (1981a), reprinted here as chapter 11. And empirical investigation, without the guidance of nonempirically testable *a priori* categories, has the danger of producing a deluge of nonorderable and logically inconsistent data (Haskill 1940: 15).

Rationalist methodology in ecology was strengthened in the 1920s by Alfred Lotka's (1956) and Vito Volterra's (1926) logistic models of competition, and in the 1930s by the competitive-exclusion principle that George Gause (1964) derived mathematically from the Lotka-Volterra equations. Rationalism continued to flourish during the 1950s and 1960s with the resurgence of mathematical ecology spawned by the brilliant work of MacArthur (1955) and his investigations with Edward O. Wilson on island biogeography equilibrium theory (MacArthur and Wilson 1963, 1967).

Powerful computers and quantitative systems ecology (e.g., Howard Odum 1957) have continued to secure the role of mathematics in ecology. "Today a background in mathematics is a requirement for the serious ecologist. Without a working knowledge of certain mathematical concepts the current ecological literature is virtually meaningless" (Vandermeer 1981: 1). For ecosystem ecologists, the centrality of arithmetic in modeling the structure and function of ecosystems holds the promise for a unifying paradigm (Margalef 1963).

If some claim mathematics is indispensable for ecology, others find it utterly useless (e.g., Levandowsky 1977). Even ecologists such as Lawrence Slobodkin (1975), who see the need for unifying theory and recognize some relevance of mathematics for ecology, still doubt that the bulk of mathematical models have much to do with the realities of biological systems. Daniel Simberloff charges that mathematical ecologists multiply models without doing any empirical testing: "Ecology is awash with . . . untested (and often untestable) models, most claiming to be heuristic, many simply elaborations of earlier untested models. Entire journals are devoted to such work, and are as remote from biological reality as are faithhealers" (1981: 52). At times the mathematics simply overwhelms the investigative procedure and ends up being counterproductive. Charles Elton complains of Lotka: "Like most mathematicians he takes the hopeful biologist to the edge of the pond, points out that a good swim will help his work, and then pushes him in and leaves him to drown" (quoted in Kingsland 1985: 249; McIntosh 1985: 176).

Ignoring *ad hominem* quips about plastic shirt-pocket pen protectors and calculator belt holsters, the contribution of rationalist methodology to the development of ecology cannot be overlooked.

Inductive Empiricism in Ecology

The ecosystem model has gone a long way in providing a unifying model for ecology (E. Odum 1964), but a nagging problem persists: a single, totalizing theory has not yet been discovered. Many ecologists and philosophers of ecology are skeptical that any unifying *a priori* theory is possible and see inductive empiricism as a viable alternative.

For Canadian biologist Robert Peters, ecology is a "weak" science in that it lacks a coherent organizing paradigm. In *A Critique for Ecology* (1991: 1), Peters argues that in weak sciences, "goals and criteria are poorly enunciated, less accepted, and more sporadically applied. These sciences are less coherent, they contain many constructs of dubious merit, and their growth is lethargic. It is my thesis that much of contemporary, academic ecology belongs with the latter group." To revive ecology, ecologists must quit wasting their time on questions that cannot be answered empirically. Theories should not be the center of ecologists' attention. Models and theories are like wrenches; you reach in the toolbox and pull out whichever one fits the situation. Such a "pluralistic" philosophy of science is consonant with the method of "multiple working hypotheses" advocated by geologist Thomas Chamberlin (1965) in 1890.

The consequence of inductive empiricism is pluralistic methodology. In "Community Ecology, Population Biology, and the Method of Case Studies" (1994),[2] reprinted here as chapter 12, philosopher Kristin Shrader-Frechette and biologist Earl McCoy also advocate inductive empiricism over *a priori* theory building—what they call a "bottom-up" rather than a "top-down" approach. In *Method in Ecology: Strategies for Conservation* (1993: 1), Shrader-Frechette and McCoy state that "insofar as ecology is required for solving practical environmental problems, it is more a science of case studies and statistical regularities, than a science of exceptionless, general laws. Insofar as ecology is an applied endeavor, it is more a science that moves from singular to theoretical explanation, than one that proceeds from theoretical to singular explanation." The inductive "bottom-up" approach uses a pluralistic methodology rather than organizing observation by only one theory.

Conclusion

Currently, pluralism is in and grandiose theory building is out (Hull 1993: 473). The lack of a unifying paradigm does not mean that the role of theory in ecology is shrinking, or that the science is weak or sickly. Rather, as historian of biology Joel Hagen (1989) correctly remarks, the lack of a unifying paradigm for ecology is not so much a result of its immaturity as a sign of its comprehensiveness and complexity.

Common sense suggests that the proper methodology for scientific ecology must consist of both rationalist and empiricist approaches. Theories are important because they direct and organize empirical observation: observation must be goal-directed. In turn, observation and empirical testing ensure that the scientific endeavor does not become disconnected from phenomena. If it does, there is a danger that it will lose its connection with real biological systems. Earl Werner and Gary Mittelbach make the point well in a discussion of optimal foraging theory: "It is critical, however, that the theory now develop in concert with empirical work. This area, as many, runs the risk of compiling large amounts of trivial theory unguided by attention to what animals are actually capable of doing and therefore what the important problems are[;] polemics over the usefulness or 'validity' of the theory degenerates into rather scholastic exercises" (1981: 827).

Notes

1. According to the Verifiability Criterion of Meaning (Ayer 1952: 9), statements are either: (a) true or false (i.e., meaningful) by observation; (b) true or false (i.e., meaningful) by definition; or (c) neither true nor false (i.e., meaningless) by observation or definition. Notoriously, subjects like aesthetics, ethics, and religion fall into the third category (see ibid., chapter 6, "Critique of Ethics and Theology"). A logical problem for logical positivists is to show that the Verifiability Criterion of Meaning itself does not necessarily fall into the third category as well.

2. This paper was originally published under the title "Applied Ecology and the Logic of Case Studies."

CHAPTER 10

The Bucket and the Searchlight:
Two Theories of Knowledge

Karl R. Popper

The purpose of this paper is to criticize a widely held view about the aims and methods of the natural sciences, and to put forward an alternative view.

I shall start with a brief exposition of the view I propose to examine, which I will call "the bucket theory of science" (or "the bucket theory of the mind"). The starting point of this theory is the persuasive doctrine that before we can know or say anything about the world, we must first have had perceptions— sense experiences. It is supposed to follow from this doctrine that our knowledge, our experience, consists either of accumulated perceptions (naïve empiricism) or else of assimilated, sorted, and classified perceptions (a view held by Bacon and, in a more radical form, by Kant).

The Greek atomists had a somewhat primitive notion of this process. They assumed that atoms break loose from the objects we perceive, and penetrate our sense organs, where they become perceptions; and out of these, in the course of time, our knowledge of the external world fits itself together [like a self-assembling jigsaw puzzle]. According to this view, then, our mind resembles a container—a kind of bucket—in which perceptions and knowledge accumulate. (Bacon speaks of perceptions as "grapes, ripe and in season" which have to be gathered, patiently and industriously, and from which, if pressed, the pure wine of knowledge will flow.)

Strict empiricists advise us to interfere as little as possible with this process of accumulating knowledge. True knowledge is pure knowledge, uncontaminated by those prejudices which we are only too prone to add to, and mix with, our perceptions; these alone constitute experience pure and simple. The result of these additions, of our disturbing and interfering with the process of accumulating knowledge, is error. Kant opposes this theory: he denies that perceptions are ever pure, and asserts that our experience is the result of a process of

From *Discovering Philosophy*, ed. Matthew Lipman. Englewood Cliffs, N.J.: Prentice-Hall, 1977, pp. 328–34. Translation of a lecture delivered in German in August 1948. Edited by Lipman. Textual additions to the translation are in square brackets. Notes omitted.

assimilation and transformation—the combined product of sense perceptions and of certain ingredients added by our minds. The perceptions are the raw material, as it were, which flows from outside into the bucket, where it undergoes some (automatic) processing—something akin to digestion, or perhaps to systematic classification—in order to be turned in the end into something not so very different from Bacon's "pure wine of experience"; let us say, perhaps, into fermented wine.

I do not think that either of these views suggests anything like an adequate picture of what I believe to be the actual process of acquiring experience, or the actual method used in research or discovery. Admittedly, Kant's view might be so interpreted that it comes much nearer to my own view than does pure empiricism. I grant, of course, that science is impossible without experience (but the notion of "experience" has to be carefully considered). Though I grant this, I nevertheless hold that perceptions do not constitute anything like the raw material, as they do according to the "bucket theory," out of which we construct either "experience" or "science."

In science it is *observation* rather than perception which plays the decisive part. But observation is a process in which we play an intensely *active* part. An observation is a perception, but one which is planned and prepared. We do not "have" an observation [as we may "have" a sense experience] but we "make" an observation. [A navigator even "works" an observation.] An observation is always preceded by a particular interest, a question, or a problem—in short, by something theoretical. After all, we can put every question in the form of a hypothesis or conjecture to which we add: "Is this so? Yes or no?" Thus we can assert that every observation is preceded by a problem, a hypothesis (or whatever we may call it); at any rate by something that interests us, by something theoretical or speculative. This is why observations are always selective, and why they presuppose something like a principle of selection. . . .

Let us now return to the problem of observation. An observation always presupposes the existence of some system of expectations. These expectations can be formulated in the form of queries; and the observation will be used to obtain either a confirming or a correcting answer to expectations thus formulated.

My thesis that the question, or the hypothesis, must precede the observation may at first have seemed paradoxical; but we can see now that it is not at all paradoxical to assume that expectations—that is, dispositions to react—must precede every observation and, indeed, every perception: for certain dispositions or propensities to react are innate in all organisms whereas perceptions and observations clearly are not innate. And although perceptions and, even more, observations, play an important part in the process of *modifying* our dispositions or propensities to react, some such dispositions or propensities must, of course, be present first, or they could not be modified.

These biological reflections are by no means to be understood as implying my acceptance of a behaviorist position. I do not deny that perceptions, observations, and other states of consciousness occur, but I assign to them a role very different from the one they are supposed to play according to the bucket theory. Nor are these biological reflections to be regarded as forming in any sense an assumption on which my arguments will be based. But I hope that they will help toward a better understanding of these arguments. The same may be said of the following reflections, which are closely connected with these biological ones.

At every instant of our prescientific or scientific development we are living in the center of what I usually call a "horizon of expectations." By this I mean the sum total of our expectations, whether these are subconscious or conscious, or perhaps even explicitly stated in some language. Animals and babies have also their various and different horizons of expectations though no doubt on a lower level of consciousness than, say, a scientist whose horizon of expectations consists to a considerable extent of linguistically formulated theories or hypotheses.

The various horizons of expectations differ, of course, not only in their being more or less conscious, but also in their content. Yet in all these cases the horizon of expectations plays the part of a frame of reference: only their setting in this frame confers meaning or significance on our experiences, actions, and observations.

Observations, more especially, have a very peculiar function within this frame. They can, under certain circumstances, destroy even the frame itself, if they clash with certain of the expectations. In such a case they can have an effect upon our horizon of expectations like a bombshell. This bombshell may force us to reconstruct, or rebuild, our whole horizon of expectations; that is to say, we may have to correct our expectations and fit them together again into something like a consistent whole. We can say that in this way our horizon of expectations is raised to and reconstructed on a higher level, and that we reach in this way a new stage in the evolution of our experience; a stage in which those expectations which have not been hit by the bomb are somehow incorporated into the horizon, while those parts of the horizon which have suffered damage are repaired and rebuilt. This has to be done in such a manner that the damaging observations are no longer felt as disruptive, but are integrated with the rest of our expectations. If we succeed in this rebuilding, then we shall have created what is usually known as an *explanation* of those observed events [which created the disruption, the problem].

As to the question of the temporal relation between observation on the one hand and the horizon of expectations or theories on the other, we may well admit that a new explanation, or a new hypothesis, is generally preceded in time by *those* observations which destroyed the previous horizon of expectations and thus were the stimulus to our attempting a new explanation. Yet this must not be understood as saying that observations generally precede expectations or hypotheses. On the contrary, each observation is preceded by expectations or hypotheses; by those expectations, more especially, which make up the horizon of expectations that lends those observations their significance; only in this way do they attain the status of real observations.

The question, "What comes first, the hypothesis (H) or the observation (o)?" reminds one, of course, of that other famous question: "What came first, the hen (H) or the egg (o)?" Both questions are soluble. The bucket theory asserts that [just as a primitive form of an egg (o), a unicellular organism, precedes the hen (H)] observation (o) always precedes every hypothesis (H); for the bucket theory regards the latter as arising from observations by generalization, or association, or classification. By contrast, we can now say that the hypothesis (or expectation, or theory, or whatever we may call it) precedes the observation, even though an observation that refutes a certain hypothesis may stimulate a new (and therefore a temporally later) hypothesis.

All this applies, more especially, to the formation of scientific hypotheses. For we learn only from our hypotheses what kind of observations we ought to make: whereto we ought to direct our attention; wherein to take an interest. Thus it is the hypothesis which becomes our guide, and which leads us to new observational results.

This is the view which I have called the "searchlight theory" in contradistinction to the "bucket theory." [According to the searchlight theory, observations are secondary to hypotheses.] Observations play, however, an important role as *tests* which a hypothesis must undergo in the course of our [critical] examination of it. If the hypothesis does not pass the examination, if it is falsified by our observations, then we have to look around for a new hypothesis. In this case the new hypothesis will come after those observations which led to the falsification or rejection of the old hypothesis. Yet what made the observations interesting and relevant and what altogether gave rise to our undertaking them in the first instance, was the earlier, the old [and now rejected] hypothesis.

In this way science appears clearly as a straightforward continuation of the prescientific repair work on our horizons of expectations. Science never starts from scratch; it can never be described as free from assumptions; for at every instant it presupposes a horizon of expectations—yesterday's horizon of expectations, as it were. Today's science is built upon yesterday's science [and so it is the result of yesterday's searchlight]; and yesterday's science, in turn, is based on the science of the day before. And the oldest scientific theories are built on prescientific myths, and these, in their turn, on still older expectations. Ontogenetically (that is, with respect to the development of the individual organism) we thus regress to the state of the expectations of a newborn child; phylogenetically (with respect to the evolution of the race, the phylum) we get to the state of expectations of unicellular organisms. (There is no danger here of a vicious infinite regress—if for no other reason than that every organism is born with some horizon of expectations.) There is, as it were, only one step from the amoebae to Einstein.

Now if this is the way science evolves, what can be said to be the characteristic step which marks the transition from prescience to science[?]

I hope that some of my formulations which at the beginning of this lecture may have seemed to you far-fetched or even paradoxical will now appear less so.

There is no road, royal or otherwise, which leads of necessity from a "given" set of specific facts to any universal law. What we call "laws" are hypotheses or conjectures which always form a part of some larger system of theories [in fact, of a whole horizon of expectations] and which, therefore, can never be tested in isolation. The progress of science consists in trials, in the elimination of errors, and in further trials guided by the experience acquired in the course of previous trials and errors. No particular theory may ever be regarded as absolutely certain: every theory may become problematical, no matter how well corroborated it may seem now. No scientific theory is sacrosanct or beyond criticism. This fact has often been forgotten, particularly during the last century, when we were impressed by the often repeated and truly magnificent corroborations of certain mechanical theories, which eventually came to be regarded as indubitably true. The stormy development of physics since the turn of the century has taught us better; and we have now come to see that it is the task of the scientist to subject his theory to ever new tests, and that no theory must be pronounced final. Testing proceeds by taking the theory to be tested and combining it with all possible kinds of initial conditions as well as with other theories, and then comparing the resulting predictions with reality. If this leads to disappointed expectations, to refutations, then we have to rebuild our theory.

The disappointment of some of the expectations with which we once eagerly approached reality plays a most significant part in this procedure. It may be compared with the experience of a blind man who touches, or runs into, an obstacle, and so becomes aware of its existence. *It is through the falsification of our suppositions that we actually get in touch with "reality."* It is the discovery and elimination of our errors which alone constitute that "positive" experience which we gain from reality.

It is of course always possible to save a falsified theory by means of supplementary hypotheses [like those of epicycles]. But this is not the way of progress in the sciences. The proper reaction to falsification is to search for new theories which seem likely to offer us a better grasp of the facts. Science is not interested in having the last word if this means shutting off our minds from falsifying experiences, but rather in learning from our experience; that is, in learning from our mistakes.

There is a way of formulating scientific theories which points with particular clarity to the possibility of their falsification: we can formulate them in the form of prohibitions [or negative existential statements] such as, for example, "There does not exist a closed physical system, such that energy changes in one part of it without compensating changes occurring in another part" (first law of thermodynamics). Or, "There does not exist a machine which is one hundred percent efficient" (second law). It can be shown that universal statements and negative existential statements are logically equivalent. This makes it possible to formulate all universal laws in the manner indicated; that is to say, as prohibitions. However, these are prohibitions intended only for the technicians and not for the scientist. They tell the former how to proceed if he does not want to squander his energies. But to the scientist they are a challenge to test and to falsify; they stimulate him to try to discover those states of affairs whose existence they prohibit, or deny.

Thus we have reached a point from which we can see science as a magnificent adventure of the human spirit. It is the invention of ever new theories, and the indefatigable examination of their power to throw light on experience. The principles of scientific progress are very simple. They demand that we give up the ancient idea that we may attain certainty [or even a high degree of "probability" in the sense of the probability calculus] with the propositions and theories of science (an idea which derives from the association of science with magic and of the scientist with the magician): the aim of the scientist is not to discover absolute certainty, but to discover better and better theories [or to invent more and more powerful searchlights] capable of being put to more and more severe tests [and thereby leading us to, and illuminating for us, ever new experiences]. But this means that these theories must be falsifiable: it is through their falsification that science progresses.

CHAPTER 11

The Role of Theory in Ecology

Robert M. May

Introduction

Ecological theory comes in many forms. . . . Some of the work . . . deals with mathematical models in the traditions of theoretical physics or classical applied mathematics. Such models range from simple, general models aimed at explaining some of the similarities and differences among observations on a range of organisms, to relatively complicated and detailed models aimed at putting a theoretical curve through a collection of data points. Other work borrows from systems analysis or hierarchy theory, in the hope of providing transcendent ways of describing pattern and process in complex communities, or of providing recipes for appropriate ways of aggregating variables. Not all theorizing, however, need be cast in mathematical terms; a lot of useful theory takes the form of verbal models. Such verbal models or schema range from sets of ideas giving coherence to a body of data and making sense of observed patterns, to questions and speculations that stimulate empirical studies. . . .

My own work is mainly in the idiom of applied mathematics. What follows is an eclectic set of examples, illustrating the diverse ways in which such ecological theory can be helpful in advancing our understanding of the natural world. These examples reflect my own interests, and not any absolute judgment about what problems are important; some serve merely as fables, or as points of departure for opinionated comment. The examples are described *very* briefly, because the details are available in easily-accessible journals and unnecessary repetition is undesirable.

Theoretical Ecology and Whaling Quotas

Theoretical population biology has, for many years, played an explicit role in many areas of resource management. In particular, the concept of Maximum Sustainable Yield, framed by Graham (1952) and Schaefer (1957),

From *American Zoologist* 21 (1981): 903–10, pp. 903–9.

has—at least in principle—been the basis for management of most fishing and whaling industries since World War II. . . .

Recent theoretical work is modifying the International Whaling Commission deliberations in three main ways. First, whale population densities are typically estimated from data on the catch per unit harvesting effort (CPUE), it being usually assumed that stock density is linearly proportional to CPUE. Borrowing from recent developments in the theory of insect prey-predator interactions (Hassell 1978), Beddington, Holt, Chapman, and others (International Whaling Commission 1979, 1980) have . . . shown that incorporation of the effects of finite "handling times," and of aggregation of predators (whaling ships) around clumps of prey (whales), makes for nonlinearities in the relation between stock density and CPUE; the upshot is reduced estimates of stock density, and lowered quotas. I regard this as a tangible accomplishment for ecological theory.

Second, conventional Maximum Sustainable Yield theory assumes that recruitment relations and the like are deterministic curves. But natural populations are subject to all manner of environmental fluctuations and vagaries, such that we necessarily must deal with probability distributions, not unique deterministic relationships. This points the way to the next generation of management rules, which ask questions not merely about the maximum sustainable average yield, but also about the associated level of fluctuation in stock and yield; we need to pay more deliberate attention to designing "risk averse" management strategies. (For a more full discussion see Beddington and May 1977; May et al. 1978, and references therein.)

Third, as harvesting pressures intensify, the assumption that whale species can each be treated as a single population, ignoring biological interactions with other species, increasingly breaks down. In a massive, ill-documented, and unintentional experiment in "competitive release," it is likely that minke whale populations have roughly doubled as their main competitor, the blue whale, has been virtually removed (International Whaling Commission 1978, 1979, 1980). Such multispecies considerations become even more complicated and pressing when harvesting takes place simultaneously at more than one trophic level. This is beginning to be the case for baleen whales and krill in the Southern Ocean. May et al. (1979) have used simple, general models to give a qualitative analysis of the effects of various management regimes in such multispecies situations; although admittedly abstract and lacking detailed realism, this work has, *faut de mieux*, become the basis for the principles of management included in the draft Convention of the Southern Ocean.

These practical problems differ from many other, more academic, areas of ecology in that decisions simply must be made, often (to use Clausewitz's telling metaphor) "in the fog." Well-designed procedures could reduce this fog, and enable more information to be gathered about the interplay between theory and reality. For example, the International Whaling Commission could . . . use different parameters or different decision rules in different areas of the oceans, thus having, as it were, more experimental "controls." Too often, other considerations overrule scientific good sense.

Theoretical Ecology and Fisheries

Most of the conventional work in fisheries management assumes that recruitment is essentially independent of stock density, and seeks to determine the minimum age at which fish should be harvested (set, for example, by mesh size on nets) in order to produce the maximum sustainable catch. The concern is to avoid "growth overfishing." Recently, however, with the collapse of several major fisheries, it appears that "recruitment overfishing" can occur, with overexploitation leading to a collapse in recruitment.

As a result, several people are investigating theoretical models for the recruitment process in fish stocks, paying especial attention to factors that might serve to herald the imminence of collapse in recruitment (Cushing 1973, 1977; Gulland 1979; for a review and some new ideas see May 1980). One outcome of this research could be a more quantitative understanding of why typical fish populations have "forward peaked" recruitment curves (with recruitment essentially independent of stock density around pristine population values), while typical marine mammal populations have pronouncedly "backward peaked" curves. It is even possible that one could end up with a biological basis for [population equations].

Dynamical Behavior of Single Populations

Over the past seven years or so [1973–80], mathematical studies have shown that the simple, deterministic equations proposed by practical entomologists, epidemiologists, and fishery managers as descriptions of the way particular populations change over time are capable of exhibiting an astonishingly rich array of dynamical behavior (May and Oster 1976). . . .

In detailed applications, single-species models have been successful in explaining observed data for particular field and laboratory populations. Broadly, the examples include laboratory populations of blowflies, daphnia, rotifers, weevils, collembola, and other organisms exhibiting stable points or sustained cycles (and sometimes both, as a function of temperature), along with field populations exhibiting stable four- or ten-year cycles. . . .

Interaction between Species

From the time of Lotka (1956) and Volterra (1926), mathematical models have contributed to our understanding of the interactions between species, as prey-predators, competitors, or mutualists. Because of their greater complexity and greater number of parameters, fewer such two-species models make contact with data than do single-species models.

The generic heading of "prey-predator" covers many different kinds of pairwise interactions. Some aspects of theory and observations [are] about plant-herbivore relations. . . . As reviewed by Hassell (1978), arthropod prey-predator systems, especially host-parasitoid ones, offer some simplifying biological features that permit a degree of direct confrontation between mathematical models and data (at least in the laboratory). Vertebrate prey-predator interactions, however, seem to me to have so many behavioral and ecological complexities that simple mathematical models can usually do no more than indicate general trends (see, e.g., Tanner 1975).

Theoretical Ecology and Community Ecology

In theoretical studies of entire communities of interacting plants and animals, abstract mathematical models have been used to explore general questions (e.g., the relation between "stability" and "complexity"), and more concrete models have been used to explore explicit patterns in community organization or trophic structure (e.g., the relative abundance of the various species in different types of communities, or the relation between the number of species and the physical size of their constituent individuals, or the factors governing the remarkable constancy of only around three or four trophic levels in most food webs). In many instances, including all those I just listed, such community-level models have been helpful in bringing the issues into sharp focus, sometimes overthrowing the conventional wisdom in the process. But many of these questions, again including all those just listed, still lack apodictic resolution (see May 1981b: 197–227). . . .

The Role of Theory

As stressed in the Introduction, there are many aspects, and many valid approaches, to ecological theory. Insofar as relatively simple mathematical models can be helpful, it is usually in clarifying what are (and what are not) the essential features in a complicated natural situation. An applied mathematician tries to identify these essential features, using what is often called "common sense" or "physical intuition"; they are then incorporated into a mathematical model, which makes qualitative or quantitative predictions. The predictions may be far from obvious, even in simple situations. . . . If the predictions accord with reality, our understanding is advanced; if not, we try to find what necessary ingredient was omitted. . . . Of course, ecological theory, in both verbal and mathematical forms, often runs ahead of observation. . . . [S]uch theory delineates possibilities, from among which empirical studies discriminate the actualities.

Rather than further airing my opinions about the role of theory, I end by mentioning Darwin's. Darwin's private correspondence reveals his views: "All observation must be for or against some view if it is to be of any service! . . . let theory guide your observations, but till your reputation is well established be sparing in publishing theory. It makes persons doubt your observations." (See Gruber and Barrett 1974: 123.)

Publicly, however, Darwin consistently portrayed himself as adhering to the accepted scientific pieties of his day, namely the Baconian Method, in which one first marshals the facts and then sees what conclusions emerge. Thus on the first page of the *Origin,* Darwin claims to have "patiently accumulated and reflected on all sorts of facts which could possibly have any bearing on it. After five years' work I allowed myself to speculate on the subject, and drew up some short notes." Likewise, on the first page of *The Expression of the Emotions in Man and Animals,* he says, "I arrived, however, at these three Principles only at the close of my observations." As Gruber and others have emphasized, Darwin's private notebooks tell a very different story, and one which is more familiar to a practicing scientist:

> The pandemonium of Darwin's notebooks and his actual way of working, in which many different processes tumble over each other in untidy sequences—theorizing, experimenting, casual observing, cagey questioning, reading, etc.—would never have passed muster in a methodological court of inquiry. . . . He gave his work the time and energy necessary to permit this confusion to arise, at the same time persistently sorting it out, finding what order he could. It was an essential part of this method that he worked at all times within the framework of a point of view which gave meaning and coherence to seemingly unrelated facts. (Gruber and Barrett 1974: 122)

I think this is important. Naively simple formulations of The Way To Do Science—be they the Baconian Method of the Victorians or the extreme logical positivism of Popper today—are harmless in themselves, but have unfortunate consequences when they inspire doctrinaire vigilantes to ride the boundaries of a discipline, culling the sinners. The scrabbling, nonlinear way Darwin pursued his ends is typical of most good science. Writing about him, Ghiselin (1969: 236) says: "Viewed from without, science appears to be a body of answers; from within, it is a way of asking questions. . . . The 'predictionist thesis' and 'hypothetico-deductive' model seem a bit trivial as clues to what real scientists are trying to do." Although his avowedly anarchic, "anything goes," position is clearly too extreme, Feyerabend (1975) may be closer to the actuality than are his canonized colleagues, from Bacon to Popper. Indeed, I find it paradoxical that Popper's tenets are themselves unfalsifiable: Feyerabend would find it easier to explain their current vogue than would Popper himself.

CHAPTER 12

Community Ecology, Population Biology, and the Method of Case Studies

Kristin Shrader-Frechette and Earl D. McCoy

1. Introduction

Two decades ago, Schoener (1972) warned that ecology has a "constipating accumulation of untested models," (p. 389), most of which are untestable, and Peters (1991) complained that the vast majority of models in the ecological literature do not describe the phenomena they purport to describe, or they contain internal mathematical problems, or both. One question such criticisms raise, apart from what sorts of claims are most appropriate to pure ecology, is whether applied ecology is ever likely to have any general theories or exceptionless laws to aid in problem solving. If not, then the logic and methods most appropriate for confirming general theories and exceptionless laws may not be those most suited to applied ecology. Sidestepping this issue of the types of causal claims most appropriate to pure ecology (if there is such a thing), we argue that some sort of "logic of case studies"—and an associated method—may be required in applied ecology. Although there are no sets of purely deductive inferences that one can draw from analysis of a unique, singular situation—and hence no applicable logic in the strictest sense—we argue that there is a "logic" of case studies in the sense of informal inferences (that give us a way to make sense of a situation), even though we cannot completely guarantee their soundness. Likewise we argue that there is a "method" of case studies in the sense of rules of thumb or a systematic plan for generating reliable case studies and hence for facilitating the relevant informal inferences.

2. Why We Might Need a Method of Case Studies

In community ecology we are unlikely to find many (if any) simple, exceptionless laws applicable to a variety of communities or species. One rea-

From *Philosophy of Science* 61 (1994): 228–49. Published originally as "Applied Ecology and the Logic of Case Studies."

son is that fundamental ecological terms (like "community" and "stability") are too vague to support precise empirical laws (see Shrader-Frechette and McCoy 1993, chap. 2). For example, although the term "species" has a commonly accepted meaning, and although evolutionary theory gives a precise technical sense to the term, there is general agreement in biology neither on what counts as causally sufficient or necessary conditions for a set of organisms to be a species nor on whether species are individuals (Cracraft 1983, 169–170; see also Mayr 1942, 1963, 1982b, 273–275, 1987; Simpson 1961; Sokal and Sneath 1963; Ghiselin 1969, 1987; Hull 1974, 1976a, 1978, 1988, 102ff., 131–157; Van Valen 1976; Gould 1981; Sober 1981; Kitcher 1985a; Rosenberg 1985, 182–187; Van Der Steen and Kamminga 1991). Such laws also appear unlikely because the apparent ecological patterns keep changing as a result of heritable variations and evolution (see, e.g., Simpson 1964; Ruse 1971, 1989; Mayr 1982b; Rosenberg 1985; Sattler 1986, 186ff.; Sober 1988). Moreover, neither specific communities nor particular species recur at different times and places. Both the communities and the species that comprise them are *unique* (see, e.g., Norse 1990, 17ff.; and Wilcove 1990, 83ff.). Of course, every event is unique in some respects (see Stent 1978, 219), and repetition of unique events is in principle impossible (Hull 1974, 98). Although—in terms of the covering-law model—initial conditions might be able to capture some of the uniqueness of an event, ecologists often do not have the historical information either to specify the relevant initial conditions or to know *what counts as* the unique event (see Kiester 1982, 355ff.). Consequently, instead of developing their own general theories and laws, ecologists are often forced to be content with a "user" science, a discipline based on borrowings and insights from other sciences.

Admittedly ecologists may apply useful findings about particular models to other situations, species, or communities. Nevertheless, such models are unlikely to help us develop general, exceptionless laws. One reason is that the ultimate units of ecological theory (e.g., organisms) are few in number as compared with the ultimate units in other scientific theories (e.g., molecules or subatomic particles), and they cannot easily be replicated. As a result, ecologists can rarely discount the random or purely statistical nature of events or changes; one disturbance in one key environment may be enough to wipe out a species. Model applications are also limited because we do not know the natural kinds. And if not, then perhaps the best paradigms of laws in ecology do not mention the species category at all. Eldredge (1985; see also Brandon 1990, 72ff.), for example, argues that because species are members of a genealogical hierarchy only, they do not take part in biological processes.

If exceptionless ecological laws are unlikely, and if there are problems with applying general ecological theory, given that species are not obviously natural kinds and that each individual in a population is unique, then apart from what sorts of claims are most appropriate to pure ecology (if there is such a thing), problem solving in applied ecology may require a new logic of case studies as well as a new method for helping to obtain reliable inferences. In this essay, we are more interested in describing, illustrating, and defending the methodological process leading to such inferences rather than in their epistemological status. Moreover, although there may be a variety of "logics" and associated methods able to encourage progress in ecology, there are four reasons that we are interested only in the "logic" and method characterizing case studies: (1) A discussion of the types of claims relevant to community, population, and ecosystems ecology is a difficult and massive undertaking, given the problems (already noted) with general ecological theory and concepts. (2) Others have already begun this undertaking (e.g., Van Der Steen and Kamminga 1991). (3) Our focus, instead, is primarily on the "logic" and method that might be most appropriate to the environmental problem solving of *applied* ecology. (4) Also, although a variety of logics and associated methods may be useful in applied ecology, our own field work (see McCoy et al. 1993; Shrader-Frechette and McCoy 1993), as well as the insight of a recent committee of the U.S. National Academy of Sciences (NAS) and the National Research Council (NRC) (Orians et al. 1986) suggest that case studies may provide the best approach to applied ecology. Indeed, when it was asked to assess the use of ecology in environment problem solving, the committee chose to illustrate how the practice of ecological science focused on case-specific ecological *knowledge*, rather than on the development or application of some general ecological *theory* (ibid., 1, 5). Faced with the absence of general ecological theory and laws available for environmental problem solving, the U.S. NAS-NRC committee recognized that ecology's greatest predictive success occurs in situations having weak or missing general ecological theory and involving only one or two species (ibid., 8). These situations suggest that the success might be coming from sources other than the general theory: lower-level ecological theories and the natural-history knowledge of specific organisms (ibid., 13, 16; see also Gorovitz and MacIntyre 1976). As the authors of the National Academy report put it, "the success of the cases described . . . depended on such [natural-history] information" (Orians et al. 1986, 16).

3. The Method and "Logic" of Case Studies

The vampire bat research included in the NAS report is an excellent example of the value of specific natural-history and case-study information when ecologists are interested in practical problem solving (ibid., 28). Its goal was to find a control agent that affected only the "pest" species of concern, the vampire bat. The specific natural-history information useful in finding and using a control, diphenadione, included the following facts: The bats are much more susceptible than cattle to the action of anticoagulants, they roost extremely closely to each other, they groom each other, their rate of reproduction is low, they do not migrate, and they forage only in the absence of moonlight (Mitchell 1986). Rather than attempting to apply some general ecological theory, "top down," scientists scrutinized this particular case, "bottom up," in order to gain explanatory insights (see Kitcher 1985b; Salmon 1989, 384–409). The success of the NAS case study suggests that one important method of applied ecology, focusing on case studies, might be applicable in unique situations where we cannot replicate singular events. But what "logic" and method are appropriate to case studies? In subsequent paragraphs we shall attempt to answer this question. We shall use examples from the various analyses of Northern Spotted Owl conservation in the Pacific Northwest both to develop and illustrate our claims about the method of case studies and to motivate our discussion of the "logic" of case studies.

The survival of Northern Spotted Owls has been an increasing concern over the last two decades because timber harvests during this time have removed almost all the accessible lowland old-growth forest and forced the much-reduced Spotted Owl population to exist primarily in the rugged, mountainous old-growth forest of the Pacific Northwest. The basic problem facing applied ecologists studying the taxon is to determine how to protect resident populations of the Northern Spotted Owl so as to make policy recommendations that achieve both owl protection and the multiple uses of the forest required by law. To solve this problem, applied ecologists need to determine (1) habitat characteristics required for nesting and for successful survival; (2) successful owl dispersal and distribution; (3) owl population sizes able to withstand environmental fluctuations and random demographic changes; and (4) effective population sizes able to minimize genetic depression. Over the last 23 years, ecologists studying the Spotted Owl have made some progress in understanding these four issues. Regarding (1), for example, some ecologists concluded that Spotted Owls do not breed in young, second-growth forests (Salwasser 1986, 232). Using the framework of island biogeographic theory, the Interagency Scientific Committee to Address the Conservation of the Northern Spotted Owl drew a number of conclusions (typical of problem solving, in applied ecology) regarding (1) through (4): that nesting and survival of the Northern Spotted Owl requires 191 habitat blocks of old-growth forest, each block 50 to 676,000 acres; that the blocks ought not be more than 12 miles apart, boundary to boundary; that the blocks (in Oregon, California, and Washington) need to be connected either by corridors or by "suitable forest lands" (with timber having an average diameter at breast height of at least

11 inches and with at least 40 percent canopy cover); and that habitat blocks need to contain at least 20 pairs of owls (Thomas et al. 1990). Congressional hearings (including examination of the key scientists involved in the Spotted Owl studies and recommendations), however, made it clear both that there is no confirmed general ecological theory to justify the conclusions and recommendations of the Interagency Committee and that even the best studies of the owl have been explicit neither about the logic underlying their conclusions nor about all of the methods used (U.S. Congress 1990). Instead, the most prominent researchers on the Spotted Owl filled the gaps in their limited data with appeals to untested (often untestable) general theories such as island biogeography (Thomas et al. 1990; see U.S. Congress 1990). In doing so, ecologists studying the owl came under attack for using general theories that were untested "in the real world," for employing inadequate "rigor," and for drawing conclusions unlikely to be supported by other reasonable persons (U.S. Congress 1990, 260–296).

Even though neither a case-study method nor an associated "logic" is explicit and fully defended in any owl studies, we argue that by examining, evaluating, and making explicit various inferences in the best owl studies (for example, Gutierrez and Carey 1985; Salwasser 1986; Dawson et al. 1987; Thomas et al. 1990), we can "make sense" of many Interagency Committee conclusions (Thomas et al. 1990). In subsequent paragraphs, we develop and illustrate a method of case studies and a set of informal inferences ("logic") associated with it. Although space constraints prohibit our using Spotted Owl studies to illustrate this "logic" in more detail, we believe that our subsequent discussion may help to provide the rough outlines of a framework or "recipe" for using the method of case studies and its associated "logic" in other unique situations in applied ecology. In the following paragraphs, we shall outline a method of case studies, illustrating each step with examples from owl research, and then we shall discuss the informal "logic" associated with the method.

Campbell (1984, 8) claims that the method of case studies is "quasi-experimental"—an interesting choice of terms since ecologists sometimes classify their methods as "classical experimental," "quasi-experimental," and "observational" (see Parker 1989, 199). Classical experimental methods involve manipulation, a control, replicated observations, and randomization. Observational methods may not include any of these four components. Between these two methodological extremes lie quasi-experimental approaches, like the method of case studies, that embody some manipulations but lack one or more of the four features of classical experiments.

The method of case studies is (in part) "experimental"—as opposed to merely observational or descriptive—in that its goal is specification of cause-

and-effect relationships by means of manipulating some of the variables of interest (Merriam 1988, 6–7). It is "quasi"-experimental, however, in that control of these variables often is difficult, if not impossible. In ecology, quasi-experimental methods often involve some manipulation and partially replicated observations. The interactions are complex (see, e.g., Levins 1968b, 5ff.; McEvoy 1986, 83), and there are typically uncertainties regarding subject and target systems, boundary conditions, bias in the data and results, and the nature of the underlying phenomena (Berkowitz et al. 1989, 193–194). As a result, usually it is impossible to use either classical experimental or statistical methods or even to specify an uncontroversial null hypothesis (see Parker 1989).

In general, the case-study method aims at clarifying, amending, evaluating, and sometimes testing examples or cases. Unfortunately, in investigating particular cases, no simple logic, such as hypothesis-deduction, is applicable. Instead, one must follow a method, a set of procedures and rules of thumb, that help one to confront the facts of a particular situation and then look for a way to make sense of them through a set of informal inferences ("logic"). Often one knows neither the relevant variables nor whether the situation can be replicated. Indeed, in some of the best case studies performed (cited in the NAS report) ecologists remained divided even on the issue of the relevant variables. In the study of the Spotted Owl in the Pacific Northwest (Salwasser 1986), for example, some theorists claimed that limiting genetic deterioration is the most critical variable in preserving the owl and determining minimal population sizes. Other researchers, however, maintained that demographic (not genetic) factors are the most critical variables.

As a consequence of uncertainties about the relevant variables, researchers using the method of case studies have often been forced to use a "logic" of informal causal, inductive, retroductive, or consequentialist inferences in order to "make sense" of a particular example or situation (see Grünnbaum 1984, 1988, 624ff.; Gini 1985; Carson 1986, 36; Edelson 1988, xxxi–xxxii, 237–251; Shrader-Frechette and McCoy 1993). In the Spotted Owl study from the NAS volume, for example, ecologists "made sense" of the situation by means of a number of inductive inferences based on observations about reproductive ecology, dispersal, and foraging behavior. As such, the inductive "logic" in the Spotted Owl case might be said to be "quantitative natural history" (see Salwasser 1986, 227; Ervin 1989, 86ff., 205ff.; Norse 1990, 73ff.). In using such informal inferences, the case-study analyst has two main objectives: to pose and to assess competing explanations for the same phenomenon or set of events and to discover whether (and if so, how) such explanations might apply to other situations (see Yin 1984, 16ff.). When they wrote *All the President's Men* (1974), for example, Bernstein and Woodward used a popular version of the method of case studies. They also used an informal "logic" to assess competing explanations for how and why the Watergate coverup occurred to suggest how their explanations might apply to other political situations (see Yin 1984, 24).

3.1. Five Components of the Method of Case Studies

In order to assess competing explanations of the same case, scientists must consider at least five factors: (1) the research design of the case study; (2) the characteristics of the investigator; (3) the types of evidence accepted; (4) the analysis of the evidence; and (5) the evaluation of the case study. The research design of the case study is a plan for assembling, organizing, and evaluating information according to a particular problem definition and specific goals. It links the data to be collected and the resulting conclusions to the initial questions of the study. Because the use of case studies is so new, however, no accepted "catalog" of alternative case-study research designs is available (see, however, Cook and Campbell 1979). Most research designs, nevertheless, appear to have at least five distinct components: (1) the questions to be investigated; (2) the hypotheses; (3) the units of analysis; (4) the "logic" linking the data to the hypotheses; and (5) the criteria for interpreting the findings. (See Yin 1984, 27, 29ff.; Edelson 1988, 278–308, 231ff.; Merriam 1988, 6, 36ff.)

In the Spotted Owl case study from the NAS volume, scientists addressed two main *questions:* (1) What are the minimal regional population sizes of owls necessary to ensure long-term survival? (2) What are the amounts and distribution of old-growth forests (the owls' habitat) necessary to ensure their survival? Although a given case study involves multiple *hypotheses*, one hypothesis in the Spotted Owl case was the following, "This particular Spotted Owl management area (SOMA) is supporting as many pairs of owls as expected on the basis of calculations of N_e, expected population size" (Salwasser 1986, 242). The *unit of analysis* in the Spotted Owl case study was the existing population of individual owls. The northwestern regional Spotted Owl population is currently estimated at approximately 2,000 in the U.S. In other case studies, the unit of analysis could be an individual organism. Or, there could be multiple units of analysis.

The most problematic aspect of the research design of a case study is the fourth component, the logic linking the data to the hypotheses. Essentially this "logic" is an informal way to assess whether the data tend to confirm the hypotheses. Because the auxiliary assumptions and controlling parameters in a case study frequently are not clear, and because a case study often represents a unique situation, scientists typically are unable to use hypothetico-deductive logic. Instead, they often are forced to use what Kaplan (1964, 333–335) calls "pattern" models of inferences. Pattern models rarely give predictive power and instead enable us merely to fill in and extend data to formulate some hypothesis or pattern. For example, discussing the relationship between the annual number of traffic fatalities and automobile speed limit in the state of Connecticut, Campbell (1975) illustrated "pattern matching." Each of his two hypotheses—that the speed limit had no effect on number of fatalities, and that it had an effect—corresponded to a different pattern of fatalities. Although he was not able to formulate an uncontroversial null hypothesis and to test it statistically, Campbell concluded that there was apparently a pattern of "no effects" (Campbell 1969; see Yin 1984, 33–35). He simply looked at the number of fatalities, over nine years, and determined that there was no pattern, no systematic trend.

In using informal inferences to examine whether data are patterned, of course, one can always question whether an actual inference is correct. In the Spotted Owl case study, ecologists used a number of "patterns" from theoretical population genetics and ecology, including specific formulas for factors such as F, the inbreeding coefficient. Because some of the variables in the formula for F, for example, cannot be measured in wild populations, the ecologists' informal inferences about actual F are questionable (Salwasser 1986, 236). Likewise, although Campbell claimed, for example, that his data matched one pattern much better than another, it is not clear how close data have to be in order to be considered a "match." Campbell could not use a statistical test to compare his patterns because each data point in the pattern was a single number—fatalities for a given year—and he had five data points prior to the reduced speed limit and four after. These are an insufficient basis for reliable statistical testing. If one analyzes the literature on case studies, however, one can discover a number of criteria for assessing the quality of the case-study "logic," the informal inferences and the associated research design. Yin (1984, 35ff.) and Kidder (1981), for example, suggest construct validity, internal validity of the causal inferences, external validity or applicability of the case study, and reliability.[1]

One tests the reliability of the research design, for example, by using a protocol—an organized list of tasks, procedures, and rules that are specified ahead of time and that help one take account of all relevant variables and methods. In the Spotted Owl study from the NAS report, the protocol consisted of eight steps. One early step was to perform censuses (of the owl) on all national forest land. A subsequent step in the protocol was to perform a risk analysis of the demographic, generic, and geographical results of different management alternatives. The final step was to monitor the various SOMA in order to determine whether those managing the owls were achieving their goals (Salwasser 1986, 238–242). To test the reliability of the research-design "logic" requires developing, amending, and continually improving a data-base against which the case-study findings can be reassessed. The final step in the Spotted Owl protocol, for example, described just such an updating and revision of the Spotted

Owl preservation plan and conclusions. One also assesses the reliability, in part, by determining whether another researcher, evaluating the same case and the same evidence, would draw the same inferences or conclusions.

In analyzing the evidence used in the case study, the scientist employs three general analytic or methodological strategies. The first is developing a case-study *description* that is capable of organizing the data and hypotheses. In the Spotted Owl study the description emphasized the small size of the owl populations and their vulnerability as a result of habitat destruction, chance factors, and genetic deterioration. Formulating such a description presupposes both (1) developing categories that enable us to recognize and collect data; and (2) looking for regularities in the data. A second evidential strategy is *hypothesis formation*, using an inductive or retroductive "logic" to discover patterns or possible causal explanations for the data. Hypothesis formation often can be assisted by organizing inductive events chronologically, as a basis for time-series analysis, or by data-base management programs. In the Spotted Owl study one important hypothesis assessed was that long-term protection of local populations of owls might require exchanging individuals among regional populations. A third evidential strategy, *informal testing*, consists of using an informal "logic" to compare actual empirical results (e.g., survival of the owls) with the predictions generated by the case-study hypotheses and causal explanations. Available statistical techniques are not likely to be relevant here, because each data point in the pattern is probably a single point. Nevertheless, in the Spotted Owl case, for example, ecologists have been able to "test" their models of owl-population viability by using research on the long-term viability of other taxa. (See Yin 1984, 99–120; Lincoln and Guba 1985; Salwasser 1986, 243, 232; Merriam 1988, 133ff., 13–17, 140ff., 147ff., 123ff., 163ff.)

After analyzing the evidence, the scientist can use an informal logic to draw conclusions and compose the case-study report. In the Spotted Owl case, one group of scientists concluded that internal factors (such as changes in fertility) and external stresses (such as habitat disturbances) increase the risk of extinction, that these factors can be offset only by immigration from other populations, and that regional populations of approximately 500 are necessary to protect the owls for several centuries (Salwasser 1986, 235–238). Of the main possible compositional forms of the case-study report—chronological, theory building, linear analytic, or comparative—the Spotted Owl report was a combination of theory building and linear analytic.[2] The final component of the case-study method is to assess the report and conclusions. Often this can be accomplished by using the four criteria already mentioned for evaluating the "logic" underlying the research design: construct validity, internal validity, external validity, and reliability. One can also evaluate the case study in terms of the standard explanatory values, such as completeness, coherence, consistency, heuristic power, predictive power, and so on. Often these evaluations are best accomplished through outside review (see Merriam 1988, 170ff.). In the Spotted Owl study, the ecologists have an on-going plan of monitoring and research to evaluate critical assumptions in their conclusions, metapopulation models, and protocol (see Salwasser 1986, 242; Shrader-Frechette and McCoy 1993).

3.2. Shortcomings of the Method of Case Studies

As many scholars have noted, case studies can easily be biased by the practitioner (Dalton 1979, 17; Callahan and Bok 1980, 5–62; Hoering 1980; Guba and Lincoln 1981, 377; Grünbaum 1984, 1988; Gini 1985; Carson 1986, 37; Edelson 1988, 239–243; Merriam 1988, 33ff.). Although there is no failsafe way to prevent all case-study bias, one way to deal with it is to realize that bias can enter the conduct of all science. Moreover, science does not require that scientists be completely unbiased, but only "that different scientists have different biases" (Hull 1988, 22). If they have different biases, then alternative conceptual analyses, accomplished by different scientists, will likely reveal these biases. Hence, it is important that practitioners of the method of case studies attempt to use an informal "logic" to confirm their results, in a partial way, by using independent data and other case studies. One also might avoid bias—or at least make it explicit—by developing rules for assessing similarities among system components, initial and boundary conditions, and by using multiple methods and multiple sources of evidence. (See Shrader-Frechette 1985, 68ff., 1991, chap. 4; Berkowitz et al. 1989, 195–197.)

Another response to possible bias in case studies is to realize that bias is possible only because of an asset: the flexibility of the method and its associated "logic" (de Vries 1986, 195). For example, in case-study work on the gopher tortoise, ecologists were able to discover a number of insights—such as the directional positions of entrances to tortoise burrows—even though they had neither algorithms nor a deductive logic to guide them (McCoy et al. 1993). Moreover, in certain situations that are unique and not subject to statistical testing, there is no alternative to the case method and informal logic. Because it is an *organized* means of obtaining information, because it can be criticized, and because it proceeds in a step-by-step fashion (problem definition, research design, data collection, data analysis, composition of results, and report), however, the method and its associated informal logic can be used in objective (i.e., unbiased) ways. For example, a number of scientists and philosophers of science have repeatedly argued that a given case study (a) does not illustrate what its practitioners claim; (b) does not fit the model imposed on it; (c) is factually deficient; (d) is a misrepresentation of the phenomena (see, e.g., Beckman 1971; Adelman 1974); (e) fails to take account of certain data (ibid.); or (f) has a "logic" that leads to inconsistency or dogmatism (Hoering 1980, 132–133) or that relies on faulty inferences—for example, the fallacy of affirming the consequent or the fallacy of assuming that two conjoint phenomena have a cause-effect relationship (see Edelson 1988, 255–266, 319ff.). Such criticisms indicate that, because use of the method of case studies, especially its associated "logic," is open to critical analysis and subject to revision, there are at least two ways in which it is rational and objective (in the Wittgensteinian sense of being "public"): (1) Expert practitioners are often able to distinguish a better application of the method and its logic from a worse one. (2) Following the method, and thinking that one is following it, are not the same thing (see Wittgenstein 1973, s. 243ff.; Baker and Hacker 1985, 150–185; 1986, 330–333).

In arguing that the case-study method and its associated "logic" need not be subjective in a damaging sense, let us refer to the Wittgensteinian insight (see section 4 later) that objectivity is not tied to *propositions* but to the *practices* of people. As Wittgenstein puts it: "Giving grounds . . . is not a kind of *seeing* on our part; it is our *acting*" (1969, 204). Admittedly, our more traditional accounts of objectivity are tied to seeing, to mind-independent beliefs about the world, to impersonality, and to a set of judgments or logic. Typically we do not attach praise or blame, respectively, to *judgments* that fail to be objective in this traditional sense. The newer Wittgensteinian account of objectivity, however, is tied to actions, impartiality, and a method or procedure for behaving in a way that lacks bias. Usually we do attach praise or blame, respectively, to *persons* who fail to be objective in this sense (see Newell 1986, 63, 16ff., 23, 30). If a judgment is thought to be objective in the first sense, then obviously a single counterinstance can be enough to discredit it. Objectivity in this sense is not compatible with error. However, objectivity in the second sense is compatible with error. The upshot of distinguishing these two senses of objectivity is that the method of case studies—tied as it is to actions and practices rather than rules, propositions, or a deductive logic—is not infallible although it may be objective in a scientific sense.

Another problem with the method of case studies is that its associated logic provides little basis for scientific generalization (see Yin 1984, 21). In the Spotted Owl case study, for example, generalizations about habitat requirements were problematic because the owls' needs varied from place to place. In California, each pair of Spotted Owls used 1,900 acres of old-growth forest. In Oregon, the per-pair acreage was 2,264, and in Washington, 3,800 acres per pair (Wilcove 1990, 77). While concerns about the ability to generalize are well placed, this apparent problem with the "logic" of case studies is mitigated by two considerations. First, if Cartwright (1989), Fetzer (1974a, 1974b, 1975), Humphreys (1989, 1991), and others are correct, then it may be possible to establish some reliable singular causal claims without first establishing regularities. Second, the single case study and the single experiment face the problem that both can be generalizable to theoretical propositions but not to populations or universes. Both face the problem of induction. Although the scientist must replicate a case or an experiment in order to generalize from it, mere replication is never sufficient for theorizing. The scientist's goal is not adequately accomplished merely by enumerating frequencies. Nevertheless, single cases and experiments, even in physics, are often sufficient for scientific theorizing. As Popper (1968, 28ff., 251ff.) pointed out, the severity of the tests, not mere replication, is important. Cases such as parity nonconservation (Franklin 1986, 100, 192ff.) and the Einstein–de Haas experiment (see Cartwright 1989, 349) likewise suggest that often controversies can be decided on the basis of a single convincing experiment. Hence it is not obvious that use of case studies is seriously flawed because its logic provides little basis for generalization.

The "logic" of case studies is also frequently criticized on grounds that it enables one to evaluate only those interpretations that use of the method of case studies already presupposes. However, any method of "logic" or confirmation is able only to evaluate hypotheses or interpretations that have already been discovered (see Hoering 1980, 135). Moreover, to the degree that the method of case studies, especially its associated logic, is open to critical evaluation, its conclusions are not merely begged and may, to some degree, be tested. M. Edelson and A. Grünbaum both provide insights regarding such testing. Recognizing that direct replication of a case study is typically impossible, Edelson argues persuasively that "partial replication," inference to the best explanation, and pitting a conclusion against rival hypotheses are all useful. Although Grünbaum claims that the data in an individual case cannot be used to *test* psychoanalytic propositions, for example, he maintains that one can use eliminative induction in experimental and epidemiological tests of the "logic" underlying the conclusions of a case study. One also can seek confirming instances or exclude plausible alternative explanations. Although Edelson and C. Glymour recognize that "testing" case-study conclusions is difficult, they appear unwilling to relegate the "logic" of case studies only to the context of discovery. Doing so likely would discourage rigorous argument about the relationship among hypotheses and evidence and might also force us to presuppose an account of testing that was rarely applicable in science. (See Glymour 1980; Meehl 1983; Grünbaum 1984; Edelson 1988, 120, 363, 231–265, 275–276.) Another allegation against the "logic" used in the case-study method is that, because it follows no purely deductive scheme of inference, its practitioners often fall victim to uncritical thinking (see Hoering 1980, 135; Francoeur 1984, 146), erroneous inductive inferences, or the fallacy of false cause (see Grünbaum 1984, 1988). Community ecologists, for example, have been divided recently over the causal role of competition versus random chance in structuring biological communities. As a consequence of this division, different ecologists (using the same case study) often make incompatible and controversial inductive and causal inferences regarding alleged competition data (see, e.g., Simberloff 1976; Simberloff and Abele 1976; Diamond and Case 1986). The solution to such controversies, however, is not to abandon the method of case studies, but to subject its "logic," especially its problematic inferences, to repeated criticism, reevaluation, and discussion—to seek independent evidence and alternative analyses of the same case (see Edelson 1988, 237–251, 286ff., 319ff.). Grünbaum's (1984, 1988) criticisms of many of the central causal and inductive inferences of psychoanalysis, for example, provide an alternative to the case analyses provided by doctrinaire Freudians. Likewise, in evaluating evidence of a species' shared genealogy, Sober (1987, 466; 1988) has discussed in detail which sorts of causal inferences are justified and which are not. In general, much of the literature (like Sober's) that discusses problems with the "principle of the common cause" (see Reichenbach 1956; Salmon 1984) or with inductive inferences helps us avoid questionable inferences in assessing case studies.

4. The Scientific Status of the Method of Case Studies, Especially Its Underlying Logic

Although the method of case studies typically employs a "logic" of informal inferences that are difficult to evaluate, the method has a number of assets. (1) It enables scientists to gain a measure of practical control over real-world problems, like pest management. (2) Its associated "logic," a set of informal inferences, often allows us to make rough generalizations (that often suffice for sensible explanations) for example, about a taxon's susceptibility to anticoagulants, even though exceptions cannot be treated in a systematic way (see Van Der Steen and Kamminga 1991). (3) By facilitating such inferences, the method shows us how to use descriptions of a particular case in order to study different, but partially similar, cases. Apart from practical benefits, the method of case studies is also important because its systematic procedures and logic are applicable to unique situations that are not amenable to replication statistical testing, and the traditional logic associated with hypothesis testing. The method provides an organized framework for consideration of alternative explanatory accounts, for doing science in a situation in which exceptionless empirical laws typically are not evident or cannot be had. Case studies enable us to learn about phenomena when the relevant behavior cannot be manipulated, as is often the situation in ecology (Yin 1984, 19). Case studies are also valuable for some of the same reasons that Kuhnian "exemplars" are important. They show, by example, how the scientific "job" is to be done, how one problem is *like* another (see Kuhn 1970, 187–191). A final benefit of the method is that, although its "logic" is often unable to provide information about regularities or to confirm hypotheses, it does enable us to see whether a phenomenon can be interpreted in the light of certain models or assumptions (see Mowry 1985). More generally, as the authors of the NAS study on applying ecological theory put it: "The clear and accessible presentation of the [case-study] plan . . . focuses the debate and research" (Orians et al. 1986, 247). It enables us to deal with a full range of evidence in a systematic, organized way and therefore to uncover or illustrate crucial details sometimes missed by more formal methods of science.

Despite these benefits of the method of case studies, critics are likely to object that there is no rationale for claiming that either its method or its "logic" is *scientific*. Moreover, any comprehensive defense of the "logic" of case studies would require us to defend some causal account of explanation as well as some account of the type of generality we expect in causal claims in applied ecology. We need to have an account, both of how causality operates and of how at least two independent avenues function, in causal explanation, so as to advance our understanding of phenomena. Nevertheless, no fully developed, specific account of causality is yet available (Salmon 1989, 409). Because it is not, we shall not provide much insight into the sorts of claims most appropriate to applied ecology: causal claims made on the basis of inductive and retroductive inferences and rough generalizations made on the basis of descriptions of natural history. Instead, we shall attempt to show merely that a causal account of scientific explanation—focused on the singular claims and unique events of case studies—is prima facie plausible for at least four reasons. First, pragmatically speaking, many complex situations (like those in ecology) have no obvious other "logic" that appears applicable. Second, there is no recipe for moving from singular claims to abstract regularities. Third, even general laws rely in part on scientific practice to specify how to apply them. Fourth, even general causal laws require reference to singular claims, if they are to work. One of the strongest arguments that the "logic" of case studies is scientific is pragmatic. The "logic" is more useful, appropriate, and workable than others—such as hypothesis-deduction—when dealing with unique situations in which testing and experimental controls are impossible (see Merriam 1988, 20–21). In other words, in many situations, we have no reasonable alternative to the logic of case studies.

A second, Wittgensteinian-Kuhnian justification for the logic of case studies is that it is scientific by virtue of being embodied in the practices (the relevant actions and dispositions) of the scientific community, such as its ability to "see" situations as exemplars, as like each other (see Kuhn, 1970, 191–204). As both Kuhn and Wittgenstein showed, there is room in scientific method for the rationality of *practice,* for a way of grasping, a rule that is exhibited in obeying the rule, rather than in being able to formulate it (see Baker 1986, 255).

Moreover, because this behavior involves an implicit reference to a community of persons (as Kripke 1982; Bloor 1983; Peacocke 1986; and Smith 1988 suggest), case-study practices need not be purely subjective (see Newell 1986). Another Wittgensteinian criterion for the correctness of case-study practices is whether they enable us "to go on" to "make sense" of further practices and to see likenesses among different cases (Wittgenstein 1973, 47–54, 1979, 61, 77–79; Eldridge 1987; Ackermann 1988, 131).

A third prima facie reason for believing that the method of case studies and its associated "logic" need not be dismissed as nonscientific (because it has no exceptionless rules to guide practices) is that no science relies solely on exceptionless rules. This is because, as Kripke (1982) and Cartwright (1989) point out, no rule can determine what to do in accord with it, because no rule for the application of a rule can "fix" what counts as accord. Therefore, every rule generates the same problem: how to apply it. If the practices of experts ultimately guide scientists in applying rules, then it is reasonable to believe that practices can also guide scientists in applying the "logic" of case studies (see Wittgenstein 1973; Baker and Hacker 1985, 1986; Baker 1986; Picardi 1988; and Smith 1988). Hence the "logic" of case studies may be appropriate to science if one conceives of scientific justification and objectivity in terms of method, in terms of *practices* that are *unbiased*—rather than in terms merely of a set of inferences, *propositions* that are *impersonal*. For Wittgenstein, practices are normative and not purely subjective in part because their existence requires multiple (not unique) occasions (Baker and Hacker 1985, 151). Although the "logic" of case studies is typically applied to a unique phenomenon, however, even accounts of unique events may be "tested" indirectly (see Olding 1978) on the basis of past scientific practices, heuristics, or "rules of thumb." Moreover, as mentioned in section 2, even unique events sometimes may be explained if one is able to supply appropriate initial conditions (see, e.g., Fetzer 1975; Van Der Steen and Kamminga 1991).

Fourth, general causal laws require reference to singular claims, if they are to work. All forms of inference, whether deductive or inductive, presuppose the recognition of regularities, parallel cases, and this recognition presupposes recognition of similarities and differences (see Wisdom 1965; Kuhn 1970, 1977; Dilman 1973, 115–120; Fetzer 1975, 95–96; Bambrough 1979, chap. 8; Newell 1986, 88–94). Deduction addresses all possible singular instances that support a particular claim, whereas case "logic" addresses only *some* of the singular instances, often one at a time. Both may transmit truth, but neither is capable, alone, of initiating it. Initiating truth requires some sort of tacit knowledge in the form of both ultimate premises and knowing what explains a particular case or cases. Indeed, all knowing requires some appeal to tacit knowledge.[3] Whenever one *applies* a generalization, law, or theory to a particular case, as

Kripke (1982) realized, one must use tacit, not explicit, knowledge. Likewise tacit knowledge tells a scientist what needs to be explained and what counts as a criterion for justification. As Wittgenstein pointed out in discussing the foundations of mathematics, even mathematical proofs proceed by means of analogy, by means of the tacit knowledge that one case is like another. Hence, if there is a problem with the tacit knowledge that characterizes the "logic" of case studies, then there is likewise a problem with the tacit knowledge that underlies all science. (See Polanyi 1959, 13, 26, 1964; Adelman 1974, 223; Gutting 1982, 323; Bloor 1983, 95; Newell 1986, 92–110.)

Admittedly one might object that, although tacit knowledge is necessary for all science, it is not sufficient. In other words, in contrast to deductive scientific logic which employs tacit knowledge, the "logic" of case studies bears the additional burden of being able, at best, only to show the rationality of a particular scientific conclusion, not to confirm it (see Plantinga 1974, 220–221). Merely being able to show that particular conclusions are rational, however, need not count against a scientific "logic." For the "logic" to be defective, confirmation—or *something more* than an illustration of rationality—must be possible in the situations in which it is used. As with the Spotted Owl case, it is not obvious that there are deductive methods and associated logic able to confirm hypotheses in situations in which the method of case studies and its associated logic are used.

5. Conclusion

Obviously the best way to defend a "logic of case studies" is to illustrate what it can do in a real scientific situation, such as the case of preserving the Pacific Northwest Spotted Owl or controlling the vampire bat. Both studies—although merely sketched here—showed that practical and precise knowledge of particular taxa is often important to the practice of applied science when no general ecological theory is available. This practical and precise knowledge—rules of thumb and informal inferences based on natural history—coupled with the conceptual and methodological analysis typical of the case study, is an important departure from much earlier ecological theorizing based on untestable principles and deductive inferences drawn from mathematical models. Moreover, informal inferences based on natural history are often more capable of being realized in contemporary community ecology than are hypothetico-deductive inferences based on exceptionless general laws. Important ideals for ecological method and its associated "logic," classical testing and use of null models sometimes fail to address the uniqueness of many ecological phenomena and the ambiguity of many ecological concepts. Hence, in addition to a logic of justification, applied ecology—and perhaps other areas of science—need a "logic" of case studies.

Acknowledgments

We are grateful to Greg Cooper, Reed Ross, Michael Ruse, and Dan Simberloff for criticisms of earlier drafts and to the National Science Foundation for grants BBS-86-159533 and DIR-91-12445, which supported work on this essay. Remaining errors are our responsibility.

Notes

1. One evaluates construct validity by employing multiple sources of evidence, attempting to establish chains of evidence and having experts review the draft of the case-study report. One tests for internal validity or causal validity by doing pattern matching, as exemplified in the Campbell case already noted, and attempting to provide alternative explanations (see Grünbaum 1984, 1988). One tests for external validity by attempting to replicate the case-study conclusions in other situations. To the extent that the case is wholly unique, however, replication will be impossible. Nevertheless, the findings of the case study may be valuable if they exhibit heuristic power.

2. The chronological form consists of a case history in temporal order. Theory-building case studies provide an account of the case, usually in the form of causal claims. Linear-analytic case-study reports follow the format of research reports and grant requests. They cover a typical sequence of topics, such as the problem being studied, methods used, findings from the data, and significance and applications of the conclusions. Comparative case-study reports assess alternative descriptions or explanations of the same case (see Yin 1984, 126–135; Edelson 1988, 278–308; Merriam 1988, 185ff.).

3. Tacit knowing is required, for example, when we discover novelty, reorganize experience, understand symbols, distinguish what is meant from what is said, and understand in a gestalt or wholistic way (Polanyi 1959, 18–29). Tacit knowledge is also required when we value something, know reasons and not merely causes, understand subsidiary rather than focal points, grasp unique things, and make methodological value judgments in science—e.g., "reagent x is the best one for the next test" (ibid., 38–93).

Reductionism and Holism

Under headings as various as "reductionism and holism," "analysis or synthesis," "internal versus external relations," "emergent property," and "gestalt-thinking," another epistemological debate in the philosophy of science bears squarely on the proper method for ecology (and, in an important sense, is behind the merological versus holological paradigm schism—that is, between autecology and synecology). The question is: Can an ecological entity be understood through an analysis of its biotic and abiotic components (reductionism), or must any ecological entity be explained by treating it as a unitary entity with unique characteristics (holism)? Part 4 is devoted to this question.

The question is an epistemic one with ontological underpinnings: the conditions for knowing entities are contingent upon the essence of the entities. In the case of a composite entity, such as an ecosystem or community, reductionists assert that the essence of the entity is a function of its parts, and thus knowledge of the parts is adequate for knowledge of the whole. Holists assert that some entities, such as ecosystems and communities, have special ("emergent") properties that are not properties of the parts, and thus knowledge only of the parts is not knowledge of the whole.

Reductionists maintain that parts are *externally* related; holists maintain that parts are *internally* related. Two things are internally related if the property of one is essential to the essence of the other, for example, the property of rationality to humanness. Two things are externally related if the property of one is not essential to the essence of the other, for example, the property of being tall or short to humanness (see Rorty 1972).

The reductionism/holism debate, of course, is not black and white. David Blitz (1992: 178) discerns several shades of gray by identifying five epistemic/metaphysical methodologies along a reductionism/holism continuum (see table 3). In terms of epistemology, at one extreme, reductionists assert that knowledge of the parts is both necessary and sufficient for knowledge of the whole. Holists take a polar position: knowledge of the parts is neither necessary nor sufficient for knowledge of the whole. Emergentists take a middle position, holding that knowledge of the parts is necessary but not sufficient for understanding the whole. Mechanists interpret reductionism in terms of causal relationships, and organicists emphasize the point that the parts cannot be known independently of the whole (Blitz 1992: 176–77).

TABLE 3.
A comparison of methodological approaches to ecology.

Methodology	Ontology	Epistemology
Reductionism	Properties of wholes are always found among the properties of their parts	Knowledge of the parts is both necessary and sufficient to understand the whole
Mechanism	Properties of wholes are effects of the parts and their structure	Knowledge of the kind or type of the cause suffices to understand the type of kind of the effect
Emergentism	Some properties of wholes are not the properties of any of their parts	Knowledge of the parts is necessary but not sufficient to understand the whole
Organicism	Parts cannot exist independently of a whole	Knowledge of the whole is necessary in order to understand the parts and vice-versa
Holism	The basic unit is that of the whole—wholes are independent of parts	Knowledge of the parts is neither necessary nor sufficient to understand the whole

Source: From Blitz 1992: 178.

Reductionism and Its Critics

For good or ill, reductionism has been the distinguishing feature of Western science. From the pre-Socratics to the modern mechanists, philosophers and scientists have sought to explain all natural phenomena by a single theoretical apparatus.

The Milesian philosopher Thales (fl. c. 585 B.C.E.) is an early example. He hypothesized that the totality of nature is to be understood in terms of different forms of water (Aristotle 1979: 17, 983b20ff), perhaps after seeing hail melting into water and boiling water becoming steam; perhaps based on his knowledge of the fact that life is associated with moisture (Guthrie 1962: 54–72). The reductionist project was refined in the mechanistic materialism of Johannes Kepler, Francis Bacon, Galileo Galilei, William Harvey, Thomas Hobbes, René Descartes, Isaac Newton, and others, who saw nature as a grand machine composed of smaller parts which operated according to strict deterministic physical laws. Mathematical laws, applied to physical entities, can explain the motion of the entirety of physical nature.

As we have just seen, reductionism is both ontology and epistemology. Since science, strictly speaking, is not so much concerned with metaphysics as with the empirical tasks of explanation and verification, epistemic (or theory) reduction is most proximately relevant to the philosophy of science (and thereby the philosophy of ecology). Even so, it is crucial to bear in mind that epistemic reduction presupposes ontological reduction.

Ontological reduction is the notion that the cosmos is a composite built out of smaller, simpler units: atoms (or now, quarks, mesons, and gluons). Ontological reduction is the thesis that the properties of any entity may be understood by knowing the properties of its parts, because nothing can be explained about the entity without reference to the parts. For example, it has been argued that the properties of a cell may be understood by the properties of molecules (the reduction of biology to physics and chemistry), or consciousness by neurophysiology.

Epistemic reduction is the situation in which one scientific theory is absorbed by some other more inclusive theory (Nagel 1961: 336–37). Epistemic reduction has had some brilliant successes in Western science, for example, Copernicus's elegantly simple heliocentric explanation of planetary motion over Ptolemy's cumbersome earth-centered model of nested crystalline spheres.

The method of reduction generates critics in proportion to proponents. Faultfinders say that reduction is unsound on both metaphysical and epistemological grounds because wholes are not *merely* composites of parts. A whole is a synergy; its essence is "emergent" in the sense that it has properties that are not describable or predictable from an analysis of the structure and function of the parts.

Philosopher Arthur Lovejoy defines *emergence* or *epigenesis* as "any process in which there appear effects that . . . fail to conform to the maxim that 'there cannot be in the consequent anything more than, or different in nature from, that which was in the antecedent'" (1927: 22).[1] Lovejoy identifies five modes of emergence (ibid.: 26–27), the first epistemic and the remaining four ontological: (1) emergence of instances that cannot be predicted by old laws (e.g., the laws of physics and chemistry not predicting the evolution of life), (2) emergence of new qualities (e.g., secondary qualities), (3) emergence of new types of entities (e.g., vertebrates), (4) emergence of new kinds of event or process (e.g., stochastic process), and (5) emergence of a larger quantity of instances within the system (e.g., increasing number of species).

The phenomenon of emergence underscores the importance of trying to see the "big picture." As Philosopher Ervin Laszlo puts it, "reductionism generates a multiplicity of limited-range theories, each of which applies to a small domain of highly specific events but says nothing about the rest" (1971: 57). If life is an emergent property of matter, then life cannot be explained only in terms of physics and chemistry (Polanyi 1968); if mentality is an emergent property of neural processes, then mentality cannot be explained only in terms of brain physiology.

Defenders of reduction are quick to lay bare the potential problems with the doctrine of emergence. For instance, if pushed to its logical conclusion, the doctrine seems to entail the conclusion that wholes are not fully comprehensible because, by definition, emergent properties are entirely *novel*! Philosopher Samuel Alexander makes the point clear. Life, he says,

> is at once a physico-chemical complex and is not merely physical and chemical. . . . The higher quality emerges from the lower level of existence and has its roots therein, but it emerges therefrom, and it does not belong to that lower level, but constitutes its possessor a new order of existent with its special laws of behaviour. The existence of emergent qualities thus described is something to be noted, as some would say, under the compulsion of brute empirical fact, or, as I should prefer to say in less harsh terms, to be accepted with the "natural piety" of the investigator. It admits no explanation. (1927: 46–47)

Not so, reductionists counter—the doctrine of emergence is actually an impediment to scientific progress. Carl Hempel and Paul Oppenheim contend that emergence needs to be stripped of its "unfounded connotations": "emergence of a characteristic is not an ontological trait inherent in some phenomena; rather it is indicative of the scope of our knowledge at a given time; thus it has no absolute, but a relative character; and what is emergent with respect to the theories available today may lose its emergent status tomorrow" (1948: 150–51). According to reductionists, Alexander is wrong: no unexplained properties should be accepted with "natural piety" (ibid.: 150).

The reductionism/holism debate has been one of the most interesting discussions in the philosophy of science. What bearing does the debate have on the philosophy of ecology? A variety of positions are identifiable.

Merological Ecology

The vanguard of reduction in the life sciences is molecular biology, which attempts to explain biological processes in terms of physical-chemical processes. Detractors of reduction in biology were muted when James Watson and Francis Crick, building on the work of Rosalind Franklin and others, discovered the double helix structure of deoxyribonucleic acid in 1952 (Watson and Crick 1953a, 1953b). And since, as Kenneth Schaffner maintains, genetics has been reduced to physics and chemistry (1969: 342), the "outcome of . . . the development of molecular genetics . . . is to warrant as a working hypothesis a *biological principle of reduction*. This principle, it seems, holds not only for genetics, but also for other biological theories. The principle can be stated as follows: given an organism composed out of chemical constituents, the present behaviour of that organism is a function of the constituents" (ibid.: 345–46; Schaffner's italics). Full reduction becomes possible if it becomes possible to explain all biological processes in terms of physical and chemical processes.

The issue of reduction in ecology revolves around the question: Does an ecological entity, such as an ecosystem or community, have properties that are not explainable or predictable by an analysis of its biotic and abiotic components? Or, to put it more colloquially, is an ecosystem more than the sum of its parts? Merological ecologists, typically with backgrounds in evolutionary biology, tend not to see organization above the level of the individual and do not pay much attention to the physical environment (E. P. Odum, personal communication, October 22, 1997). Consequently their answer is no: significant properties do not "emerge" on a systemic level.

Reductionist ecologists have used the Hempel-Oppenheim argument against emergent properties in terms of ecosystems and communities. For instance, Michael Edson et al. (1981: 594) assert that the difference between reduction and emergence is really the difference between predictability and unpredictability; and the difference between predictability and unpredictability is the difference between having a sufficient or insufficient model. Claims of emergence contribute nothing to the advance of ecological science: "continued use of the term *emergent property* . . . can serve only to divert limited time, attention, and resources from the real focus of ecological inquiry" (ibid.: 595).

In "Mechanistic Approaches to Ecology: A New Reductionism?" (1986), reprinted here as chapter 13, population ecologist Thomas Schoener argues that there is a perfect hierarchy of ecological subdisciplines, and that higher-order subdisciplines of the hierarchy are reducible to lower-order ones (although not all ecological subdisciplines are members of this hierarchy: evolutionary ecology does not fit into it). The subdisciplines comprising the hierarchy are community ecology, population ecology, and individual ecology. According to Schoener, community ecology, which deals with groups of populations, is reducible to population ecology, which deals with populations of single species. Likewise, population ecology is reducible to individual ecology, which deals with individual organisms.

The thrust of reductionist arguments is significant: if the nonbiotic component of ecosystem ecology is reducible to inorganic chemistry via geology, and if the biotic component is reducible to genetics via population and community biology, then the science of ecology itself might be appropriated by chemistry and molecular biology!

Holological Ecology

One of the most interesting narratives in the story of scientific ecology is the history of holistic ideas. The issue of holism is especially important for ecology because the entities of ecological investigation are complex amalgam-ations of biotic and abiotic components.

Ecologists have developed radically different holological interpretations of the ecological community. As we have already seen, Frederic Clements (1916 [reproduced in this volume], 1936) defined both individuals and communities as organisms—the latter being *superorganisms*. Thus, for Clements, "succes-sion was not just analogous to ontogeny; it was ontogeny" (Hagen 1992: 82). South African ecologist John Phillips defended Clements (Phillips 1931, 1934, 1935a, 1935b) by extending the holistic social philosophy of Jan Christian Smuts (1926) to ecology.

Phillips's sustained advocacy of Clementsian ecology prompted Alfred Tans-ley (1935 [reproduced in this volume]) to repudiate Phillips's teleological or-ganismic holism with a nonteleological mechanistic holism. By defining the ecological community as a complex system of biotic and abiotic parts in physi-cal equilibrium, Tansley initiated a new direction for ecology—*ecosystem ecol-ogy*. Raymond Lindeman (1942) advanced the ecosystem concept by interpret-ing it in terms of energy flow. For example, the structure of a pond ecosystem can be understood in terms of energy flow between trophic levels.

The subsequent genesis of ecosystem ecology through the work of Howard Odum (1957), Bernard Patten (1975), Eugene Odum (1962), Frank Golley (1972), and others is the working out of holistic ideas as they relate to the over-all structure and function of ecological communities. Ecosystem ecologists typically have backgrounds in physiological sciences and tend to see overall plans and structures (in contrast with evolutionary ecologists, who tend not to see organization above the level of the individual). These ecologists believe that ecosystems are results of self-organization. Ecosystems can only be understood as unique wholes with emergent properties (E. P. Odum 1986; Golley 1993). According to Odum, philosophers and ecologists "have for a long time main-tained that, at levels of organization above the organism, the whole is more than a sum of the parts. In other words, as components are combined to pro-duce larger and more complex systems, new properties, often called 'emergent properties,' appear" (1986: 3). To illustrate, Odum says that "the fabulous pro-ductivity and diversity of coral reefs and rainforests involve emergent proper-ties that are only found at the level of the community" (ibid.)—in other words, the net metabolic output of coral reef and rainforest ecosystems is not predict-able from the biotic and abiotic components of these systems. For this reason, George Salt (1979: 147) sees the concept of emergent properties as indispensable for the delineation of trophic structure in theoretical population ecology.

To sum up: On the one hand, the fleshing out of merological and holological themes has given twentieth-century ecology its flavor and character. On the other hand, the debate between the opposing schools has been, to a large degree, intransigent. The current challenge is to adjudicate the two views.

Alternative Methodologies

Creative minds have dealt with the diametric tendencies of merological and holological ecology in a variety of ways. Eugene Odum has been at the forefront of ecologists eager to deal with the problem of analysis versus synthesis. Good science, Odum observes in "The Emergence of Ecology as a New Integrative Discipline" (1977), reprinted here as chapter 14, should do both: "It is self-evident that science should not only be reductionist in the sense of seeking to understand phenomena by detailed study of smaller and smaller components, but also synthetic and holistic in the sense of seeking to understand large components as functional wholes." Ecosystem ecology, according to Odum, meets the requirements for good science: ecosystems are at once analyzed in terms of constituent subsystems, but also examined for emergent properties on a gestalt level.

In "'Reductionist Holism': An Oxymoron or a Philosophical Chimera of E. P. Odum's Systems Ecology?" (1995), reprinted here as chapter 15, Donato Bergandi questions Odum's claims by pointing out the problems in the contention that ecosystem ecology is "reductionistic holism." According to Bergandi, reductionism and holism are mutually exclusive, and in claiming to hybridize the two methodologies, Odum does not remain faithful to the ontological and epistemological presuppositions of holism. "Reductionistic holism" is a chimera that obfuscates the need for a truly holistic ecology.

Richard Levins and Richard Lewontin's answer to the reduction/holism question is to reject both the reductionism of mechanistic materialism and the idealism of holism—namely, Clementsian ecology. The correct theory, they say, is the intermediary "dialectical ecology" or "dialectical materialism" (see figure 12). Lewontin explains: "What characterizes dialectics is its rejection of the terms of the argument that places all questions somewhere on a line between explanation by properties of part and explanation by emergent properties or wholes. It is not that a whole is more than the sum of its parts, but that the parts themselves are redefined and recreated in the process of their interaction" (1983: 37). In "Dialectics and Reductionism in Ecology" (1980), reprinted here as chapter 16, Levins and Lewontin answer Simberloff's (1980 [reproduced in this volume]) widely discussed attack on holism by elaborating the method of dialectical ecology.

Another influential answer to the reductionism/holism debate is the "hierarchy theory" pioneered by T. F. H. Allen and others (vide Allen and Starr 1982

Merological ecology		Dialectical ecology		Holological ecology
(mechanistic materialism)	⟷	(Dialectical materialism)	⟷	(Dialectical idealism)

Figure 12. Dialectical ecology

[reproduced in this volume]; O'Neill et al. 1986; Allen and Hoekstra 1992; Ahl and Allen 1996). Hierarchy is a theory of the observer's role in organizing complex systems for the purpose of scientific study (see part 1). The goals of the investigation dictate the scope of observation, for example, the nutrient cycling of a pond or the meteorological repercussions of El Niño. One must focus on certain patterns and processes at the exclusion of others.

Hierarchies are either nested or nonnested. In a nested hierarchy, processes are present at each level as one ascends the hierarchy. Pickett et al. (1994: 20) propose the following nested hierarchy (cf. figure 2):

Biosphere
Ecosystem
Community
Population
Organism
Organ
Tissue
Cell
Subcellular structure
Molecule

Certain processes (such as electron energetics at the molecular level or the Krebs cycle at the cellular level) are present at each level, and the processes of a certain level are predictable, to a degree, from the processes of the levels below it. On the other hand, in a nonnested hierarchy, higher-level events are not inclusive of lower-level events. Social systems are examples of nonnested hierarchies: in a university, graduate students are not found on the level of tenured faculty.

"Hierarchy: Perspectives for Ecological Complexity" (1982), reprinted here as chapter 17, summarizes T. F. H. Allen and Thomas Starr's interpretation of emergent properties in light of the nested/nonnested dichotomy. Consonant with Hempel and Oppenheim's claim, they dispute the presence of emergent properties in nested hierarchies. In contrast, nonnested hierarchies have some properties (such as the personality of the general in a military hierarchy) that are not explainable in terms of the constituents of lower levels, and hence are considered by some to be genuinely emergent.

Samuel Scheiner et al. (1993: 20) argue that at any one level, *all* properties are emergent because they are the result of ecological process at that level: "An emergent property is one that appears at one level in a hierarchy without being a simple summation of the properties at a lower level. In our schema, all properties are emergent since all are determined by the processes at a given level."

Conclusion

For many philosophers of ecology, thinking ecologically *means* thinking in terms of wholes and hierarchies of organization. As Arne Naess puts it, "unecological thinking is thinking in terms of sectors and aspects rather than in terms of hierarchies of wholes" (1981: 10). Yet others champion the reductionist paradigm and claim that all ecological properties will be reduced ultimately to chemical and physical mechanisms. Within the science both viewpoints coexist and the tension between them is sometimes creative and productive, and at other times destructive and counterproductive.

Note
1. Here Lovejoy is referring to the scholastic maxim that there cannot be more "reality" or "perfection" in an effect than in its cause. Anselm (Hakim 1997: 222–25) and Descartes (1989: 105–10) famously used this maxim in their "ontological arguments"—as Kant (1965: 500) would later call them—for the existence of God.

CHAPTER 13

Mechanistic Approaches to Ecology: A New Reductionism?

Thomas W. Schoener

Community ecology is chronically among the most tumultuous and at the same time alluring of ecology's subdisciplines. Presently there is as much controversy over method as over fact (see symposia volumes by Price et al. 1984; Strong et al. 1984a). During the past decade or so, a rather distinct methodology in community ecology, called the mechanistic approach, has been slowly coming into prominence. This approach can be most simply defined as the use of individual-ecological concepts—those of behavioral ecology, physiological ecology, and ecomorphology—as the basis for constructing a theoretical framework with which to interpret the phenomena of community ecology. The approach contrasts with the "descriptive" one, in which community phenomena are represented by models with no lower-level derivations, but which have one or both of descriptive prowess and mathematical convenience. Of course, any given work may contain both mechanistic and descriptive elements. Hence in principle there exists a continuum of mixtures, although the distribution of actual studies may be bimodal. Somewhat informally, I will use "mechanistic" to refer to studies primarily employing the mechanistic approach....

Reducibility of Community Ecology

Formal Conditions for the Reduction of One Area of Science to Another
While most philosophers (review in Wimsatt 1980a) now advocate a broader concept of reductionism than did Nagel (1961) in his classic work, Nagel's formalism, because of its precision and primacy, is an attractive place for our discussion to begin. In general, "a reduction is effected when the experimental laws of the secondary science (and if it has an adequate theory, its theory as well) are shown to be the logical consequences of the theoretical assumptions . . . of the primary science" (Nagel 1961: 352). By experimental

From *American Zoologist* 26 (1986): 81–106.

law, Nagel means generalizations about the phenomena of the science in question, whether they be absolute or statistical. An example of an absolute experimental law is "lead melts at 327°C." Nagel distinguishes "theoretical laws" or "theories" from experimental laws (32) as statements "whose basic terms are not associated with definite experimental procedures for applying them . . . so that a theory cannot be put to direct experimental test" (85). An example of a theoretical law would be a statement dealing with the atomic theory of matter. Nagel admits the distinction is in some ways not hard and fast, and given the statistical and often weakly verified nature of "laws" in ecology, it is even looser here than it is in physics and chemistry.

A problem with reduction can be that the laws of science to be reduced may contain terms not found in the reducing science. Under such circumstances, reduction can be said to be effected formally when the following two conditions are met (Nagel 1961: 353–94). The first is *connectability*, in which terms not in the to-be-reduced science must be related to terms in the reducing science. These relationships can take three forms: (1) logical connections between established meanings; (2) conventions, or definitions; and (3) facts, i.e., relations established empirically. The second condition is derivability, in which all laws of the to-be-reduced science, including those with terms not in the reducing science, must be derivable from the theoretical principles of the reducing science, using where necessary the connectability relationships.

Various difficulties with Nagel's criteria have been summarized by Wimsatt (1980a; see also Hull 1976b). On the one hand, the use of approximations makes a literal execution of Nagel's conditions problematical (in what sense are they logical deductions?) and jeopardizes most conceivable realizations in actual science. On the other hand, if such difficulties can be overcome, the in-principle satisfying of Nagel's conditions may generally be possible, even though the in-practice realization may be too complex to be useful to science and may merely satisfy a philosophical desire for ontological simplification. Because scientists seem often to show reductionist behavior, broadly construed, Wimsatt (1976b, 1980a) argues that rather than emphasizing Nagel's idealization, we should adopt a broader characterization which corresponds more closely to the actual practice of science. Toward this end, Wimsatt (1976a, 1976b) proposes "explanatory" reduction, in which one has "an explanatory relation between a lower-level theory or domain of phenomena and a domain . . . of upper-level phenomena" (Wimsatt 1976b: 220). Further, one "attempts to identify or explain the upper-level whole and its properties with or in terms of a configuration of lower-level parts and their known monadic or relational properties" (ibid.: 208). A consequence of Wimsatt's modification is that practical aspects of reductionism, e.g., "research strategies," are emphasized over formal aspects; if the lower-level characterization becomes too intricate or cumbersome, it is no longer explanatory.

My tactic in this section is to evaluate how closely Nagel's conditions might be met by a mechanistic approach to community ecology. I will try to show that, while of course community ecology has not been reduced *in toto,* some prototype reduction exists for certain of its aspects that appear explicitly to satisfy Nagel's conditions in most major ways. In so doing, I will be trying out the notion that to the degree Nagel's conditions are met by extant mechanistic approaches without excessive postulation of links and deductions not now existing, and without excessive complexity, the reduction is usefully explanatory. The existence and success of such prototype reductions, I will argue, makes a much broader reduction at least plausible.

Levels of Ecology

To attempt application of Nagel's conditions, we must first decide what the various subdisciplines of ecology are; each of these will then be considered a separate "science" in the above terminology. Moreover, as reduction proceeds from "higher" to "lower" sciences in some sense (Medewar 1974), we need to arrange the subdisciplines, inasmuch as possible, into a hierarchy of levels. Hierarchies are perhaps best defined with respect to objects, such as organisms or populations, which are in fact the "parts" composing the various levels. I am going to assume in what follows that these "parts" are the objects of *principal* focus with respect to a particular subdiscipline, rather than being any kind of object that is mentioned in the phenomenology or theory of the subdiscipline. Hence, *term* is obviously a more inclusive label than *part.*

Beckner (1974) has stated the formal conditions for a perfect hierarchy as follows:

a. every part P_i is assigned to exactly one level L_i;

b. every part P_i (except those of the highest level) is a part of exactly one part at each level above L_i; and

c. every part P_i (except those of the lowest level) is exhaustively composed of parts at each level below L_i.

I will now attempt to show that while a perfect hierarchy exists for a subset of ecological subdisciplines, not all such subdisciplines are members of a perfect hierarchy. This is despite the fact that one of the latter subdisciplines is often considered to deal with the highest level of ecology.

The three disciplines of ecology that, when narrowly enough constructed, do form a perfect hierarchy, are *individual ecology* (the parts are individuals); *population ecology* (the parts are single populations, defined as those individuals in some place belonging to a single species); and *community ecology* (the parts are collections of populations occurring in some place). Individual ecology is itself decomposable into physiological, behavioral, and functional-morphological ecology; all such disciplines focus on the individual but have somewhat different objectives and theoretical structures. Population ecology deals with single-species populations one at a time. Phenomena of interest include kinds of items in the diet, behavioral and physiological thermoregulation, territory size, and mating strategies. The phenomena of interest in population ecology are aggregate properties of the individuals composing the population; examples are age structure, sex ratios, growth rates, and reproductive schedules. Community ecology deals with a group of populations in some place. Here the aggregate properties of interest concern the various species populations: abundance distributions, species diversity, species-turnover rates, and so on. These three kinds of ecology satisfy all of conditions (a)–(c), at least ideally. That condition (a) is satisfied follows from the definition of the subdisciplines. That (c) is satisfied follows from the definitions of populations and communities. Condition (b) is satisfied if communities comprise mutually exclusive populations. Because communities are in practice rather arbitrarily and dissynchronously defined, this need not be the case, but one might imagine that a study by a consortium of researchers would designate communities in this fashion, and if the world's communities were ever catalogued, that they would consist of nonoverlapping populations.

A fourth subdiscipline of ecology, often considered the highest level, is ecosystem ecology. The parts are ecosystems; an ecosystem is defined as a community or communities plus the physical environment (Whittaker 1975). Once the physical environment is brought into the picture as a part, condition (c), that each level is exhaustively composed of parts from lower levels, is clearly violated. One might try to argue that physiological ecology also has "parts," e.g., environmental input such as solar radiation, that are purely physical; if so, however, then individual ecology would violate condition (b) and the perfect hierarchy would be destroyed from below. For reasons expressed in the following paragraphs, I prefer not to think of physiological ecology this way, and I would argue that ecosystem ecology is skewed aside from the hierarchy from individuals to communities.

For the sake of completeness, a final subdiscipline of ecology is often recognized—evolutionary ecology. It can be defined as that portion of the larger science of evolutionary biology which is relevant to ecology. (One might simply recognize evolutionary biology as the "science" in question, but I am trying to relate terms used here to those commonly found in the literature.) I see evolutionary ecology as arranged almost entirely laterally with respect to ecology's other subdisciplines, and certainly not in any perfect hierarchy. It is not so clear what the "parts" of evolutionary ecology are. If they are alleles and/or genotypes, and if evolutionary ecology were the lowest level, then condition (c) is violated. If its "parts" also include individuals and populations, then condition (a) is violated, and either or both of conditions (b) and (c) are also violated. Less formally, some ecologists are interested in evolutionary phenomena at various levels: the individual, the population, and the community (e.g., Lewontin 1970; Wilson 1980). All this may mean that evolutionary ecology is better considered a "perspective" (in Wimsatt's (1976b: 254) sense; for more on identifying boundaries of sciences, see Darden and Maull (1977)).

As mentioned, the rationale for proposing a hierarchy is that typically reduction is attempted from higher-level sciences to lower-level sciences. However, the fact that one science stands higher than another in a perfect hierarchy is neither a necessary nor sufficient condition for reduction of the higher to the lower science. Levels are defined with respect to parts; reduction is defined with respect to theories. Hence reduction may or may not be possible, given a perfect hierarchy (Beckner 1974). It is also false to argue that if sciences are not in a perfect hierarchy, one cannot be reduced to another. One might be misled by making the argument that parts in the higher-level science may not be exhaustively composed of parts in the lower-level science; therefore connectability is violated. But connectability refers to terms, not parts; the former is a more inclusive category than the latter, so that even if there are "parts of parts" in a higher-level science not occurring as parts in the lower-level science, they may still occur as terms at the lower level, or at least be relatable to terms at the lower level.

A related complication is that the theoretical structure of a particular ecological subdiscipline may include as "terms" those predominantly used at higher levels, or even include the "parts" of higher levels. For example, in models of the costs and benefits of territorial defense, a major conceptual issue in individual ecology, the term "rate of intrusion" is necessary. In turn, this term is strongly related to "number of individuals in a population," a term of major explanatory focus in population ecology. But the fact that there exists another subdiscipline of ecology devoted to explaining such properties as population number need not destroy the integrity of lower-level subdisciplines using those terms. In individual ecology, population number might simply be considered an input parameter in the same sense as radiant energy or some other physical quantity whose etiology is unnecessary for some kind of explanation in that subdiscipline. On the other hand, there might exist some "metatheory" spanning several disciplines such that, for example, understanding population phenomena may contribute to predictive power at the individual level. . . . More generally, the subdisciplines that we showed stand in a perfect hierarchy on the basis of their parts may not have entirely discrete bodies of theory. Moreover, if in some sense theories can be said to be hierarchical, a perfect hierarchy defined with respect to parts may not imply the equivalent sort of hierarchy with respect to theories or descriptions (Wimsatt 1974). . . . In short, our analysis of levels has left entirely open the question of the reducibility of ecology's subdisciplines. We now attack that issue directly.

Formal Reducibility of Community Ecology

This section considers the possible reductive relationships between community ecology, population ecology, and individual ecology. I will first attempt to show that the population-dynamical approach to modeling community-ecological phenomena makes a reduction of community to population ecology plausible. I will then attempt to show how much if not all of the theoretical model-structure of population ecology might be reduced to individual ecology.

Community phenomena such as numbers of species and their abundance distributions can in principle be understood by analyzing a set of differential (or difference, or hybrid) equations, each having the abundance of a component species as the dependent variable. For each such equation, independent variables may include population abundances of other species on the same trophic level and/or on different trophic levels and physical quantities such as the supply of some nutrient. The equations represent changes in abundances through time, as these are affected by the independent variables and parameters through births, deaths, immigrations, and emigrations. Such equations in fact compose a large part of the theoretical machinery of the subdiscipline population ecology. Certain phenomena of interest there, e.g., population growth, are directly representable by such equations; others, such as age structure, are representable by a more extensive set of equations whose output can be combined to give changes in total number of individuals in a population.

So commonplace is the use of such equations in community ecology that it is easy to miss that this usage in fact may automatically constitute a reduction of that subdiscipline to population ecology. To be convinced that this interpretation makes sense, it is helpful to imagine theoretical approaches in community ecology that do not involve population-dynamical equations. A number of prominent ones exist, including the early MacArthur (1960) broken-stick models for species abundances and the MacArthur-Wilson (1967) theory of island biogeography. The fact that the first has been declared obsolete by its founder (MacArthur 1966) and the second by Williamson (1983) may indicate a general trend of declining popularity of such models. Although many, myself included, are far from ready to write off the second as yet, replacement of MacArthur-Wilson dynamics (where the dependent variable is number of species) with population-dynamical models is certainly conceivable. All this illustrates that the reduction of community to population ecology is far from complete although it is plausible that it will eventually become entirely or nearly so.

A second difficulty for the just-proposed course of reduction is that a theory having to do with evolution in communities (e.g., character displacement) may not be representable using models whose variables are of the kind listed above.

Because the same problem arises for the reduction of population ecology, I discuss the two together below.

A much stronger kind of reduction of community ecology to population ecology would occur were there to exist no experimental laws or valid theoretical laws that in any major way entail interaction between species populations. Something like this view has been favored by ecologists from Gleason (1926) to Simberloff (1983; see Simberloff 1980 [reproduced in this volume] for the weakest of disclaimers). The latter writes, for example, "We are asking if species' individual responses to the physical environment suffice to explain their distributions" (Simberloff 1983). If it can be shown that neither vertical (e.g., predation) nor horizontal (e.g., competition) connections in the food web are very important, then the ecology of single-species populations is sufficient to explain the phenomena of focus in community ecology—species diversity, relative abundances of species, species turnover, and so on. (Notice that almost certainly predation and competition will be shown either both important or both unimportant, as they march in logical lock-step. That is, significant resource competition at one level implies significant predation at the level of the resources and vice versa (Hairston et al. 1960).) In a very literal sense the community then becomes a whole which is entirely the sum of its parts, and the limits that one places on the set of species composing the community become totally arbitrary. This does not mean that community-level phenomena would necessarily disappear from ecological consideration. But it does mean that the theoretical explanation for such phenomena would at best involve large-number concepts such as the central limit theorem—e.g., the log-normal distribution of species abundances results from many independent effects acting multiplicatively on independent populations (May 1975).

A lot of the controversy in present-day community ecology can thus be viewed [as being] about the strong reducibility of community to population ecology, i.e., whether or not species populations are additive or conjunctive. If Simberloff and colleagues can show that species interactions are minor, then community ecology will cease to exist as an interesting theoretical discipline. As is well known among ecologists, many persons, myself included, strongly oppose the view that anything like this has been shown or even that recent research results are headed in that direction (e.g., Schoener 1982; Connell 1983; Quinn and Dunham 1983; Roughgarden 1983; Schoener 1983a; Paine 1984). The only reason the issue is not yet settled is that so little research has been done, relative to the number of existing systems, to make a statistically valid generalization. . . .

In summary of this section, I have argued that a nonevolutionary community ecology is in principle reducible to a nonevolutionary individual ecology via a reduction through population ecology, and that an evolutionary community (and population) ecology is probably in principle reducible to either a nonevolutionary individual ecology plus evolutionary ecology or to an evolutionary individual ecology by itself. Furthermore, I have argued that for certain aspects of the upper-level theory, the reduction is practical and/or has already taken place. Notice that there is nothing but complexity barring the way from systems of equations with many variables being reduced in the same way as [a] logistic equation. . . . For example, we could have many species in a food web rather than a few, and we could be interested in modeling indirect effects (those passing through intermediate species) rather than direct ones—the reduction could in principle still go forward. Moreover, the behavioral complexity focused upon by the non-population-dynamical mechanistic approach could in principle be incorporated into population-dynamical equations. Finally, when there is one equation per species, the sometimes unsatisfactory assumption that all individuals are equal could be taken care of by replacing it with a set of equations for each species, distinguishing (as is often done) age classes, or size classes, or sexes. The resulting complexity, while not being a formal impediment, could of course be a major practical impediment, so that the reduction would contribute little to understanding. We shall return to this possibility. . . .

The Mechanistic Approach as a Research Strategy

Wimsatt (1980a) has written that "the in principle claim of the reductionist is seldom in dispute," and that in the fields he is familiar with, "the issue between scientists who are reductionists and holists is not over the in principle possibility of an analysis in lower-level terms but on the complexity and scope of the properties and analyses required." Whether this is true or false for community ecology (I doubt community ecologists have thought much about it until recently), Wimsatt is certainly correct that, given that reduction is in principle possible, its execution may not be worth the trouble in terms of insights gained or research facilitated. Toward evaluating this possibility, I now discuss the pros and cons of the mechanistic approach to community ecology as (to use Wimsatt's phrase) a "research strategy."

Many advantages of the mechanistic approach have been mentioned above and discussed. . . . For myself, its chief advantage is that it allows a theoretical understanding of how variation in individual-ecological properties—those of behavioral and physiological ecology—affect population and community structure. As Wimsatt (1976a) points out, this would not be so vital were there few exceptions to laws at the macrolevel, or were exceptions homogeneous when translated into microlevel terms. In fact, variation at the community (and population) level is extensive, so much so that rather than exceptions to a few "laws," it appears that community ecology is a genuinely pluralistic field, with many different "laws," each restricted to a rather narrow domain (Schoener 1985).

There are at least three major consequences of the advantage just discussed. First[,] qualitative predictions about how behavioral and physiological properties affect community and population dynamics and equilibrium become possible. For example, does an energy-maximizing predator stabilize or destabilize a predator-prey relationship? What behavioral traits would result in a population with a leptokurtic utilization distribution? How does metabolic rate affect population growth rate and stability of species interactions?

Second, from an array of possible submodels available for a community (or ecosystem) model, the mechanistic approach suggests which to select. The most appropriate model would not be an issue were all models with roughly similar qualitative properties to behave the same way. But it is becoming obvious that in some major cases, and perhaps in many, they do not. Two relevant examples from community ecology stand out. In the first, Gilpin and Justice (1972) showed that, depending upon whether the zero-isocline of a competition model were linear (Lotka 1956; Volterra 1926) or concave, two *qualitatively* opposite predictions would be made about the outcome of competition in an actual system, two species of *Drosophila*. In fact, the isoclines were in reality concave, and the Lotka-Volterra model gave the incorrect prediction. At a more general level, Turelli (1981) showed that which of three qualitatively similar population-growth functions were used in a stochastic model determined the degree *and direction* of the effect of environmental variation on community stability: one gave a positive effect, one a negative effect, and one no effect! Ecosystem ecologists are also becoming aware of the problem. After citing some examples, Watt (1975: 140), in a spate of disillusionment with ecosystem modeling, wrote:

> What is the meaning of the phrase "a general function which describes this curve is . . . ?" Does it mean that the function was plucked out of thin air as being reasonable, or that it was tested against various sorts of ecological data to ensure that it described reality reasonably well, or that it was the product of some type of deductive process which will be outlined at some later time so as to be completely intelligible? Particularly where the function is new in ecological writings, and the explanation for its origin is not given, the critic is basically trapped in a guessing game with the author.

My suspicion, as also voiced above, is that the first of Watt's alternatives is almost always true, and unfortunately, it appears that this may be no longer good enough. The above examples make me less than optimistic, contra Levins

(1966; see also Wimsatt 1980b), about the robustness of community-ecological models, even regarding small details, much less at the scale he is talking about.

If choice of model or submodel makes a difference, how is one to choose? The answer may well lie with the mechanistic approach: select an appropriate mechanistically derived model, rather than one that is arbitrary or at best purely descriptive. And as a coda, do not hesitate to change models or submodels when the situation changes.

The third consequence is perhaps the most ambitious in its claims, but some outstanding examples of its success exist. It is that the mechanistic approach allows *quantitative* predictions to be made about community structure from behavioral and physiological considerations which can be tested with independently gathered macrolevel data. Four studies illustrating this advantage . . . are briefly reiterated here. First, Belovsky (1984, 1986) fitted by nonlinear regression population data describing competition between moose and hare. The "best-fit" population parameters were then translated into behavioral parameters, via a mechanistic model, and those estimates were compared with independently derived estimates of the behavioral parameters obtained from behavioral-ecological (feeding-strategy) considerations. The two were found to be very close, greatly increasing our confidence in the theory.

Second, Tilman (1976, 1977, 1986) used Michaelis-Menten growth considerations to predict quantitatively the values of nutrient ratios that determine different kinds of competitive outcomes. These predictions were verified with experiments. Third, Abrams (1981) checked his estimates of a competition "coefficient" obtained from a model of shell dynamics with observations of marked shells in the field. Again, the two were very close. Fourth, Spiller (1986) evaluated a mechanistic competition-coefficient formula with field observations, then performed field experiments to measure the coefficient directly. Again, agreement was very good. All four of these studies are extremely powerful, in that they allow two independent assessments of a community-ecological theory. When the two are in agreement, our confidence in the theory is greatly increased.

One might wonder from the rosy picture I have just painted why ecologists have not all boarded the mechanistic bandwagon. I think the basic caution of the dissenters is that this approach may portend an extraordinary degree of complexity when many-species interactions are considered. The complexity could arise in two kinds of places in the theory. First, any particular model, if it is to incorporate enough behavioral or physiological variation, may have to be so complex as to be analytically opaque. Already, a tradeoff in this area is detectable within the mechanists themselves. Those who delete population dynamics from their approach can incorporate more behavioral variation than those who do not. . . . Second, even if individual models are manageable, too many models, each with a very narrow application, may render the entire theory so massive and arcane that community ecology will become an impossibly esoteric field, unteachable to undergraduates and run mainly by experts in information retrieval. Worse, a theory too composed of special cases may be untestable, at least without intergalactic travel, as the earth may not contain sufficient communities to provide adequate statistical power.

Some ecologists are probably willing to give up a lot of precision and linkage to lower levels if these things can be avoided. Moreover, as Wimsatt (1980a) points out, advocacy of a reductionist approach coincides with emphasizing internal, rather than external, factors when simplification is necessary. Thus mechanistic people will stress behavioral and physiological detail at the expense of, say, food-web detail. Two-species systems rather than many-species systems, and direct rather than indirect effects, will be emphasized. This is already to some extent happening (see contrasts in Diamond and Case 1985, for example).

As pointed out above, it is not that in principle the mechanistic approach is unable to handle phenomena involving numerous population or community-level variables, e.g., numerous species. It is just that in practice, this may be too overwhelming. The hope that computer technology can make any "in principle" actual is dashed by reading Boyd (1972) and Wimsatt (1980a). For example (Wimsatt 1980a), there are approximately 10^{130} possible chess games of one hundred moves, larger by about forty-one orders of magnitude than the number of elementary particles in the universe and by about eight orders of magnitude of the number of physical events between such particles since the "big bang." So there have not been enough actual states to represent the chess game even if the universe since its inception were a computer! Those who have ever contemplated a very microreductionist approach to community ecology, e.g., following the fate of *each* individual (rather than representative individuals) in a set of interactions potentially very much more complicated than chess should be sobered by these calculations. That which is in principle possible may in fact not be physically possible. The mechanistic people, of course, are not advocating such an approach, and their hope is that reduction (in the way I have described it) may actually sometimes lead to meaningful simplification, not greater complexity.

Even for the same degree of complexity (as measured, say, by the number of free parameters in a model), the descriptive approach may be more suitable than the mechanistic one *if description is an end in itself.* That is, it is conceivable that the most descriptive model for a particular case is nonmechanistic, or more likely, that the single model describing a set of cases better than any other is nonmechanistic. Because, as stressed above, ecological phenomena *in toto* rarely fit any single model well, the latter is in my opinion not so likely; an example is found in my own work on habitat shift (Schoener 1974b). This is also why the role of upper-level generalizations in "winnowing out" inappropriate lower-level representations (as suggested by Wimsatt 1976b, footnote 11) is not likely to be conspicuous for population and community ecology even if it were looked for carefully.

A related advantage for nonmechanistic models, especially linear ones, is their typically intimate association with statistical estimation. Again, however, estimation is possible with nonlinear models; it is just more cumbersome. Moreover, if the assumptions of the estimation (e.g., linearity) are far from true, reliability of the estimation is compromised, and a more complicated estimation procedure (or no procedure) may be preferable.

Finally, of course, reduction has to stop somewhere along its downward path. While I have argued that it may often be practical to reduce community and population ecology to individual ecology, would it be sensible to go farther? That is, should we use physiological laws such as the metabolic-rate-to-body-weight function in their simple descriptive form, or should we use a probably more complicated mechanistic version were one available? And if the answer is yes, should we continue through biochemistry, physical chemistry, and physics? If this *reductio ad absurdum* (or *ad nauseum*!) were possible in principle, it would be strangulating in practice. Scientists will place bounds on a train of reductions that are in principle possible when the sequence becomes too long to have explanatory power (Wimsatt 1976b).

Despite occasional bursts of ambitious pronouncement, we are not going to know for a very long time how the balance of advantages and disadvantages will finally fall. But it is amusing, not very risky, and perhaps even a bit inspirational to speculate, which I now do.

A Mechanistic Ecologist's Utopia

What if the mechanistic program realized its wildest aspirations? What would ecology be like then? Here I imagine the characteristics of a mechanistic community ecologist's utopia. I distinguish six such characteristics.

First, the macroparameters of community ecology will be deemphasized. Less use will be found for concepts like "niche overlap," "niche breadth," and indeed even "niche." "Niche overlap," for example, might be represented by an array of more specific concepts, such as Abrams's (1980) competition ratio, Schoener's (1974a) competition coefficient, and so on.

Second, theoretical models will have proliferated, and each will have a rather specific domain. A pluralistic theory will have replaced an attempted universal one. Pluralism will involve specificity at both the organismic and environmental levels, i.e., with respect to the biological traits of the type of organism being considered (e.g., generation time) and the environmental traits of the community's location (e.g., degree of spatial fragmentation). Elsewhere (Schoener 1985), I have suggested a first list of such traits.

Third, arbitrary models whose sole virtue is mathematical convenience will no longer be acceptable. In order to be used, a model will have to be mechanistically justifiable. It may be that manipulation of such models will require a great deal of mathematical skill with approximations and so on, and perhaps a lot of computer time as well.

Fourth, in both observational and experimental approaches, a greater emphasis will be placed on discovering the mechanism of an interaction or process, not just its existence and strength. The ingenuity required to get at such mechanisms will probably be much greater than that to design the removal or introduction experiments that most of us do today.

Fifth, individual-ecological terms, e.g., those from behavioral and physiological ecology, will commonly appear in designations of kinds of ecological communities. Thus we might have ectothermic communities, semelparous communities, or long-generationed communities.

Sixth, population and community-level hypotheses will be framed in much more precise and obviously testable terms than is presently the case. Perhaps Beckner's (1974) application strategy involving event reduction (see above) will be realized: the "revision of higher-level theory in a manner that facilitates event reduction; that is, the introduction of higher-level descriptions with an eye toward the lower-level explanation of events under those descriptions." The use of quantities and units from behavioral and physiological ecology may bring testability of population- and community-ecological models on a par with that currently possible for, say, feeding-strategy models (e.g., Krebs et al. 1983).

Notice that nothing in this scenario suggests *replacement* of community ecology by individual ecology as a science, despite the prospect of reduction. The phenomena of community ecology will still be of interest[,] although as stressed above, there will be a good deal more unity between the subdisciplines than presently exists. (In this regard, I am supporting Wimsatt's (1976b: 222) view of "interlevel" reduction.)

Of course, as already noted, actualization of the mechanistic program could well falter on complexity and unwieldiness. Exactly what will happen remains to be seen, but we may ask in closing about the effect this and other philosophical analyses might have on the development of community ecology. Will a philosophically self-aware science pursue a different path than one that is philosophically ignorant? Philosophers are sometimes surprisingly self-effacing on this question (e.g., Beckner 1974), and in fact [it] is probably unanswerable; we are participants in an experiment without a control.

CHAPTER 14

The Emergence of Ecology
as a New Integrative Discipline

Eugene P. Odum

It is self-evident that science should not only be reductionist in the sense of seeking to understand phenomena by detailed study of smaller and smaller components, but also synthetic and holistic in the sense of seeking to understand large components as functional wholes. A human being, for example, is not only a hierarchical system composed of organs, cells, enzyme systems, and genes as subsystems, but is also a component of supraindividual hierarchical systems such as populations, cultural systems, and ecosystems. Science and technology during the past half century have been so preoccupied with reductionism that supraindividual systems have suffered benign neglect. We are abysmally ignorant of the ecosystems of which we are dependent parts. As a result, today we have only half a science of man. It is perhaps this situation, as much as any other, that contributes to the current public dissatisfaction with the scientist who has become so specialized that he is unable to respond to the larger-scale problems that now require attention. There is a rich literature on hierarchical theory and philosophy which deserves to be read by today's specialists (Koestler and Smythies 1969; Whyte et al. 1969; Pattee 1973). As expressed by Novikoff (1945), there is both continuity and discontinuity in the evolution of the universe. Development may be viewed as continuous because it is never-ending, but also discontinuous because it passes through a series of different levels of organization.

An important consequence of hierarchical organization is that as components, or subsets, are combined to produce larger functional wholes, new properties emerge that were not present or not evident at the next level below. Feibleman (1954) has theorized that at least one new property emerges with each new integrative level of organization. Whatever the emergent rate, we can conclude that results at any one level aid the study of the next level in a set but never completely explain the phenomena occurring at that higher level, which

From *Science* 195 (1977): 1289–93. Notes omitted. Odum's unconventional use of *hierarchal* has been replaced with the more common *hierarchical*.

itself must be studied to complete the picture. The old folk wisdom about "the forest being more than just a collection of trees" is indeed the first working principle for ecology. For example, intensive research at the cell level has established a firm basis for the future cure and prevention of cancer at the organism level, and perhaps for genetic engineering at the population level, should we ever choose to experiment in this direction. However, cell-level science will contribute very little to the well-being or survival of human civilization if our understanding of supraindividual levels of organization is so inadequate that we can find no solutions to population overgrowth, social disorder, pollution, and other forms of societal and environmental cancer. This is not to say that we abandon reductionist science, since a great deal of good for mankind has resulted from this approach, and some of our current short-range problems can perhaps be solved by this approach alone. Rather, the time has come to give equal time, and equal research and development funding, to the higher levels of biological organization in the hierarchical sequence. It is . . . the properties of the large-scale, integrated systems that hold solutions to most of the long-range problems of society. Again, Novikoff (1945) expressed it well when he wrote, "Equally essential for the purposes of scientific analysis are both the isolation of parts of a whole and their integration into the structure of the whole. . . . The consideration of one to the exclusion of the other acts to retard the development of biological and sociological sciences."

The New Ecology

The rise of what I have previously called the "new ecology" (E. P. Odum 1964) is—in part, at least—a response to the need for greater attention to holism in science and technology. Since the word *ecology* is derived from the Greek root *oikos* meaning "house," it is an appropriate designation for the study of the biosphere in which we live. However, until quite recently, ecology as an academic subject had a much more limited scope than the name indicated. When I first came to the University of Georgia as a young instructor in 1940, my suggestion that a course in ecology be included in a core curriculum for majors received an exceedingly cold reception. My colleagues of those days confused ecology with natural history and voiced the opinion that no new ideas or principles were likely to be revealed in an ecology course that had not already been covered in courses in taxonomy, evolution, physiology, and other subjects considered to be more basic. As a partial result of this rebuff I decided to write a textbook that would emphasize unique principles that emerge at the supraindividual levels of organization. The first edition of *Fundamentals of Ecology* (E. P. Odum and H. T. Odum 1953; see also 1959, 1971), written in collaboration with my brother, Howard T. Odum[,] was revolutionary in two respects: (i) principles were presented in a whole-to-part progression with consideration of the ecosystem level as the first rather than the last chapter, and (ii) energy was selected as the common denominator for integrating biotic and physical components into functional wholes. As the book passed through two more editions and was translated into other languages, these approaches and viewpoints became generally accepted, not only by professionals, but by the public at large. As the environment-awareness movement began to emerge in 1968, some professional ecologists actually resented the public's use of "their" word, but we welcomed it as a long overdue recognition of holistic concepts. Although *ecology* is frequently misused as a synonym for *environment*, popularization of the subject is having the beneficial effect of focusing attention on man as a part of, rather than apart from, his natural surroundings.

A joint research study on a coral reef by my brother and me in 1954 (H. T. Odum and E. P. Odum 1955) can perhaps serve as an illustration of how ecosystem-level study can reveal emergent properties which tend to be missed in piecemeal study. At Eniwetok Atoll we measured the metabolism of the intact reef by monitoring oxygen changes in the water flow. We also did a detailed trophic analysis as a means of charting major energy flows, and were able to construct an energy budget for the whole system. It became evident from the latter that corals and associated algae were much more closely linked metabolically than had previously been supposed, and that the inflow of nutrients and animal food from surrounding ocean waters was inadequate to support the reef if corals and other major components were functioning as independent populations. We theorized that the observed high rate of primary production for the reef as a whole was an emergent property resulting from symbiotic linkages that maintain efficient energy exchange and nutrient recycling between plant and animal components. Our work created considerable controversy and stimulated a number of investigations. Teams of researchers with expertise in chemistry, microbiology, invertebrate zoology, and other fields descended onto the reefs, but remained loosely united in their interest in testing directly or indirectly basic hypotheses about the reef as an ecosystem. Some theories were verified, others refuted, with the result that today there is a rather good understanding of coral-algal relationships and mineral cycling mechanisms in reef systems (Johannes et al. 1972). We like to think that setting up radical but testable hypotheses at the beginning had much to do with this progress. Scientists work together best when motivated by some common idea, even if—or perhaps, especially if—that idea is controversial.

Do these coral reef discoveries have any significance for urban industrial man? Perhaps they do. The Pacific coral reef, as a kind of oasis in a desert, can stand as an object lesson for man who must now learn that mutualism between autotrophic and heterotrophic components, and between producers and consumers in the societal realm, coupled with efficient recycling of materials and use of energy, are the keys to maintaining prosperity in a world of limited resources.

Since the study of ecosystems is best carried out by teams of investigators who are united in their objective of seeking to discover the emergent properties of the whole but have different skills and secondary interests, I realized early that it would be necessary to establish some kind of organization to promote such teamwork. At the University of Georgia we established the Institute of Ecology for this purpose and, with the help of outside financial support, we carry out long-term studies. The Sapelo Research Foundation has provided continuous support that enabled us to mount an unhurried study of the Georgia salt marsh estuaries. A long-term contract with the Atomic Energy Commission (now the Energy Research and Development Administration) has provided a similar opportunity for population and ecosystem-level study of terrestrial and freshwater environments on a large area set aside for atomic research along the Savannah River, an area recently designated as the nation's first national environmental research park.

The complex of Georgia salt marshes and estuarine channels belongs to a general class of ecosystems which we have designated fluctuating water-level ecosystems. They are pulse-stabilized by tidal flows which act as energy subsidies that enhance productivity as much as ten times over that which would be achieved without this natural use of tidal power. Because we could document the work potential and, therefore, the value of these estuaries, our findings have been widely used as a basis for formulating laws and other measures to protect the U.S. coastal zone from insidious alterations (Gosselink et al. 1974). This work, along with parallel investigations of other natural landscapes, has led to the recognition of an important class of ecological systems which from the holistic viewpoint may be termed subsidized, solar-powered ecosystems. Human agriculture belongs to this class. We have much to learn from natural systems of this sort since most of our agroecosystems lack stability and tend to behave in a boom-and-bust manner, perhaps because we do not yet understand the network of feedback energy flows necessary to maintain continuous high productivity.

While much of the work at the Savannah River area has of necessity involved piecemeal studies on the effect and environmental fate of radionuclide and thermal discharges from atomic reactors, we did at the onset select ecosystem development as our central or unifying focus. The locale provided an unusual opportunity to observe and experiment with the process of natural ecological succession, and to study the impact of artificial reforestation as well, because hundreds of fields were taken out of cultivation when the atomic energy facilities were constructed in 1952. We theorized that new systems properties emerge in the course of ecological development, and that it is these properties that largely account for the species and growth form changes that occur

(E. P. Odum 1969). The idea that there is a holistic strategy for ecosystem development remains controversial. An alternate theory that species aggregations do not interact as a whole, and that ecological succession can be adequately explained on the basis of competitive exclusion and other species-level processes, has also been vigorously promoted (Drury and Nisbet 1973; Horn 1974). Again, controversy is welcome since disagreement on the "big ideas" is certain not only to generate useful knowledge, but to promote the art and science of both the experimental and analytical approaches.

The somewhat disappointing performance of the U.S. effort [in] the International Biological Program (IBP) can, in hindsight perhaps, be attributed to the fact that unifying theories or concepts were not set up for testing on the onset. Hundreds of investigators with widely different training and expertise were funded and were expected to work together as a team without a clearly defined common denominator (Mitchell et al. 1976). The biome program as a major part of the U.S. effort under the IBP was conceived in a holistic vein, and the idea of studying the totality of major solar-powered natural ecosystems such as grasslands, forests, deserts, tundras, and so forth was a uniquely American concept. But in practice there was a shortfall of integration. For example, in the grassland studies, which received the first and largest funding, there was never any "grassland theory" for the reductionists to rally around. A prodigious effort by a handful of systems ecologists did manage to link some of the fragmented data into something approaching an ecosystem-level model, but even the most sophisticated mathematical models cannot compensate for inadequate planning, uncoordinated data gathering, or, most of all, the lack of [a] central theme.

The new ecology, then, is not an interdiscipline, but a new integrative discipline that deals with the supraindividual levels of organization, an arena that is little touched by other disciplines as currently bounded—that is, by disciplines with boundaries established and strongly reinforced by professional societies and departments or curriculums in universities. Among academic subjects, ecology stands out as being one of the few dedicated to holism. But I do not mean to imply that ecology is emerging by default; other disciplines, including perhaps even economics, as I will note later, are striving to climb upward on the hierarchical ladder.

The Link with Social Sciences

From another context, the new ecology links the natural and the social sciences (E. P. Odum 1975). In the bottom of the Great Depression of the 1930s sociologists began to shift from the dictum that the proper study of man is man (alone) to the idea that the proper study is man in environment. For example, my father, the late Howard W. Odum, directed a major effort within the Institute for Research in Social Sciences at the University of North Carolina toward development of the concept of regionalism, which he viewed as an approach to the study of society based on the recognition of distinct differences in both cultural and natural attributes of different areas, which, nevertheless, are interdependent (Odum 1936; Odum and Moore 1938). Regional study of social science was widely misinterpreted in those days as being merely an inventory device designed to upgrade "backward" regions so they would contribute to, rather than detract from, the total economy of the nation. Rather, Howard W. Odum envisioned the real goal as the integration of regions, and he hoped that the concept would provide an antidote to divisive sectionalism, which was then spawning bitter economic and political warfare between sections of the nation. To a remarkable extent the philosophy of regionalism did help smooth social and economic transitions in the Southeast, but as a major theory of sociology the concept stalled because there was no appropriate linkage with natural science (applied ecology had not yet emerged to this level of thinking) and because statistical methods of the day were totally inadequate to cope with the mountains of data collected by social science researchers. Now, advances in systems science and electronic data processing promise to alleviate the data processing problem, and we have the new ecology as an improved link with the rest of science.

At the national level, I believe, the current effort to mount a program of research and management for the coastal zone may be the first major test of whether we are yet ready to combine the best of reductionist and synthesis science as a basis for rational decisions. Experience in mounting team research at the ecosystem level suggests that one or more major theories or paradigms that can be tested (and refuted, if possible) must provide a focus if the coastal zone effort—which of necessity must involve local, state, regional, and federal groups—is to be a truly scientific enterprise and not just another series of expensive and frustrating inventories. The tidal subsidy theory and the concept of regionalism stand as two unifying focuses for productive team research in this area.

Since the kind of sectional conflicts which for so long hampered our national development are now appearing on a truly frightening scale in the confrontations between so-called "advanced" and "backward" nations, even a partial success at coastal zone management could have a favorable global impact by demonstrating that action based on holistic values and properties is a viable alternative to development on the basis of competitive exclusion alone.

Technological and Environmental Impact Assessments

The need to raise thinking and action to the ecosystem level is especially evident as the practices of technology assessment and environmental impact analysis assume increasingly important roles in decision-making, especially with regard to public works and energy and industrial development. As a member of the advisory committee of the Office of Technology Assessment (OTA) established by Congress, I can vouch for the fact that there is a serious dichotomy of thinking between those who urge that OTA restrict its work to piecemeal assessment of new technology on the grounds that greater precision can be achieved in this manner, and those who argue for broadening the studies to include environmental, social, and economic aspects on the grounds that a holistic level of assessment is more realistic (Shapely 1973; Decker 1975). The point is that the questions and answers can be quite different depending on the level of assessment. For example, a thoroughly competent study restricted to the technical performance of a fission nuclear reactor could well show that this method of power generation is reasonably safe. Since technologists have stressed safety as *the* limiting factor and the public has logically followed in making safety *the* issue, then a favorable technology assessment of the safety problem would become a powerful signal for government and industry to launch a massive development of this form of atomic energy. Yet a total assessment that includes economic and environmental components (and covers the whole chain of events from mining to waste disposal) might show that as a first-generation attempt to utilize atomic power the light-water fission reactor is badly flawed technology, and thus not yet ready to play a major role in power production, especially where alternatives are available.

In the pages of *Science* the writing of environmental impact statements, as required by the National Environmental Policy Act, has been denounced as a "boondoggle" (Schindler 1976) and defended as a necessary step in the right direction (Peterson 1976). In my opinion, most impact statements, as now almost mass produced, are inadequate because they focus on the wrong level— often, for example, on the species or factor level when the questions and decisions clearly involve the ecosystem level. In other words the practice of environmental impact assessment is not so much bad or inadequate applied science as it is wrong-level applied science, a viewpoint that has not been emphasized in published discussion on the subject. For example, most would agree that important chemical factors such as oxygen concentration and basic biological characteristics such as species composition should be included in a baseline assessment of a body of water. However, a table of dissolved oxygen measurements and a long list of species present, as often included in current impact statements, provide little useful information for assessing the impact of a projected perturbation such as a thermal discharge from a power plant. With little, if any, additional effort in terms of time and money, one can increase the useful information content manyfold by assessing more functional and integrative properties that relate to oxygen and species. By measuring dissolved oxygen over diurnal cycles, the balance between the two major metabolic processes, photosynthesis and respiration, can be determined; systems-level information of this sort is usable in judging the potential impact of a procedure that might change the temperature of the water, alter the input of organic matter, and so forth. Likewise, arranging species data into a diversity profile reveals how numbers and kinds interact and gives a clue as to the developmental status of the community, thus providing a far better basis for impact assessment than would a mere list of species. Accordingly, environmental impact assessment, as well as technology assessment in general, should move from mere component analysis, wherein factors and organisms are treated as if they were independent entities, to more holistic approaches wherein interactive, integrative, and emergent properties are also included. The "new ecology," of course, must provide the basic theory for this necessary evolution in practice. In the meantime, there is much to be said for a procedure that combines a few carefully selected systems-level properties that monitor the performance of the whole, with selected "red flag" components such as a game species or a toxic substance that, in themselves, have direct importance to the general public (E. P. Odum and Cooley 1977).

Economics and Ecology

The ultimate in a holistic approach to preparing environmental impact statements must involve integration of economic and environmental values. In the real world monetary values are always going to weigh heavily in any decision regarding man's use of his environment. Regrettably, environmental and economic assessments usually are made by different teams or individuals. Not only do these teams rarely communicate with one another, but each also tends to restrict evaluation to its own preconceived narrow world of the natural or man-made environment, respectively, ignoring the fact that it is the interaction between these systems that is of paramount importance. Environmental and economic assessors should work together, or at least the results of study by different groups of specialists should be integrated. There are ways to scale economic and ecologic values which at first might seem incapable of comparison. For example, where alternate choices are involved one can set up a numerical scaling system in which the maximum quantitative value for each component is set at one (or one hundred), and all other values scaled accordingly. These scaled values can then be weighted according to a Delphi or other technique based on consensus of knowledgeable assessors (E. P. Odum et al. 1976). Another approach involves using energy as a common denominator for man and nature (Gosselink et al. 1974; H. T. Odum and E. P. Odum 1976).

If subjects were organized according to the literal derivation of their names, then ecology and economics would be companion disciplines since the words are derived from the same Greek root, with ecology translating as "the study of the house" and economics as "the management of the house." The disciplines remain poles apart on college campuses as well as in the minds of the general public as long as each restricts itself to only a part of the house, nature's and man's part, respectively. As ecologists have begun to take an interest in the man-made environment a few economists, notably Kenneth Boulding, Nicholas Georgescu-Roegan, William Nordhouse, James Tobin, and Herman Daley, have begun talking about an emerging "new economics" that is more attuned to natural laws and that includes a more equitable valuation of the natural environment (see especially Georgescu-Roegen 1971).

A closer liaison between ecology and economics makes good sense because in so many cases actions which benefit the general environment also benefit the general economy in the long term. It is in the short term, especially the two- to four-year electoral cycle of political action, where the most recalcitrant conflicts occur. Take, for example, strip-mining for coal or other earthbound resources. In the short-term view government should encourage and subsidize rapid exploitation of the resource and place few, if any, restrictions on strip-mining on the theory that the national economy would quickly benefit from a substitution of domestic coal for imported oil. In the long-term view just the opposite would be indicated, namely, that careful, well-planned mining, including mandatory land rehabilitation, is in the best interest of both the economy and the environment. Land valuable for food production, recreation, and life support would be preserved or restored. The steady, moderate production of coal would ensure that energy conservation would be pursued while air pollution and other threats to health would be less likely to get out of hand. New industry and towns would have a long life in contrast to local booms and busts that generally accompany unrestricted mining. Water, which is required in huge quantities to use or process coal, would be less likely to be squandered. Otherwise, dry-country cities such as Los Angeles and Phoenix might have to use all the newfound energy to obtain water, with a probability of net economic loss. As the ecologist might say, it is the secondary impacts that will get you if you do not consider the whole. Best of all, the long-range scenario would ensure that the nation has coal long after the easily obtainable Mideast oil is gone, thus making us less dependent and more secure in terms of national defense.

Politics and Ecology

Finally, there is yet another divided world, the scientific and the politico-legal spheres of action, where holistic thinking might help. In a recent editorial in *Science* Gerald Edelman expresses pessimism that these two disciplines will ever intersect, and states that we are left with "two extreme ideological positions—scientism and anti-scientism" (Edelman 1976). As long as students and practitioners of both disciplines insist on fragmenting their subjects, rigidly adhering to their own way of thinking and calling each other derogatory names, adversary interaction will continue to predominate. I am much more optimistic about the integration of these spheres because I have found that a meeting of minds in study panels and public commissions begins with the general acceptance of the idea that large-scale problems and issues might have common denominators that could be assessed along with the more narrowly defined scientific, political, or legal aspects. If hierarchical theory is indeed applicable, then the way to deal with large-scale complexity is to search for overriding simplicity. Sometimes, it appears, this turns out to be old-fashioned common sense. As noted, the dichotomy inherent in short and long time spans imposes a major stumbling block in acting on common sense judgment.

In summary, going beyond reductionism to holism is now mandated if science and society are to mesh for mutual benefit. To achieve a truly holistic or ecosystematic approach, not only ecology, but other disciplines in the natural, social, and political sciences as well must emerge to new hitherto unrecognized and unresearched levels of thinking and action.

CHAPTER 15

"Reductionist Holism": An Oxymoron or a Philosophical Chimera of Eugene Odum's Systems Ecology?

Donato Bergandi

Introduction

The epistemological issue "holism-reductionism" affects every level of integration: from physics to chemistry, from biology to psychology and even sociology. At each of these levels, one is faced with the same type of questions: Is it possible to understand an event or an object from any given integrative level by dissecting it in ever greater detail, or is it necessary to *respect* its structure and functions as much as possible by studying it in its proper context? Can a given integrative level be *reduced*—that is to say *explained* or *predicted*—from a basis of the laws or theories of a "lower" level of integration, or can each integrative level only be explained by means of its own laws and theories? Reductionists believe that ever more sophisticated molecular research will be able to reveal the essential structure of biophysical objects. Some go on to suggest that from this basis it will be possible to understand if not all, then at least a significant proportion of psycho-sociological phenomena. Holists, on the other hand, maintain a more phenomenological worldview, which is characterized by a greater respect for the objects of study and by an attempt to take into account the complexity of the spatiotemporal interconnections which give them their character (Nagel 1961; Koestler and Smythies 1969; Ayala and Dobzhansky 1974; Hoyningen-Huene and Wuketits 1989).

The reductionist paradigm, or what one might more accurately term the atomist-analytical-reductionist paradigm, is well structured and has already been proven to function at the most different levels of integration, but increasingly often, it is encountering anomalies which undermine it. Its epistemological base is derived from determinism. The holist paradigm, on the

From *Ludus Vitalis: Journal of Philosophy and Life Sciences* 3 (1995): 145–78. Notes omitted.

other hand, with its rich history of philosophical development, is engaged in a search for a theoretical and operational unity which it has not yet attained. The issue is rendered more complex by the fact that the proponents of the two approaches (especially in scientific fields) misunderstand their actual impact. Moreover, the epistemological evolution of whole disciplines, such as ecology—the holistic science *par excellence* (E. P. Odum and H. T. Odum 1971; H. T. Odum 1982; Ramade 1992; Patten 1993a, 1993b; E. P. Odum 1993)—has been founded on misunderstanding and misinterpretation.

From time to time, in the development of a scientific discipline, some texts emerge which rapidly become and remain a fundamental work of reference. This is due to their ability to function as a "point of attraction" for the discipline, and above all, to provide a new and useful perspective. Eugene P. Odum's *Fundamentals of Ecology* belongs to this small group of texts which one can define as "paradigmatic," since they provide the starting point and the material for specialists of the discipline to progress in their research. The book was first published in 1953, and then in 1959 and 1971. It seems to me that a comparative analysis of the three editions is a necessary preliminary not only to understand the text in all its subtleties but also to grasp better the importance of the systemic ecology which is presented within it.

Odum's contribution has been essential in determining the scientific status of modern ecology—its holistic concept of the ecosystem has enabled us to focus ecological research by giving it an original epistemological foundation. It has not, however, resolved the methodological problems which accompany the analysis of "complexity" in ecology. Neither has it enabled us to move beyond the historical debate between the followers of reductionism and those of holism. . . .

There is a mutual relationship between the terms *holism* and *emergent properties:* we cannot mention one without referring to the other. However, when we speak of emergent properties, we inevitably come up against the problem of their existence in nature and that of their explanatory power in science (Salt 1979; Edson et al. 1981). . . .

If one proceeds to a comparative analysis of the three editions, one cannot fail to notice that certain analysis, certain concepts, have survived whereas others have disappeared. In particular, the approach which Odum calls "holistic" undergoes major modifications in the last two editions while remaining the main line of the text. My aim is to define the holistic approach proposed by Odum and to highlight the ways in which it is modified in the development of the work. I will further attempt to establish whether the holistic vision of the world proposed by Odum is in harmony with the methodological tools he uses, i.e., whether there is a gap between his philosophical assumptions and the research methods which he advocates.

"Holism": An Emergent Scientific Philosophy

A significant proportion of the key concepts which form the basis of the approaches encompassed by the terms *holism, globalism, systemics,* and *organicism* have their origins in philosophical thought. If to nothing else, one must refer in this context to the Anaxagorian and Platonic concepts of nature as an integral whole (Anaxagoras, 6th Fragment; Plato, *Meno*), and the Aristotelian idea of the state as an ontological entity which takes precedence over the family and the individual (*Politics*).

A brief historical review of the scientific disciplines which have felt the influence of holism will help us to better understand the scope and significance of the issue. In the second half of the nineteenth century, holistic thought, principally by means of Hegelianism, was introduced into and spread in the developing field of cultural anthropology (Tylor 1871). Tylor's conception of culture has a symbolic significance in this respect. Culture is perceived for the first time as a "complex whole" whose various elements (laws, customs, art, economy, and so on) interact. In psychology, holistic thought finds its clearest expression in the extremely well-known phrase "the whole is greater than the sum of its parts," an axiom proposed by Ehrenfels (1890), who was among the first people to concern themselves with the holistic properties of cognitive phenomena, and with the semantically similar principle of "creative synthesis" (Wundt 1912). In the twentieth century, Gestalt psychology (Koehler 1929; Koffka 1935; Wertheimer 1945), a vital transitional phase in the process of "scientification" of holistic thought, came to represent the natural development of this trend.

In biology, an initial organicist wave with radical tendencies (Montgomery 1882; Haldane 1884; Russell 1924; Smuts 1926) was succeeded by a second, more moderate one (Woodger 1932; Needham 1936, 1937; Bertalanffy 1952, 1969; Goldstein 1963; Weiss 1967). If one excepts the very definition of ecology as a scientific discipline—i.e., a global science of the relationships between organisms and their environment (Haeckel 1866a)—holism first makes its appearance in ecology with the works of Clements (1916 [reproduced in this volume]) and, providing the necessary refinements, Allee (1949), E. P. Odum and H. T. Odum (1953, 1959, 1971), [and] H. T. Odum and E. P. Odum (1955). . . .

The Confrontation between Two Scientific Philosophies

Before moving on to the analysis of instance, which is archetypal as far as the misunderstanding and misuses of epistemological meaning in the holism-reductionism debate are concerned, we shall highlight the different positions taken by the two paradigms on the ontological, methodological (strategies of research), and epistemological (the relationships which exist between the theories and laws of different levels of integration) levels. This distinction between the various semantic fields, which was used by Ayala and Dobzhansky (1974; see also Mayr 1982b, 1988) to clarify nuances of the reductionist paradigm, I shall apply to the holist paradigm. It will soon become clear that in reality there is no such thing as "reductionism" or "holism" per se, but rather that there are several forms of both reductionism and holism.

The total contrast between the two paradigms applies in every semantic field. On the ontological level, reductionism and holism share a materialist worldview, but differ in their suppositions concerning those entities which make up reality. A current of philosophical atomism generally underlies reductionist suppositions, whereas holism has a relational and continuist view of reality. According to holism, the basis of reality does not consist of discrete entities, but is rather formed by a "network" of events and relationships which cannot be broken down. This network is the primary reason for the existence of emergent properties which characterize each individual level of integration: properties which symbolize all that is new in each integrative level, and which signal the increase in its complexity.

As far as methodological reductionism is concerned, it may take either a radical or a moderate form. The radical form maintains that it is possible to *predict* the properties of a level of integration by studying its constituent parts. The moderate form, on the other hand, limits itself to stating that it is possible to *explain* those properties on the basis of a study of the level below. In any case, they are both structured around additional analysis—which holds that the whole is equivalent to the sum of its parts (Amsterdamski 1981). A typical example of methodological reductionism may be found in Simon (1969), at the point where he puts forward the method which he paradoxically calls "pragmatic holism." Bertalanffy (1952) and Bunge's (1983) positions are similar. Curiously, all these methods are presented as being part not of a reductionist, but of an emergentist approach. Only Somenzi (1987), who proposes a radical reductionist approach to atomic physics, gives it its proper name.

Methodological emergentism, on the other hand, maintains that it is necessary to consider several levels of integration: according to Feibleman (1954), this means at least three. It stands in contrast to the belief that only levels of integration lower than the one being studied can supply explanations. The radical form of methodological emergentism does not accept additional analysis (Russell 1924), whereas the moderate form, although still rejecting this particular technique, is more open to the analytical method. It is acceptable to study lower levels of integration as long as one does not claim to be able to predict the properties of higher levels solely on the basis of these studies (Koehler 1929).

As far as epistemological reductionism is concerned, we are referring to the classic model of reduction, i.e., heterogeneous or interlevel reduction, as was proposed by Nagel (1961) and commented on and developed by Schaffner (1967, 1986, 1993) and Wimsatt (1986). In the realm of science, some fields are, they maintain, more "fundamental" than others. Biological levels of integration may be "reduced" or explained by reference to the laws and theories of chemistry and physics.

Epistemological holism, in contrast, gives precedence to the idea that the general tendency of science is toward synthesis (Bertalanffy 1952) and not reduction. According to Quine (1961), no one field is any more significant than another. The importance of science is as a global system, and therefore a change in any given area of science reverberates upon the entire system. Piaget (1970), for his part, refutes the idea of "one-way" reduction, and leans toward a reciprocal assimilation of disciplines. If a higher level of integration may be derived from a lower level, this lower level will become enriched to the extent that its structure will be radically altered.

Fundamentals of Ecology from 1953 to 1971: The Evolution of Odum's "Holistic" Thought

The key work in the oeuvre of E. P. and H. T. Odum is the text *Fundamentals of Ecology*. The "epistemological manifesto" of this work will form the object of our analysis, an illustration which will help us to better understand the sharp contrast between the holist and reductionist paradigms.

Because it is so much younger a science than more established disciplines such as physics, for example, ecology is in a position where it is reliant on the basis of the dominant scientific paradigm (i.e., reductionism), while all the time proclaiming the opposite: ecology is referred to as the "holistic" science *par excellence* (E. P. Odum and H. T. Odum 1971; Ramade, 1992; Patten 1993a, 1993b; E. P. Odum 1993). The world of ecological discourse does not always coincide with the world of experience or methodological practice.

There is a "radical difference" between holism and reductionism which is not always perceived as such. A typical example of this misinterpretation is to be found in the protracted scientific labors of the Odum brothers (H. T. Odum and E. P. Odum 1955; H. T. Odum 1957; E. P. Odum 1969; E. P. Odum and H. T. Odum 1971), which helped to structure the IBP (International Biologic Program). The cornerstone of what we might in future call the "Odumian paradigm" is the concept of the ecosystem.

Traditionally, the authorship of the term *ecosystem* is attributed to Tansley (1935 [reproduced in this volume]). . . . We should not forget, however, that concepts belonging to the same semantic family both preceded and followed this definition. Terms such as *microcosm* (Forbes 1925), *holocoen* (Friederichs 1930), and *biosystem* (Thienemann 1939) share a conceptual core with the term *ecosystem:* that is, they all represent the definition of a unity of reference, an object of study where it is possible to link biotic and abiotic factors formally. This unity is identifiable by its relative autonomy, by the structures which are typical of its levels—spatial (repetitive and homogeneous surfaces), specific, and trophic—and by its characteristic functions (the flow of matter and energy).

Tansley's aim was to bring into relief the organicist concept of Clements's "complex organism" (1916 [reproduced in this volume]), which had been championed several times by Phillips (1931, 1934, 1935a, 1935b). Tansley's concept is fundamentally antiholistic insofar as, for him, the problem of emergentism is a nonissue, and in that he uses physics as his explanatory field of reference. And here, suddenly, is an epistemological paradox. Tansley's concept of the ecosystem was employed by E. P. Odum in his explicit proposal of a genre of ecology which, in its intentions at least, is holistic (E. P. Odum and H. T. Odum 1953, 1959, 1971).

The first chapter of *Fundamentals of Ecology,* in all three editions, represents the exposition of the core of major concepts which enable one to extract phenomenological reality from a given point of view, to choose certain methodologies, and to develop specific scientific theories.

For the third edition, Odum deliberately modified the grouping of the chapters, which probably reveals an evolution in his way of thinking. Each edition begins with an analysis of the ecosystem; while analysis of the organization in terms of "population" precedes that of "community" in the first and second editions, the opposite approach has been chosen for the third edition. It would appear that following the publication of the first two editions, the importance given by Odum to the principle of wholeness was such that it logically became necessary to deal with the "community/whole" before the "population/part." The preeminence of the whole is such that, even in the way the arguments are presented, the author chooses systematically a "downward" itinerary in preference to an "upward" one. It is easy to affirm that the epistemological basis of the first edition is much less thoroughly detailed and developed in contrast to the second and third editions, which appear much more structured. It is simply necessary to compare certain key passages of the three editions for this to become abundantly clear. In the first edition, Odum asserts that: "Because ecology is concerned as much with the biology of *groups* of organisms (that is populations and communities) as with individual organisms, if not more, it would perhaps be better, and more in keeping with the modern emphasis, to define ecology as the study of the structure and temporal processes of populations, communities, and other ecological systems, and of the interrelations of individuals composing these units" (1953: 4; Odum's italics). In the second and third editions, the individual organisms disappear and are replaced by "functional process." We will cite the whole of the corresponding passage in order to better convey the profound significance of the transformation: "Because ecology is concerned especially with the biology of *groups* of organisms and with *functional* processes on the lands, in the oceans and in fresh waters, it is more in keeping with the modern emphasis, to define ecology as the study of the structure and function of nature (it being understood that mankind is a part of nature)" (1959: 4; 1971: 3; Odum's italics). Odum's approach, at least at the level of its intentions, is not limited to an analysis which would separate out successive levels of integration, but on the contrary, brings out the links between the parts and the whole system, considering that the organisms "are intimately linked functionally in ecological systems, according to well defined laws" (1959: 9). By extending Lindeman's (1942) trophic-dynamic vision, Odum is seeking "clarification of the basic energy relationships of the ecosystem as a whole" (1953: 89; 1959: 147–48), emphasizing the laws which govern the flow of energy between the different compartments of an ecosystem.

In the first edition the "epistemological manifesto" concentrates on the links which may exist between ecology and other scientific disciplines, biological or not. The complementary nature of ecology and genetics is underlined: since "the organism [is the] result of interaction of heredity and environment," the research should try to define the various influences of "heredity mechanisms" and "ecological factors" (1953: 5) in studied phenomena. Odum also underlines the main affinities between ecology and physiology as "both deal with functions," while pointing out the essential methodological differences between the two disciplines:

> As an illustration of the difference in approach, let us consider the heart. The physiologist is primarily concerned with the mechanism of its contractions and with the nervous, endocrine, and other factors controlling its beat and rate. The ecologist, on the other hand, would be primarily interested in the heart as a possible "physiology-of-the-whole" indicator. That is, the ecologist might wish to use the heart rate as an index of the way in which the organism as a whole responds to some environmental factor, for example, temperature. (Ibid.: 6)

Finally, he indicates a strong and declared propensity toward physics and chemistry: "In common with all of biology, ecology leans heavily on the physical sciences. Developments in chemistry and physics continually provide new techniques and influence ecological theory" (ibid.). In the last two editions, the analysis of the links between ecology and other scientific fields and the declaration of the physicalist tendency completely disappear. This analysis is replaced by an explicit presentation of the theory of the levels of organization accompanied by a graphic missing from the first edition (1959: 6–7; 1971: 4–5). The theory of the levels of integration is naturally present in the first edition, but the awareness of its importance and its explicit representation emerged gradually. However, the idea of the arbitrary and instrumental nature of this theory is present even in the first edition: "There are no sharp boundaries between any of these subdivisions which represent ways of looking at ecological problems rather than cut and dried scientific fields. It is merely convenient and profitable to approach the study of ecology from different levels of complexity" (1953: 7). The author expresses himself in the same way in the second and third editions on what he regards as the arbitrary character of the theory of levels of integration:

> It is important to note that no sharp lines or breaks were indicated in the above "spectrum," not even between the organism and the population. Since introductory biology courses usually stop abruptly with the organism, and since in dealing with man and higher animals we are accustomed to think of the individual as the ultimate unit, the idea of a continuous spectrum may seem strange at first. However, from the standpoint of interdependence, interrelations and survival, there can be no sharp break anywhere along the line. The individual organism, for example, cannot survive for long without its population any more than the organ would be able to survive for long without its organism. Similarly, the community cannot exist without the cycling of materials and the flow of energy in the ecosystem. (1959: 6–7; 1971: 5)

Thus, Odum's attempt to create a holistic approach (a term which will only appear in the third edition) materializes above all within his assertion that levels of integration corresponding to the concepts of ecosystem, of community, and of population are functionally interlocked. The actual definition of the concept of ecosystem evolves between the first and third editions. The ecosystem is initially defined in the following way by Odum:

> Living organisms and their nonliving (abiotic) environment are inseparably interrelated and interact upon each other. Any entity or natural unit that includes living and nonliving parts interacting to produce a stable system in which the exchange of materials between the nonliving parts follows circular paths is an ecological system or ecosystem. The ecosystem is the largest functional unit in ecology, since it includes both organisms (biotic communities) and abiotic environment, each influencing the properties of the other and both necessary for maintenance of life as we have it on the earth. A lake is an example of an ecosystem. (1953: 9; 1959: 10)

While the second edition offers virtually the same definition, if we discount the fact that the ecosystem is no longer defined as "stable system" (by no means an insignificant modification), the third edition explicitly introduces the concepts of energy flow and trophic structure. Thus the idea that the "whole" exists as a structured entity is put forward:

> Living organisms and their nonliving (abiotic) environment are inseparably interrelated and interact upon each other. Any unit that includes all of the organisms (i.e., the "community") in a given area interacting with physical environment so that a flow of energy leads to clearly defined trophic structure, biotic diversity, and material cycles (i.e., exchange of materials between living and non living parts) within the system is an ecological system or ecosystem. . . . The ecosystem is the basic functional unit in ecology, since it includes both organisms (biotic communities) and abiotic environment, each influencing the properties of the other and both necessary for maintenance of life as we have it on the earth. (1971: 8)

In this way, Odum proposes a far more highly developed definition of the ecosystem than that of Tansley. There is undoubtedly a large area of convergence in that both authors claim to define the basic units or entities of nature, but Odum views things from a clearly emergentist point of view, or at least intends to, while Tansley's concept of the ecosystem is explicitly anti-emergentist. Indeed, when Odum analyzes the concepts of "population" and "community" in the chapter "Introduction to Population and Community Ecology," by using the typically holistic metaphor of the forest, he relates it to the theory of emergence: "The important point to stress is that the population and community are real entities, even though one cannot usually pick them up and put them in the collecting kit as one would collect an organism. They are real things, because these group units have *characteristics additional to the characteristics of the individuals composing them*. The forest is more than a collection of trees. The whole is not simply a sum of the parts" (1953: 88; 1959: 146; 1971: partially integrated into the first chapter; Odum's italics).

However, while doing so, the author is setting up a serious logical contradiction, which will lead him to confuse the concept of "collective properties" with that of "emergent properties," as his definitions of the concept[s] of population and community show. Odum defines population as follows: "The population, which has been defined as a collective group of organisms of the same or closely associated species occupying a particular space, has various characteristics which, although best expressed as statistical functions, are the unique possession of the group and are not characteristic of the individuals in the group. Some of these properties are: density, natality (birth rate), mortality (death rate), age distribution, biotic potential, dispersion, and growth form" (1953: 91; 1959: 149; 1971: 172). The definition of community follows the same logical structure: "A biotic community is any assemblage of populations living in a prescribed area or physical habitat; it is a loosely organized unit to the extent that it has characteristics additional to the individual and population components" (1953: 181; 1959: 245; 1971: 140).

In these two definitions, the properties which concern the group (system, wholeness, integrative higher levels) and which do not concern individuals (components, parts, integrative lower levels) which make it up, are mentioned. Odum does not make it clear whether he considers these properties to be "collective" or "emergent." It is, however, an essential difference, and one which Odum does not perceive: by definition, the emergent properties of a given integrative level cannot be deduced from the study of its components, while collective properties, even when they are not the characteristics of individuals with statistical functions, can be inferred (deduced) from the combined characteristics of all the individuals of a population or community. . . .

Moving on to an analysis of the methodological approach advocated by Odum, which one might expect would invite research methods on broadly holistic ontological lines, we should point out that the author, in the chapter "Systems Ecology: The Systems Approach and Mathematical Model in Ecology" of the third edition, puts forward a form of analysis of systems as follows: "As the formalized approach to holism, systems ecology is becoming a major science in its own right for two reasons: (1) extremely powerful new formal tools are now available in terms of mathematical theory, cybernetics, electronic data processing, and so forth, and (2) formal simplification of complex ecosystems provides the best hope for solutions to man's environmental problems that can no longer be trusted to trial-and-error, or one-problem-one-solution, procedures that have been chiefly relied on in its past" (1971: 276). Thus Odum makes cybernetics his dominant theme. His aim is to propose a holistic model—that is, a model that takes into consideration "the ecosystem as a whole" (1953: 89; 1959: 148). However, by using the cybernetic model, he paradoxically puts himself in a position which contradicts his own basic epistemological arguments. . . . Odum believes that it is possible for cybernetic models to respect the complex nature of hierarchical organization, while allowing a certain economy of effort in analysis. Odum's assessment is in error, however, for his reference to Feibleman (1954) can only make sense if the proposed models are able to take into account and to explain emergent properties.

It is certainly true that cybernetic models consider the feedback loops which exist between subsystems on the same level and the hierarchical interdependence of feedback loops which link different levels of organization, but these models do not make clear which integrative levels need to be taken into consideration in order to explain an emergent property observed on a given level. By side-stepping this issue, Odum is pushing all truly holistic concerns aside.

Furthermore, I should underline the fact that, in cybernetic models, the inputs taken into account at each level of organization are the results of an *a priori* choice which risks being arbitrary. Odum . . . asserts that: "Contrary to the feeling of many skeptics when it comes to modelling complex nature, information about only a relatively small number of variables is often a sufficient basis for effective models because 'key factors' often dominate or control a large percentage of the action" (1971: 7). A "key factor" is not an ontological characteristic determined once and for all, but rather a contextual characteristic. Only by halting the variable time and by interrupting the network of interaction can we abstract an element from the system and *a priori* define it as a key factor or otherwise.

In this respect, Odum's thought has undergone a subtle but absolutely fundamental shift. Indeed, by implying that a complex whole can be properly understood simply by taking into account basic components with respect to their dominance in the trophic network or to their control function, Odum would seem to be suggesting that an approach is holistic simply because it does not need to analyze all the components of a system in order to understand it in its entirety. It is therefore legitimate to suppose that Odum is likening reductionism to exhaustivity. But in reality, what he is proposing is a simplification. If there is no doubt that a model is necessarily a simplification, it follows that its value can either be reductionistic or holistic.

In the end, cybernetic models, with their potentially global significance with respect to the assessment of the variables of ecological systems, are in fact used to an end which is epistemologically reductionistic. Models of this type are used to describe the trophic networks of an ecosystem from the point of view of energy, and from the perspective of cycles of matter. The prototype of this form of analysis goes back to Hutchinson; Lindeman formalized and extended the form, and Patten (1959) and H. T. Odum (1982) developed it to its highest level of sophistication.

This type of analysis, legitimate when we wish to consider the dimension of energy and that of the flow of matter inside ecosystems, becomes completely illegitimate, reductive, and obscurantist when it is perceived as the one true form of analysis of ecosystems. In practice, this exclusivist tendency goes so far as to consider as relatively unimportant the fact that it destroys the structural and functional specificity of the levels under consideration.

One might think that this whole discussion is out of date, since the systemic approach in biology and ecology has decided on its tools. In our opinion, however, this is not so. The state of the methodologically holistic approach to the discipline of ecology has recently been very well described by Loehle (1988a: 100–101):

Unfortunately, holism means at least the following four things in ecology:
    1. The view that ecosystems are integrated, interconnected systems with their own laws and organizational principles. It is not necessarily denied that reductive expla-

nations are possible, but merely that they are so impractical that the higher level system should be studied phenomenologically.

2. The practice of embedding a problem in larger context. Thus when designing a waste-treatment pond, holism might mean keeping in mind that migratory ducks might use the pond and be killed. It is this type of holism that ecologists complain is lacking in civil engineers and urban planners.

3. A black box approach which includes questions such as: What is the nutrient loss response of a whole watershed to acid rain? The watershed is treated as an input-output system without detailed concern for mechanisms or particular species. This is an empirical approach that does not necessarily seek laws or principles (in contrast to 1).

4. Detailed systems analysis such as ecosystem models that are mechanism-oriented. This approach is holistic because it includes all components and processes.

Strictly speaking, only the first definition coincides to any degree with the theory of emergence as it is proposed by Feibleman (1954) and Novikoff (1945). The second definition can be accepted as holistic, but only in a very broad sense. Unfortunately, neither the first nor the second definition can be developed into a real systemic methodology. Indeed, the first limits itself to a warning as to the limits of epistemological reductionism, whereas the second is, methodologically speaking, completely trivial. The third definition coincides with the cybernetic models proposed by Odum and the fourth confuses holism with an approach which seeks exhaustivity (an impracticable theoretical ideal).

Conclusion

In Odum's work the two spirits present in the history of ecology exist side by side, at least to a certain extent: the systemic, globalistic, holistic component prevalent in the work of Clements, Phillips, Allee, as well as in that of other authors of the Chicago group, and the reductionistic component of ecology represented by Tansley, Hutchinson, Lindeman, and Margalef. Any mediation between these two visions of the world, their methodologies, and their epistemologies is very difficult, and perhaps even impossible, to effect. Odum's mistaken interpretation of the collective properties of community and population as emergent is the most obvious evidence of this theoretical aporia.

The ecosystem is mainly interpreted from the perspective of energy, as Lindeman conceived it, using a holistic ontology which has at its core the concept of emergent properties. Thus it is this "emergentist" notion that the reductionistic approach disowns, whether at an ontological, a methodological, or an epistemological level.

According to reductionism in the ontological field, the ultimate essence of any given integrative level is made up of individual units from a level lower than the one under consideration. From a methodological point of view, the properties of a given integrative level can be foreseen and/or explained from its components: this is the case with the evaluation of energy of ecosystems derived from the methods of Lindeman, who uses an approach based entirely on additional analysis. From an epistemological point of view, the laws of integrative levels relating to biology should be replaced by laws belonging to more fundamental levels, that is, ultimately, laws of biophysics or even chemistry and physics. On the other hand, according to ontological holism, each level of organization has a value in itself, for it is characterized by the acquisition of one or more new emergent properties which increase the complexity of the system. From a methodological point of view, holism cannot limit itself to a form of additional analysis of the lower integrative level, but rather must consider at least three integrative levels. Lastly, from an epistemological viewpoint, there are no fields of science, to which other fields should be reduced.

Odum's position can be defined as crypto-reductionist for, while his vision is holistic on ontological and epistemological levels, the author proves to be purely reductionist in his methodology. It is truly paradoxical that an author as strongly involved in epistemological holism as Odum confines himself to understanding the ecosystem essentially in terms of energy. This means that, even if the physicalist tendency is no longer explicitly stated in the second and third editions, a permanent trace of it remains, throughout. In this approach, cybernetics is the tool used to oversimplify the ecosystem, by reducing it to the assessment of components providing the greatest transfer of energy and to some supposed key factors. . . .

Reductionism and holism represent two opposing approaches which are not always perceived as such. A lack of understanding of their distinctive characteristics and, at the same time, more or less conscious attempts to find a middle point between these two worldviews have led to profound aporia. To illustrate the existence of this epistemological obstacle we have used an example from the field of ecology as an archetypal instance. However, the epistemological clash between holism and reductionism affects every scientific discipline. Some are naturally inclined toward reductionism, while others lean toward holism. Systemic ecology has made its choice: "reductionist holism." This latter forms a kind of scientific paradox which is not recognized for what it is.

One should not conclude, however, that this is an isolated instance, and that at other levels of integration "holism" may be presented as a truly "emergentist" approach. It might appear that to speak of "emergentist holism" would be tautological and semantically redundant, but is by no means the case.

At any given integrative level, a reductionist approach is completely coherent. Its ontology, methodology, and epistemology are linked in a way which is almost "organic" (an irony of science . . .). An emergentist approach, however, runs the risk on methodological and epistemological levels of forgetting its ontological presuppositions and metamorphosing into a hybrid form: "reductionist holism." The crux of the concept of emergence disappears and becomes secondary to attempts to explain a given level of integration by means of laws and theories from lower levels.

In conclusion, one might ask the question: Is this hybrid approach a kind of oxymoron—a fine nuance of thought rendered explicit in the form of an apparent contradiction—which may soon become a philosophical and scientific truism, or is it simply a chimera (χιμαιρα) of philosophy?

Our answer would be as follows: "Reductionist holism" must be recognized as a sort of "epistemological monstrosity"—not one to be slain by a new epistemologist Bellerophon, naturally, but one which must nonetheless be unmasked. If this is not done, proponents of this kind of approach, with their problematic similarities with an emergentist worldview, will continue to fail to appreciate that they are unwitting reductionists. . . . The failure to recognize a confusion on this scale makes the possibility of constructing a genuinely emergentist approach ever more remote.

Acknowledgments
I am grateful to Patrick Blandin (Muséum national d'histoire naturelle, Paris), Roberto Cordeschi (Università di Salerno), Jean Gayon (Université Paris 7–Denis Diderot), Laurent Palka (Muséum national d'histoire naturelle, Paris), and Vittorio Somenzi (Università "La Sapienza" di Roma) for comments and helpful discussion on this manuscript. I am very thankful to Stanley N. Salthe (City University of New York) for his remarks and recommended bibliography.

CHAPTER 16

Dialectics and Reductionism in Ecology

Richard Levins and Richard C. Lewontin

The philosophical debates which have accompanied the development of science have often been expressed in terms of dichotomous choices between opposing viewpoints about the structure of nature, the explanation of natural processes, and the appropriate methods for research.

Are the different levels of organization such as atom, molecule, cell, organism, species, and community only the epiphenomena of underlying physical principles or are the levels separated by real discontinuities? Are the objects within a level fundamentally similar despite apparent differences or is each one unique despite seeming similarities? Is the natural world more or less at equilibrium or in constant change? Can things be explained by present circumstances or is the present simply a reflection of the past? Is the world causal or random? Do things happen to a system mostly because of its own internal dynamic or is causation external? Is it legitimate to postulate hypothetical entities as part of scientific explanation or should science stick to observables? Do generalizations reveal deeper levels of reality or destroy the richness of nature? Are abstractions meaningful or obfuscatory?

As long as the alternatives are accepted as mutually exclusive, the conflict remains one between mechanistic reductionism championing materialism, and idealism representing holistic and sometimes dialectical concerns. It is also possible to opt for compromise in the form of a liberal pluralism in which the questions become quantitative: How different and how similar are objects? What is the relative importance of chance and necessity? Of internal and external causes (e.g., heredity and environment)? Such an approach reduces the philosophical issues to a partitioning of variance and must remain agnostic about strategy.

When we attempt to choose sides retrospectively, we find that it is not possible to be consistent: we side with the biologists who oppose theological idealism in insisting upon the continuity between our species and other animals or between living and nonliving matter. But we emphasize the discontinuity between human society and animal groups in opposition to various "biology is destiny" schools.

From *Synthese* 43 (1980): 47–78.

As long as we accept the terms of the debate between reductionism and idealism, we must adopt an uncomfortably ad hoc inconsistency as we see now one side, now the other advancing or holding back science. Simberloff's essay (1980 [reproduced in this volume]) seems to us to embody the false debate by being based on three fundamental common confusions. These are: the confusion between reductionism and materialism, the confusion between idealism and abstraction, and the confusion between statistical and stochastic. As a result of these confusions, Simberloff, in his attempt to escape from the obscurantist holism of Clements's "superorganism," falls into the pit of obscurantist stochasticity and indeterminism. For if one commits oneself to a total reductionist program, claiming that *in fact* collections of objects in nature do not have properties aside from the properties of the objects themselves, then failures of explanation must be attributed ultimately to an inherent indeterminism in the behavior of the objects themselves. The reductionist program thus simply changes the locus of mystification from mysterious properties of wholes to mysterious properties of parts.

We will discuss these three confusions, and some subsidiary ones, in order to outline our disagreements with Simberloff, but also to develop implicitly a Marxist approach to the questions that have been raised. Dialectic materialism enters the natural sciences as the simultaneous negation of both mechanistic materialism and dialectical idealism, as a rejection of the terms of the debate. Its central theses are that nature is contradictory, that there is unity and interpenetration of the seemingly mutually exclusive, and that therefore the main issue for science is the study of that unity and contradiction rather than their separation either to reject one or to assign relative importance.

Reductionism and Materialism

The confusion between reductionism and materialism has plagued biology since Descartes' invention of the organism as a machine. Despite the repeated demonstrations in philosophy of the errors of vulgar reductionism, practicing biologists continue to see the ultimate objective of the study of living organisms to be a description of phenomena entirely in terms of individual properties of isolated objects. A recent avatar is Wilson's (1978) claim that a scientific materialist explanation of human society and culture must be in terms of human genetic evolution and the Darwinian fitness of individuals.

In ecology, reductionism takes the form of regarding each species as a separate element existing in an environment that consists of the physical world and of other species. The interaction of the species and its environment is unidirectional: the species experiences, reacts to, and evolves in response to its environment. The reciprocal phenomenon, the reaction and evolution of the environment in response to the species, is put aside. While it is obvious that predator and prey play both the roles of "organism" and "environment," it is often forgotten that the seedling is the "environment" of the soil in that the soil undergoes lasting evolutionary changes of great magnitude as a direct consequence of the activity of the plants growing in it, and in turn feeds back on the conditions of existence of the organisms. But if two species are evolving in mutual response to each other or if plant and soil are mutually changing the conditions of each other's existence, then the ensemble of species[,] or of species and physical environment, is an object with dynamical laws that can only be expressed in a space of appropriate dimensionality. The change of any one element can be followed as a projection on a single dimension of the changes of the n-tuple, but this projection may show paradoxical features including apparent lack of causality, while the entire ensemble changes in a perfectly regular manner. For example, a prey and predator will approach an equilibrium of numbers by a spiral path in the two-dimensional space whose axes are the abundances of the two species. This path is completely unambiguous in the sense that given the location of a point in two-dimensional space at one instant of time, a unique vector of change can be established, predicting its position at the next instant. Each of the two component species, however, is oscillating in abundance so that given only the abundance of the predator, say, it is impossible to know whether it will increase or decrease during the next interval. The description of change of the n-dimensional object may then itself be collapsed onto some new dimension, for example, distance from the equilibrium point which again may behave in a simple monotonic and predictable way. The rule of behavior of the new object is not an obscurantist holism but a rule of the evolution of a composite entity that is appropriate to that level of description and not to others. In the specific case just given, neither the prey nor the predator abundances converge monotonically to their final equilibria, and the monotone behavior of the pair object is not predictable from the separate equations of each species. Moreover, the separate behavior of each species is not itself predictable from the form of their separate equations of motion, since neither of these equations is intrinsically oscillatory and the damp oscillation of the two species is a consequence of their dynamic coupling.

The Clementsian superorganism paradigm is indeed idealist. Its community is the expression of some general organizing principle, some balance or harmony of nature. The behavior of the parts is wholly subordinated to this abstract principle, which causes the community to develop toward the maximization of efficiency, productivity, stability, or some other civic virtue. Therefore, a major priority would be to find out what does a community maximize.

Having correctly identified the Clementsian superorganism as idealist, Simberloff then lumps with it all forms of "systems modeling." But the large-scale computer models of systems ecology do not fit under the heading of "holism" at all. Rather they are forms of large-scale reductionism: the objects of study are the naively given "parts"—abundances or biomasses of populations. No new objects of study arise at the community level. The research is usually conducted on a single system, a lake, forest, or prairie, and the results are measurements and projections for the same lake, forest, or prairie without attempting to find properties of lakes, forests, or prairies in general. It requires vast amounts of data for its simulations, and much of the scientific effort goes into problems of estimation. We agree with Simberloff that this approach has been generously supported and singularly unproductive.

Idealism and reductionism in ecology share a common fault: they see "true causes" as arising at one level only with other levels having only epistemological but not ontological validity. Clementsian idealism saw the community as the only causal reality with the behaviors of individual species populations as the direct consequence of the mysterious organizing forces of the community. One might *describe* the community for some purpose by a list of species abundances, but that description was of epiphenomena only. Reductionism, on the other hand, sees the individual species, or ultimately the individuals (or cells, or molecules, for there is no clear stopping place in the reductionist program), as the only "real" objects while higher levels are again descriptions of convenience without causal reality. A proper materialism, however, accepts neither of these doctrinaire positions but looks for the actual material relationship among entities at all levels. The number of barn owls and the number of house mice separately are important causal factors for the abundance of their respective competitors and are material realities relevant to those other species, but the particular combination of abundances of owls and mice is a new object which is a material cause of the volume of owl pellets and therefore of the abundance of habitat for certain bacteria.

The Community as a Dialectical Whole

Unlike the idealist holism which sees the whole as the embodiment of some ideal organizing principle, dialectical materialism views the whole as a contingent structure in reciprocal interaction with its own parts and with the greater whole of which it is a part. Whole and part do not completely determine each other.

The community in ecological theory is an intermediate entity, between the local species population and biogeographic region, the locus of species interactions. The region can be visualized as a patchwork of environments and a continuum of environmental gradients over which populations are distributed. A local community is linked to the region by the dynamics of local extinction and colonization. Local extinction depends on local conditions affecting the populations in question. Colonization depends on the number of propagules (seeds, eggs, young animals) the local population sends out, which depends on the population size achieved locally. It also depends on the behavior of these propagules, their ability to cross the gaps between suitable habitats, their tolerance of conditions along the way, and their capacity to establish themselves (anchor on the . . . new substrate, grow under the shade of established trees, defend an incipient ant nest). These properties are biological characteristics of the individual species which are not directly responsible for abundance and survival in the local community. Finally, colonization depends on the pattern of the environmental mosaic—the distance between patches and whether the patches are large or small, the structure of the gradients (whether different kinds of favorable conditions are positively or negatively associated). These biogeographic properties are not implicit in the dynamics of the local set of species.

The whole ensemble of species of a region depends on the origin of the biota, the extinction of species in the whole region, and the processes of speciation. Therefore, the biogeographic level gives us a dynamic of extinction, colonization, and speciation in which the parameters of migration and extinction are givens, partly dependent on local dynamics but not contained therein.

Below the community are the component species populations. They enter the community at a rate which depends on their abundance in other communities, in the region as a whole. But once in the locality their abundance, persistence, variability, and sensitivity to environmental variability depend on their interactions with other species and on the parameters of their ecology— birth rate, food and microhabitat preferences, mobility, vulnerability to predators, physiological tolerances which come from their own genetic makeup. The genetic makeup in turn is a consequence of the processes of selection, mutation, drift, and gene exchange with other populations of the same species, which form the domain of population genetics and reflect past evolutionary history. The other members of the community affect the direction of natural selection within the community and therefore influence these parameters, but they are not deducible from the general rules of community ecology. Thus the claim that the ecological community is a meaningful whole rests on its having distinct dynamics—the local demographic interactions of species against a background of biogeographic and population genetic parameters. . . .

[A] dialectical approach to the ecological community allows for a greater richness than the reductionist view. It permits us to work with the relative autonomy and reciprocal interaction of systems on different levels, shows the inseparability of physical environment and biotic factors, and the origins of correlations among variables, makes use of and interprets both the many-to-one relations that allow for generalization and the one-to-many that impose randomness and variation. Where particular techniques are unsatisfactory, the remedy is likely to be not a retreat from complexity to reductionist strategies but a further enrichment of the theory of complex systems.

Abstraction and Idealism

It is Simberloff's view that abstractions are a form of idealism and that the materialism in science necessarily overthrows abstractions and replaces them with [some] sort of "real" entities which are then each unique, because of the immense complexity of interacting forces on each and because of the underlying stochasticity of nature. However he cannot really mean that all abstractions are to be eliminated or else nothing would remain but chronicles of events. If any cause explanations are to be given, except in the trivial sense that a historically antecedent state will be said to be the cause of later ones, then some degree of abstraction is indispensable. There is no predictability or manipulation of the world possible except that events can be grouped into classes, and this grouping in turn means that unique properties of events are ignored and the events are abstracted. Thus, we can hardly have a serious discussion of a science without abstraction. What makes materialist science is that the process of abstraction is explicit and recognized as historically contingent within the science. Abstraction becomes destructive when the abstract becomes reified and when the historical process of abstraction has been forgotten so that the abstract descriptions are taken for descriptions of the actual objects.

The level of abstraction appropriate in a given science at a given time is a historical question. No ball rolling on an inclined plane behaves as an ideal Newtonian body, but that in no way diminishes the degree of understanding and control of the physical world that we have acquired from Newtonian physics. Newton was perfectly conscious of the process of abstraction and idealization that he had undertaken, and he says in the *De Motu Corporum:* "Every body perseveres in its state of rest, or of uniform motion in a right line, unless is it compelled to change that state by forces impressed thereon." Yet Newton points out immediately that even "the great bodies of the planets and comets" have such perturbing forces impressed upon them and that no body perseveres indefinitely in its motion.

On the other hand, the properties of falling bodies that have been abstracted away are replaced when necessary; Newton himself, in later sections of the *Principia,* considered friction and other such forces. Landing a space capsule on the moon requires the physics of Newtonian ideal bodies moving in vacuums for only part of its operation. For other parts, an understanding of friction, hydrodynamics, and aerodynamics in real fluid media must be involved.

Finally, there are correction rockets, computers, and human minds to cope with the idiosyncrasies of actual events. No space capsule could land on the moon without Newtonian abstractions nor solely with them. The problem for science is to understand the proper domain of explanation of each abstraction rather than becoming its prisoner.

The argument given by Lewontin (1974) that Darwin and Mendel represented a materialist revolution in biology was not based on the assertion that they overthrew *abstractions* but that they overthrew *Platonic ideals*. Darwin's and Mendel's works are filled with abstractions (species, hereditary factors, natural selection, varieties, etc.). The error of idealism is the belief that the ideals are unchanging and unchangeable essences that enter into actual relationships with each other in the real world. Ideals are abstractions that have been transformed by fetishism and reification into realities with an independent ontological status.

Moreover, idealism sees the relationships entered into by the ordinary objects of observation as direct causal consequences, albeit disturbed by other forces, of the actual relations between the essences. Marx, in discussing the fetishism of commodities in Chapter I of *Capital*, draws a parallel with "the mist-enveloped regions of the religious world. In that world the productions of the human brain appear as independent beings endowed with life, and entering into relations both with one another and the human race." In a similar way idealistic, pre-Darwinian biology saw the actual organisms and their ontogenetic histories as causal consequences of real relations among ideal, essential types, as opposed to the materialistic view that sees the actual physical relations as occurring between actual physical objects with any "types" as mental constructs, as *abstractions* from actuality. The precise difficulty of pre-Darwinian evolutionary theory was that it could not reconcile the actual histories of living organisms, especially their secular change, with the idea that these histories were the causal consequences of relationships among unchanging essences. The equivalent in Newtonian physics would have been to suppose (as Newton never did) that if a body departed from perfectly rectilinear, unaccelerated motion, there nevertheless remained an entity, the "ideal body," that continued in its ideal path and to which the actual body was tied in some causal way. What appears to be the patent absurdity of this view of moving bodies should make clear to us the contradictory position in which pre-Darwinian evolutionists found themselves.

In ecology the isolated community is an abstraction in that no real collection of species exists which interacts solely with its own members and which receives no propagules from outside. But the total isolation of a group of species from all interactions with other species is not a requirement of the usefulness of the community as an analytical tool. . . . To put the matter succinctly, what distinguishes abstractions from ideals is that abstractions are epistemological consequences of the attempt to order and predict real phenomena, while ideals are regarded as ontologically prior to their manifestation in objects. . . .

Conclusions

Biology above the level of the individual organism—population ecology and genetics, community ecology, biogeography, and evolution—requires the study of intrinsically complex systems. But the dominant philosophies of Western science have proven to be inadequate for the study of complexity:

1. The reductionist myth of simplicity leads its advocates to isolate parts as completely as possible and study these parts. It underestimates the importance of interactions in theory, and its recommendations for practice (in agricultural programs or conservation and environmental protection) are typically thwarted by the power of indirect and unanticipated causes rather than by error in the detailed description of their own objects of study.

2. Reductionism ignores properties of complex wholes; the effects of these properties are therefore seen only as noise; this randomness is elevated into an ontological principle which leads to the blocking of investigation and the reification of statistics, so that data reduction and statistical prediction often pass for explanation.

3. The faith in the atomistic nature of the world makes the allocation of relative weights to separate causes the main object of science, and makes it more difficult to study the nature of interconnectedness.

Where simple behaviors emerge out of complex interactions, it takes that simplicity to deny the complexity; where the behavior is bewilderingly complex it reifies its own confusion into a denial of regularity.

Both the internal theoretical needs of ecology and the social demands that it inform our planned interactions with nature require an ecology that makes the understanding of complexity the central problem: it must cope with interdependence and relative autonomy, with similarity and difference, with the general and the particular, with chance and necessity, with equilibrium and change, with continuity and discontinuity, with contradictory processes. It must become increasingly self-conscious of its own philosophy, and that philosophy will be effective to the extent that it becomes not only materialist, but dialectical.

CHAPTER 17

Hierarchy: Perspectives
for Ecological Complexity

T. F. H. Allen and Thomas B. Starr

Hierarchical Structure

Ecology, even most theoretical ecology, can be seen as largely a formalization of common knowledge, as institutionalized walks through the woods. This aspect of the discipline's character comes from the essential tangibility of the research material; apparently, it is difficult for ecologists to look beyond phenomena that are in tune with familiar rhythms. Sophistication may lie in the methods of quantification of field data and in the formalization of those numbers into ecological models; but the underlying common-knowledge base, it could be argued, sets limits upon the level of sophistication which can be achieved. All too often, ecological subject matter is determined by a reductionism whose level of resolution is arbitrarily set so as to map easily onto casual perception through largely unaided eyes and ears. But that is not to say that data collection is casual or proceeds through the unaided senses; we refer here only to the model, the research goal, and the criteria adjudged as sufficient for explanation. Even when computers and fancy measuring devices are employed, some might dismiss the whole discipline as mere natural history.

Then there are those who would reply that there is nothing wrong with being a natural historian, asserting that natural history is not limited. Such an observer also might be unhappy at the state of ecology, but not for the reasons expressed above. He might look to the last century and see natural historians of Herculean stature. Any reading of Darwin and many others humbles the contemporary biologist, field man and theoretician alike. By comparison, a dip into the most prestigious contemporary journals comes up with something which is somehow a little thin. We see ecology at this juncture constrained by its origins on one hand, but not living up to them on the other. Perhaps this situation could be turned around so as to reverse a trend where articles become more arcane about the more mundane to one where baroque ecological struc-

From *Hierarchy: Perspectives for Ecological Complexity.* Chicago: University of Chicago Press, 1982, pp. 3–42; pp. 3–4 and 37–42 are reprinted here.

tures are factored so they can easily be seen whole. More detailed specification of more local circumstances so that more exotic mathematics can be applied will only make things worse. On the other hand, a retreat from quantification will not lead back to a golden Victorian age. We do, however, see some hope in trying to formalize, through the power of quantification, the sort of richness of understanding that was common to last century's country vicars. We must seek to model somehow the passing of a butterfly along with the growing of a tree.

With the gradual encroachment of ecology into larger and smaller realms (ecosystem analysis and microbial ecology) ecological insights do begin to be won outside the commonplace human experience. There are changes of perceptual scale taking place that will allow the discipline to transcend the limits imposed by the naked senses, and new perceptual patterns are likely to widen conceptual horizons. It is the aim of this contribution to deal with the scale problem in its own right and so work toward tools and general models that may encourage ecology in its escape from limitations that have historically prevailed. Any ecological theory that applies only to the time constants associated with everyday encounters is likely to be rather local in its application. That part of ecological theory which will survive with significance a translation to larger and smaller scales is the body of concrete achievement in ecology thus far. At the moment we can only guess what that significant subset might be, but formal treatment of scale should permit its identification.

The ideas collected here come from that part of general systems theory which is beyond the mechanistic cybernetic approach. Those who share our disillusion with lumbering simulations should not reject all modes of systems analysis. There is a better way than stringing together five hundred differential equations, a perspective which is cognizant of the intrusion of the humanity of the scientist into his observations. This approach views systems theory not just as a tool for solving problems already defined, but as a conceptual framework within which one might develop new ideas about biology. . . .

Nesting and Derivability: Nonnesting and Emergence

We have defined a hierarchy to be a system of communication where holons with slow behavior are at the top while successively faster-behaving holons occur lower in the hierarchy. The characteristic time for behavior of a holon is determined by the cycle time of endogenously driven behavior, also called the natural frequency (May 1973b). The endogenous cycle time determines the time taken for a holon to return to its equilibrium behavior after being influenced by an external signal. This is sometimes called the relaxation time, an alternative criterion for arranging holons in hierarchies. . . .

Lower holons in the hierarchy are to various extents constrained by higher holons with which they communicate. Higher holons are to various extents the environment of lower holons. While constraint does not determine the endogenously driven behavior of lower holons, it does give that behavior a context and boundary conditions. This, then, is our conception of hierarchy, and we see it as having sufficient definition. Other conceptions will often include additional criteria, such as higher levels containing lower levels, and we shall now put these other criteria in the context of our necessary and sufficient conditions for hierarchy stated above.

Nested and Nonnested Hierarchies

The notion of hierarchical arrangement is central to biology and even has an Aristotelian origin. The two standard ways of organizing biological systems, that is, into taxonomic units and structural relationship, both represent nested hierarchies. A nested hierarchy is one where the holon at the apex of the hierarchy contains and is composed of all lower holons. The apical holon consists of the sum of the substance and the interactions of all its daughter holons and is, in that sense, derivable from them. Individuals are nested within populations, organs within organisms, tissues within organs, and tissues are composed of cells. In taxonomy a family consists of its constituent genera and their component species. These two hierarchical arrangements are ubiquitous in modern biology to such an extent that these arrangements are commonly used to order the material in introductory textbooks.

The biologist encounters this pair of hierarchies so frequently that he is wont to consider nested hierarchies as being more properly hierarchical than their nonnested counterparts. In our more general approach to hierarchies, however, it is profitable to view the nested hierarchical condition as only a special and restricted case rather than a perfect hierarchy. Nested hierarchies meet all the criteria of the more general hierarchical condition. The higher holons in the system are associated with slower time constants of behavior (i.e., longer cycling times and relaxation times), and, if manifested in space, they occupy a larger volume. This is common to all hierarchical construction. Also, the higher holons in a nested system constrain the behavior of the lower holons; where in the world a heart shall beat is determined by the movement of the whole organism. However, these are properties common to nonnested hierarchies such as social hierarchies. Nonnested hierarchies relax the requirement for containment of lower by higher holons and also do not insist that higher holons are derivable from collected lower holons. It is profitable to view the nesting as a restriction, for if we are limited to just the nested condition, then generality is lost because the many nonnested ecological structures must be viewed as a separate class of phenomena.

Also relevant to comparisons of nested and nonnested systems is the nature of emergent properties. Margalef (1968) discusses such properties and labels them as "macroscopic." These are properties of higher levels in the system that are not obvious from the properties of the parts. In fact a clear view of the parts may preclude observation of the emergent property. An example here could be the properties of macromolecules which cannot even be seen, let alone given account, at a scale of observation that focuses on only one or a few atoms at a time. A bond cannot be seen until a level of resolution is used which allows both parties to be seen at once. By this definition the property emerges because of a change in the way observations are made. But there are other considerations.

With appropriate observer scale as a given, one type of emergence only indicates that the property arises from incomplete specification of the parts or an insufficient understanding of their interactions. Such properties are emergent only because they were unexpected when they were found—a simple matter of observer inadequacy. Again, with appropriate observer scale as a given, another type of emergence relates to properties of the whole that are innately irreducible to the parts; they are not reduced because they cannot be reduced. These pertain to linguistic aspects of the whole; linguistic relationships involve signs and symbols whose relationships to that which they represent are arbitrary. Therefore they cannot be reduced by application of universal laws. . . . Returning to the comparison of nested and nonnested hierarchies, emergent properties of higher-level properties are seen more easily in the nonnested case as being something more than mere addition of the parts, although such emergence also applies to higher-level properties of nested structures.

Koestler's (1967) example of a hierarchical military command is a useful nonnested example here. It is particularly illustrative because a nested equivalent may be developed with ease. Clearly the general at the apex of the military command does not contain and is not composed of his subordinates. Nevertheless he does exhibit the characteristic behavior of a top-level holon in that he makes decisions and utterances to the men that he controls much less frequently than do his low-ranking officers. In spatial terms the influence of the general is felt throughout the area over which his army is deployed. A platoon leader makes frequent decisions and directly influences only the limited area occupied by his small group.

The alternative nested military hierarchy has an apex which is the entire army itself. Here the top holon is indeed composed of all the individual soldiers which make up its parts. A contrast between the general as the top holon of the military command and the army as the top holon of a military force is helpful in identifying the special properties of a nested hierarchy. In order to determine the behavior of the entire army it is only necessary to have an account of all the individual soldiers and commanders. Nested hierarchies are, in this sense, determinable in their behavior. Emergent properties here could be of the type of emergence that arises from incomplete observation. Inadequate knowledge of the placement of soldiers or an incomplete specification of their interaction could lead to emergence. This alone without any mystical emergence of the army could make the hierarchy appear indeterminate, although with sufficient specification the type of determinacy that we attribute to nested hierarchies would emerge. There is, however, an indeterminacy of even nested hierarchies that relates to why the parts behave as they do and to the role of the whole in constraining the parts. This indeterminacy relates to the other type of emergence mentioned above, the one that we said pertains to linguistic aspects of the whole.

In the nonnested condition, the behavior of the highest holon, a general, is not derivable from even the most complete account of the lower levels. In nonnested hierarchies emergent properties must remain uncertain until the system is actually let run for a while. The nested hierarchical model is the one which generally facilitates a reductionist model of scientific investigation. Holistic science, on the other hand, can always operate perfectly well with the nonnested condition. In the nonnested case the individual holons are taken at face value as quasi-independent wholes which are part of a hierarchical system of communication. When Howard Odum deals with ecosystems using energy as a *lingua franca,* he is generally increasing the nestedness in his ecosystem models. Odum's approach has more than a touch of reductionism to it. The incorporation of conservation principles in the construction of complex models such as those for ecosystems also tends to make the system nested. This allows large ecosystem models to be built in the reductionist mode. Conservation principles are often reasonable, but provide simple cast-iron constraints for the whole model. If, however, the system to be modeled has subtle self-organizing forces, the conservation constraints do not allow their emergence in the model. Prescribed conservation masks evolved emergent conservation. Thus the large reductionist ecosystem models may tell something of the *how* of ecosystems but lose much of the *why*. They focus on system dynamics rather than rate-independent system constraints (Pattee 1978).

The general at the apex of the military command is himself a nested hierarchy including personality, past experience, and physiology. . . . The interconnections within his personal nested hierarchy are much stronger than those in his role as the top holon in a military command. It is in his personal nested hierarchy that he gains the extra degrees of freedom which make him non-derivable from the lower levels of his military command. The interactions in his personal nested hierarchy occur much faster than his orders travel down the military command. The general could be modeled with a difference equation with one time unit between each command. The processes in his personal nested hierarchy are equivalent to a large r, the growth rate in difference equations discussed by May and Oster (1976) in the context of populations. May (1976) points out that if r in a simple difference equation becomes at all large, then the equation enters a realm of chaos where its behavior, while deterministic, is subject to summary only in statistical terms, like that of the military general.

Levins (1974) includes long-term oscillations as a type of instability and says: "Instability must not be confused with lack of persistence. In cellular metabolism and in the central nervous system it is rather interpretable as spontaneous activity." It is from the tightness of the integration of his personal hierarchy that the general gains his spontaneous behavior and inscrutability with respect to the rest of the military command. At the top of the command hierarchy the general displays emergent properties which cannot be determined even with the most perfect knowledge of the rest of the nonnested hierarchy. That is not to say that the general's personal nested hierarchy may not allow the derivation of his behavior, but emphasis upon the personal nested hierarchy of all the men in the army from the general to the soldier is a very inefficient account of the nature of military organization and our particular general and his specific army. Worse than this, the significance of the nonnested hierarchy would be lost in the noise of overspecification. The military command would disappear in a mountain of personal histories for each soldier. Knowing all the nucleotide placements in fly does not tell you much about its flyness.

> In fact, if we should actually achieve a microscopic rate-equation description of the measurement constraints for the system we are "explaining," we would find that not only the measurement but the system we originally had in mind would disappear, only to be replaced by a new system with an immense number of new initial conditions requiring new measurements. One can recognize this process as related to what we call reductionist explanation where one has explained away the original system. . . . The essence of the measurement problem at the quantum mechanical level is that reductionism apparently "explains away" the measurement itself. (Pattee 1979)

There is an inescapable presence of the observer which is seen particularly clearly in nonnested hierarchies. In every measurement there is present a choosing to make the measurement in the first place, and in that contextual choosing the observer defines the form of the nonnested hierarchy.

Ecology and Evolution

Ecological science and evolutionary law are tightly linked. Ecology involves evolution, as the interactions of the biota with each other and with the abiotic environment depend on the evolved properties of organisms. Evolution involves ecology, as the properties of organisms are a function of their interactions with each other and with the environment. (As we shall see, the degree to which the traits are a function of interactions with the environment is hotly debated.)

Part 5 addresses the ways evolutionary law contributes to scientific ecology. Evolution, natural history, and systems science form the core of scientific ecology. Evolution, in turn, is grounded in genetics, which provides a mechanism to understand heritable variation as well as the way the environment acts upon the genome and results in natural selection. In our survey of the connection between ecology and evolution, we will ponder the metaphysical implications of evolution, Darwin's argument for natural selection, the issue of teleology, Darwinism and molecular genetics, the adaptation debate, and, finally, evolutionary ecology.

The Evolution of Metaphysics

The idea of evolution has had an enormous effect on the Western tradition, from the humanities to natural science. The idea of evolution is not merely a biological theory; it indicates *a fortiori* a shift in worldview. The catalyst for this sea change was the English biologist Charles Darwin and the 1859 publication of *On the Origin of Species by Means of Natural Selection, or The Preservation of Favored Races in the Struggle for Life.*

For Darwin's part, as he sailed to the Galápagos Archipelago on H M S *Beagle,* he was far from ready to accept the idea that species change. At Cambridge, Darwin was most impressed by the theologian William Paley's work (see Darwin 1988: 155; 1991: 388). In a famous argument, Paley (1825) compares a watch and a stone. If you are walking along, Paley argues, and happen to find a watch lying on the ground, you assume that the watch had an intelligent maker, because the parts are related to each other in such a way as to exhibit *design.* Cogs, springs, axles, and arms function in concert with each other to create a beautifully synchronized whole. But the situation is different if you are walking along and happen to kick a stone: you assume that the stone was not made for some intentional purpose, like the watch, because it lacks intelligent design. Then Paley draws an analogy between artificially manufactured objects like watches, and natural objects, like eyes. Eyes, like watches, exhibit design. As a watch is designed and constructed by an intelligent artificer, so must be natural objects. The intelligent designer of natural objects is God. *Nature,* in effect, is a magnificent corporeal clockwork designed by God.

The implication of Paley's argument for biology is that species do not change, as ontic change is inconsistent with the idea that God created nature according to a perfect, divine plan. Change implies imperfection. Natural theology was the *Zeitgeist* of nineteenth-century Europe and squarely in line with the elegant teleology of Occidental thinking—a theme woven throughout the fabric of this book. For the young Darwin en route to South America, the amazing plenitude of life represented the creative work of a supernatural omniscient, omnibenevolent, omnipotent creator.

Many Enlightenment thinkers, however, questioned the reality of this utopian metaphysic. Voltaire (1999) notoriously chastised Gottfried Leibniz's "this is the best of all possible worlds" version. Hume, in the words of interlocutor Philo, makes the point clear: "The parts hang all together; nor can one be touched without affecting the rest, in a greater or less degree. But at the same time, it must be observed, that none of these parts or principles, however useful, are so accurately adjusted, as to keep precisely within those bounds, in which their utility consists; but they are, all of them, apt, on every occasion, to run into the one extreme or the other. . . . Rains are necessary to nourish all the plants and animals of the earth: But how often are they defective? how often excessive?" (1990: 120).

Eventually Darwin asked the same question. How could an omnipotent and omnibenevolent being design a world with so much apparent maladjustment? The degree of struggle and strife inherent in nature contradicted perfection of design; the various parts of nature often impede and destroy one another. In a candid May 22, 1860, letter to Harvard botanist Asa Gray, Darwin admits:

> I had no intention to write atheistically. But I own that I cannot see, as plainly as others do, and as I [should] wish to do, evidence of design and beneficence on all sides of us. There seems to me too much misery in the world. I cannot persuade myself that a beneficent and omnipotent God would have designedly created the Ichneumonidae with the express intention of their feeding within the living bodies of caterpillars, or that a cat should play with mice. Not believing this, I see no necessity in the belief that the eye was expressly designed. (1993: 224)

As a biologist, Darwin could no longer accept divine design-in-nature metaphysics. Darwin writes in his autobiography: "The old argument of design in nature, as given by Paley, which formerly seemed to me so conclusive, fails, now that the law of natural selection has been discovered. We can no longer argue that, for instance, the beautiful hinge of a bivalve shell must have been made by an intelligent being, like the hinge of a door by man. There seems to be no more design in the variability of organic beings and in the action of natural selection, than in the course which the wind blows" (1958: 87). And if design in nature falls short of perfection, what is the use of retaining the corollary static species concept? None.

Against the background of natural theology, it is clear that the idea of evolution is far more than biological theory: it represents a fundamental shift in worldview across the entire Occidental tradition. In "The Metaphysics of Evolution" (1967b), reprinted here as chapter 18, philosopher and historian of biology David Hull traces differing metaphysical concepts of species from the ancient Greeks to Darwin. Hull shows that a necessary condition for the theory of evolution to be taken seriously was a rejection of divine design-in-nature metaphysics. That is to say, the theory of evolution required an evolution of metaphysics!

One Long Argument: The Theory of Descent with Modifications through Natural Selection

At the end of the *Origin,* Darwin writes: "It is interesting to contemplate a tangled bank, clothed with many plants of many kinds, with birds singing on the bushes, with various insects flitting about, and with worms crawling through the damp earth, and to reflect that these elaborately constructed forms, so different from each other, and dependent upon each other in so complex a manner, have all been produced by laws acting around us" (1975: 489). If these laws are not the divine laws of natural theology, what are they? According to Darwin, they are the laws of natural selection.

The *Origin,* in Darwin's own words, is "one long argument" for "the theory of descent with modifications through natural selection" (1975: 435). In order to discern the importance of evolutionary law for ecological science, it is worth recounting the logical flow of Darwin's argument. (For a summary of the argument, see figure 13. Mayr (1991: 72) provides a similar synopsis.)

The first subargument of the main argument can be attributed to the English economist Thomas Malthus. In his 1798 *Essay on the Principle of Population,* Malthus points out that natural populations can increase rapidly (premise 1), while resource utilization can, at best, increase slowly (premise 2): "The power of population is infinitely greater than the power in the earth to produce subsistence for man. Population, when unchecked, increases in a geometric ratio. Subsistence increases only in an arithmetic ratio. . . . This implies a strong and constantly operating check on population from the difficulty of subsistence"

Premise 1:   Rate of population growth increases geometrically
Premise 2:   Rate of resource utilization increases arithmetically
Premise 3:   Actual population levels remain relatively constant
    Conclusion 1:   Because the number of offspring produced exceeds the subsistence level,
            there is competition between individuals for survival
Premise 4:   Individuals within populations differ
Premise 5:   Heritability of traits
    Conclusion 2:   The best-adapted individuals survive (natural selection)
Premise 6:   The process of natural selection takes place over long periods of time (gradualism)
    Conclusion 3:   Over generations, differential reproduction changes the overall genetic
            makeup of populations (evolution of species by natural selection)

Figure 13. The structure of Darwin's argument for natural selection

(1926). What Malthus has in mind by using the mathematical words *geometric* and *arithmetic* is that populations have the *potential* to increase exponentially (e.g., if two parents have four offspring, and the offspring breed at the same rate, the population could grow from 2 to 2,048 in only ten generations), while *potential* food production increase is linear (e.g., the amount of cultivated land can be increased only one acre at a time). The main point is that reproductive rates always outstrip carrying capacity (see figure 14).

Yet it is easy to see that actual population levels remain relatively constant over time (premise 3). Therefore, in conjunction with premises 1 and 2, there must be competition for limited resources. The penalty of losing is death (conclusion 1). As Darwin succinctly puts it: "The struggle for existence inevitably follows from the high geometrical ratio of increase which is common to all organic beings. This high rate of increase is proved by calculation. . . . More individuals are born than can possibly survive" (1975: 467). In contrast to the concordant clockwork ontology of natural theology, wherein all the parts of nature fit and function harmoniously, this picture is one of constant strife and fierce competition for survival. Nature is, in Alfred Tennyson's (1987: 373) famous phrase, "red in tooth and claw."

In addition, individual members differ observably from one another within populations (premise 4), and these variations are capable of being passed from generation to generation (premise 5). The whole purpose of breeding animals and plants, Darwin says, is to pick individuals with specific traits out of populations in order to produce progeny with the desired traits: "Man can and does select the variations given to him by nature, and thus accumulate them in any desired manner. He thus adapts animals and plants for his own benefit or pleasure. . . . This process of selection has been the great agency in the production of the most distinct and useful domestic breeds" (1975: 467).

Next, Darwin draws an analogy, and the conclusion is natural selection (conclusion 2): "There is no obvious reason why the principles which have

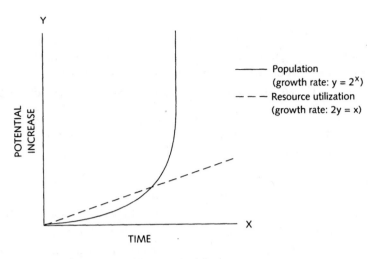

Y

POTENTIAL
INCREASE

——— Population
(growth rate: $y = 2^x$)
— — Resource utilization
(growth rate: $2y = x$)

TIME

X

Figure 14. Thomas Malthus's population principle

acted so efficiently under domestication should not have acted under nature"
(ibid.: 467). Hence, if individuals can be selected by humans for propagation
(artificial selection), then there is reason to believe that in all populations,
certain organisms are favored over other organisms in their ability to obtain
resources, withstand environmental vicissitudes, escape predators, and overcome other challenges to survival (natural selection). To put it another way,
certain individuals are *selected* for survival and reproduction because they
possess traits that allow them to adapt to particular environmental conditions
better than competing individuals.

As Darwin learned from his study of Galápagos finches and the fossil record
(and substantiated by the studies of other naturalists such as von Buch, Wag-
ner, and Wallace), there is evidence of populations of related species in various
stages of evolutionary development. Because of such evidence, Darwin asserts
that the relative frequency of new traits appearing in a population from one
generation to the next can only be slight (premise 6). Natural section occurs
slowly, in small incremental steps (gradualism);[1] there are no large leaps or
jumps in the progression of evolution (ibid.: 460).

All of the pieces are in place to draw the final inference: over many generations, differential reproduction changes the overall genetic composition of the
population (i.e., the gene pool). Although the relative frequency of traits appearing in the population will not vary greatly from one generation to the next,
over numerous generations the change may be great. This process, evolution
by natural selection (conclusion 3), explains the origin of species. Speciation
occurs at the level of populations as gene frequencies change.

A review of Darwin's argument for natural selection lays bare the ecological aspect of evolution. The organism is a node of complex social, biological, chemical, and physical interactions. The environment of the organism provides the context within which genetic differences (genotypes) are manifested (phenotypes). The struggle for existence is the acting out of these differences.

The Kind Watchmaker or the Blind Watchmaker?
The Issue of Teleology

Evolution has a pattern: the geological record reveals a slow and gradual transformation from simplistic life-forms such as *prokaryotae* to more complex life-forms like *plantae* and *animalia*. What is the reason for this pattern? Is an omniscient, omnipotent, and omnibenevolent being akin to Paley's Watchmaker behind it, guiding evolution in a planned direction? Or is evolution unplanned and unpredictable, and any pattern simply the result of the metaphysical essence of nature itself?

The evidence of evolution suggests that there is no ultimate endpoint (*telos* in ancient Greek) external to the process of evolution itself. Design in living things results from the generation of genetic variability and complex interactions between organisms and their environments. Organic design does not necessitate a wise designer. Oxford zoologist Richard Dawkins writes:

> A true watchmaker has foresight: he designs his cogs and springs, and plans their interconnections, with a future purpose in his mind's eye. Natural selection, the blind, unconscious, automatic process which Darwin discovered, and which we now know is the explanation for the existence and apparently purposeful form of all life, has no purpose in mind. It has no mind and no mind's eye. It does not plan for the future. It has no vision, no foresight, no sight at all. If it can be said to play the role of watchmaker in nature, it is the *blind* watchmaker. (1996: 5)

The stochasticity of evolution calls into question the reality of an external and unchanging *telos* orchestrating the process in a rigid and predetermined way. Organisms have a definite purpose—to live and grow and reproduce—but this purpose does not guide their pattern of growth and development. All organisms are affected by constantly changing living conditions. Animals can subvert some environmental determinism by changing their habitat. Growth and development can occur in more than one manner, along more than one vector.

Consequently, contemporary biological definitions of *evolution* generally omit any reference to *a* direction. For example, compare Peter Price's (1996) definition of evolution as "genetically based, heritable change in one or more characteristics in a population or species through time" with Webster's definition of evolution as "a process of change in a certain *direction*, a process of continuous change from a lower, more simple to a higher and more complex state, or a process in which the whole universe is a progression of interrelated phenomena" (1981: 393; emphasis added). To capture the idea that a specific *direction* is not inherent in evolution but *directionality* is, we might say that the process of evolution by natural selection is not *teleological* but *teleomatic*.

Due to the stochasticity of evolutionary process, philosophers of life science tend to agree that it is legitimate only to say that natural section *has* taken place, not that it *will* take place. Recognition of natural selection is always retrospective. Because of the unpredictability of genetic variation and environmental conditions, there is no apodictic way to say, "This phylogenic trait will allow this organism and its progeny to propagate and increase the genotypic frequency in the population." For this reason, the word *selection* may be misleading because there is no intentional, forward-looking process similar to selection in crop and livestock breeding. Using the French word *bricoleur*, which refers to a person who creates new useful objects from found materials, Paul Rabinow comments: "Nature is a blind *bricoleur*, using an elementary logic of combinations, yielding an infinity of potential differences" (1996: 108).

If evolution is not *teleological* in the sense of being guided by a *telos* external to the process of evolution itself, then there is no reason to think that *Homo sapiens* is the terminus of evolution, as paleontologist and philosopher of religion Pierre Teilhard de Chardin implied with his notion of humanness as the "Omega Point" of evolution (Teilhard de Chardin 1969; Goudge 1972: 83–84). We will likely be eclipsed by some new being of our own flesh and blood—Nietzsche's *Übermensch*?

The Modern Synthesis: Darwinism and Molecular Genetics

A necessary condition for natural selection is variation of individuals within populations, and the heritability of this variation. How is variation generated and transmitted from generation to generation?

Darwin and other biologists did not know exactly how inheritance worked (although in 1865, unknown to Darwin, Gregor Mendel (1965) had provided a mathematical description of genetic heritability). Yet, despite the lack of a correct understanding of genetics, Darwin could still plausibly argue for evolution by natural selection; all he needed to know was that genetic variability is passed on from one generation to the next. As the Harvard zoologist and philosopher of biology Ernst Mayr remarks, "On the whole Darwin treated genetic variation as a 'black box.' As a naturalist and reader of the animal breeding literature, he knew that variation was always present, and this is all he *had* to know. . . . In other words, a correct theory of genetics was *not* a prerequisite for the theory of natural selection" (1991: 82). The missing genetic ingredient was provided years later by a series of investigators who showed that deoxyribonucleic acid (DNA) is the genetic material for all cellular organisms and most viruses.

The modern theory of evolution emphasizes that variants of genes enter populations through mutation or recombination. Variety is generated in at least three ways: (1) mutation in replication, (2) mutation by environmental conditions, and (3) sexual reproduction. Mutation during the process of replication (e.g., a base-pair substitution, insertion, or deletion), though statistically rare, is one source of genetic variability. Mutation may also be caused by the breakage and/or reorganization of the chromosomes due to environmental factors such as radiation or chemical exposure. Recombination occurs in sexual reproduction during the process of meiosis. The result is that natural selection has an enormous array of genetic variation upon which to act. Selection eliminates those organisms that are unfit and favors those that can cope with the environmental circumstances at that place and time or have the capacity to move to a more hospitable place.

The identification of the structures making up the chromosome and the gene are consistent with the structure-function principle; that is, organisms carry out biological processes as functions of their adaptively evolved structure. John Campbell (1985) observes that when Darwinism and even modern neo-Darwinism were formulated, the structures of organisms and genes were so poorly understood that evolutionists used the mechanical paradigm of physics, in which inert objects move in passive response to exogenous forces pushing upon them from outside, to explain the relation between structure and function.

In this interpretation, evolution occurs mainly from changes forced upon the organism by the environment instead of being based on functions the organism is structured to carry out. "During the past decade," Campbell says, "molecular biology has advanced genetics to a structure-function science" (ibid.: 133). The organizational features of gene structure include the following: (1) genes have sufficient internal structure for their functions, (2) genes are chemical substrates upon which enzymes operate, and (3) complex genes include sensing devices to bring relevant information from the environment to the DNA molecules. The complexity of genes and their interactions with other genes or regulation by the environment allow ample room for mutations to produce variation on which natural selection can work.

The contemporary evolutionary program begins with patterns observed in nature that are discrete, or are gradients of difference. The question changes from the usual ecological questions of What is there? Where is it located? and How does it come to function that way? to Why does that particular pattern appear when there are several equally plausible alternatives? This latter question is frequently examined experimentally and then submitted to deeper analytical techniques that help to explain the genetic patterns that are related to observed ecological patterns.

The Adaptation Debate

The question of adaptation is central to the philosophy of ecology because the less adaptation determines an organism's physiology, morphology, and (in animals) ethology, the smaller the role ecology plays in the evolutionary drama.

Evolutionists usually demonstrate natural selection by describing how a certain trait allowed individuals within a population to *adapt* to environmental exigencies better than other individuals, thereby increasing the relative frequency of that trait and, in the long run, altering the gene pool of the species. This explanation for *how* evolution works is adaptationism. Here is an example: Within the population of polar bears, the allele for white hair has been selected over alternate alleles for hair color, such as brown hair or black hair, because those individuals with the allele for white hair blended into the snow and were better able to sneak up on and kill prey, such as seals. Individuals that were unable to kill sufficient prey perished without producing many offspring. As a result of this selection process, alleles for nonwhite hair have been eliminated in this species.

Evolutionists agree that natural selection by adaptation takes place, but disagree over whether it tells the full story of evolution. This point of dispute within evolutionary biology can be accurately dubbed the adaptation debate. On one side are the "orthodoxy" (e.g., Richard Dawkins and Daniel Dennett), who maintain that if any nonadaptive processes do in fact occur in evolution, they play insignificant roles. On the other side are the nonadaptation "heretics," Niles Eldredge, Stephen Gould, and Richard Lewontin. The latter point out that Darwin himself believed that "natural selection has been the main but not the exclusive means of modification" (*Origin*, 6th ed., quoted in Gould 1997: 34). Gould calls this approach "Darwinian pluralism," in contrast with the "Darwinian fundamentalism" of the orthodox adaptationists. Gould states:

> I (along with all other Darwinian pluralists) do not deny either the existence and central importance of adaptation, or the production of adaptation by natural selection. Yes, eyes are for seeing and feet are for moving. And, yes again, I know of no scientific mechanism other than natural selection with the proven power to build structures of such eminently workable design. . . . But selection cannot suffice as a full explanation for many aspects of evolution; for other types and styles of causes

become relevant, or even prevalent, in domains both far above and far below the traditional Darwinian locus of the organism. . . . These additional principles are as directionless, nonteleological, and materialistic as natural selection itself—but they operate differently from Darwin's central mechanism. In other words, I agree with Darwin that natural selection is "not the exclusive means of modification." (ibid.: 35)

Clearly, then, Gould is *not* repudiating adaptationism, as is sometimes thought; he is only trying to validate the nonadaptive mechanisms of evolution. Mechanisms that *limit* genetic variability for the sake of physiological stability, for instance, are nonadaptive (ibid.).

The debate was inaugurated with Gould and Lewontin's 1979 essay, "The Spandrels of San Marco and the Panglossian Paradigm: A Critique of the Adaptationist Program," reprinted here as chapter 19. Gould and Lewontin argue that, just as certain architectural features like *spa· .'rels*—the curvilinear space created by the intersection of two arches and a dome [2]—are accidental epiphenomena of the construction of great churches, so are some biological features accidental epiphenomena of evolution. Gould and Lewontin chastise the "adaptationist program" as a utopian (that is, Panglossian) project which holds that organisms are the way they are because they are perfectly adapted to their environments, analogous to the way Dr. Pangloss (Voltaire's parody of Gottfried Leibniz) explains catastrophes and other natural evils as being part of a "preestablished harmony" (see Leibniz 1965).

Gould and Lewontin's piece is valuable, in addition to its philosophical and literary merits, for the amount of discussion it stimulated. In "How to Carry Out the Adaptationist Program?" (1983), reprinted here as chapter 20, Ernst Mayr agrees with Gould and Lewontin that claims of adaptation need to be backed up with *evidence* if they are to be anything more than "just-so stories." But in their argument against adaptationism, Mayr contends, Gould and Lewontin are guilty of the straw man fallacy: they set up a caricature of the adaptationist program, based on "a faulty interpretation of natural selection and an improperly conducted adaptationist program," only to knock it over. In fact, says Mayr, Gould and Lewontin are themselves adaptationists. Philosopher Daniel Dennett, in his spirited attack on nonadaptationism, makes the same point: "The thesis that every property of every feature of everything in the living world is an adaptation is not a thesis anybody has ever taken seriously" (1996: 276).

There seems to be agreement that adaptationism is an extremely powerful tool in explaining how natural selection works. Because of the length of evolutionary process in contrast with the shortness of our lives, natural selection is difficult to observe directly. In turn, it is difficult to know which traits are adaptations and which, if any, are not. The extent to which adaptation is the main means of evolution remains an open and interesting quandary.

Evolutionary Ecology

Ecologists have been concerned with evolution from the time of the publication of the *Origin*. Ernst Haeckel (see figure 1) was a major supporter of Charles Darwin's ideas in Germany. In 1866, when he invented the word *ecology*, Haeckel said that "ecology is the study of all those complex inter-relations referred to by Darwin as the conditions of the struggle for exis-tence" (1879).

Thus, ecology was attached to evolutionist thinking from its founding. Al-most all ecology textbooks make evolution a central organizing theme, and some ignore most other kinds of ecology except as they relate to evolutionary ecology. Evolutionary ecology forms, with genetics, an integrated foundation for conservation biology. Evolutionary ecology also provides the context for anthropological studies of social phenomena (Steward and Murphy 1977).

James Collins's "Evolutionary Ecology and the Use of Natural Selection in Ecological Theory" (1986), reprinted here as chapter 21, provides a historical overview of the development of evolutionary ecology. At the time of Haeckel and later, ecologists were interested in evolution because of their studies on speciation and the distribution of organisms. Natural history asked questions about the presence and location of organisms. Centuries of exploration and observation led to an understanding of regional patterns of flora and fauna diversity and distribution. These observations revealed many puzzles that have intrigued biologists ever since. But interest in natural history waned in the twentieth century as ecologists became interested in population dynamics, community organization, interactions among species, and systems science. Population research rapidly developed the capacity to explain patterns and pre-dict outcomes with a relatively high degree of success (Kingsland 1985). While many ecologists continued to be involved with questions of natural history, populations, or with developments in ecosystem science, other ecologists ap-plied the findings in genetics to ecological science. In 1962 Gordon Orians first used the phrase "evolutionary ecology."

An example of developments in evolutionary ecology is the study of the role of the antlers of large ruminants such as the elk (*Cervus canadensis*). Antler size varies widely among males and is highly correlated with age and nutritional condition. Antlers must be replaced each year of the animal's life. Antler size is related to the animal's success during the breeding season and to its well-being at other seasons. Researchers have concluded that a stag's antler size reflects his overall genetic quality, which in turn affects his offspring. Kodric-Brown and Brown (1984) call this "truth in advertising" and conclude that "sexual selec-tion favors the evolution of costly phenotypically variable traits whose expres-sion reflects survivorship and vigor of males, and hence their overall genetic quality."

Over the last three decades evolutionary ecology has grown enormously. In "Evolution: The Missing Ingredient in Systems Ecology" (1988), reprinted here as chapter 22, Craig Loehle and Joseph Pechmann discuss the application of evolutionary thinking to ecology, and specifically to ecosystem ecology. They contend that the integration of evolution into ecosystem modeling overcomes many current research challenges.

The research program of evolutionary ecology is very complex, and this complexity raises a philosophical question. We might call this the "locality" question because it focuses on the evolution of the individuals as we observe them directly in nature. Each organism enters the world with genetic information that gives it the capacity to react to other organisms of like kind that may be helpful or competitive; to organisms of unlike kind that may serve as food or habitat, enemies and predators, transport agents, and so on; and to the physical-chemical environment, which may support the organism or destroy it. How are we to develop a realistic model of these interactions and relationships that does not simplify the process unrealistically or bias our understanding? We observe the organism. That is the "fact" from which our speculation begins. But how do we relate the ecological complexity of the organism's life to its genetic character? There may be a tendency to find correlations between one factor and another process. Organisms may be there simply because the structures of the environment provide a niche into which they may fit, by chance, at this particular moment. How do we distinguish selection and adaptation from other causal explanations? Ecologists do not yet have the answers.

Conclusion

How did life originate? is perhaps the most profound question that a reflective and self-aware creature can ask. Human beings, fitting this description, have given answers ranging from myth to scientific explanation.

The answers we give affect the way we live, the way we orient ourselves in our earthly home (*oikos*). The prevailing answer of the Western tradition bestows upon humans a nonnatural origin that posits a fundamental ontological divide between humanity and nature. The first Genesis human creation myth (1:26)—that human beings have a supernatural origin—implies that humans are separate from and superior to nature. (Interestingly, the second Genesis creation myth (2:7)—that humans were created from the dust of the earth— implies the opposite.) The ancient Greeks felt that a social-political structure (*polis*) was a product of human intelligence and separated civilized humans from nature—and from the barbarians outside the city walls. The answer of life science is that *Homo sapiens* is the result of evolution; humans' essence is our evolutionary history. Further, all life as we know it evolved from a common ancestor. The notion of the unity of life goes back to Darwin: "All living things have much in common, in their chemical composition, their germinal vesicles, their cellular structure, and their laws of growth and reproduction. . . . Therefore I should infer from analogy that probably all the organic beings which have ever lived on this earth have descended from some one primordial form, into which life was first breathed" (1975: 484). We know now, as Darwin did not, that all life, from bacteria to bears, shares the same double helix of DNA. If all life has a common origin, then all life is interconnected in the most ontologically fundamental way. Although we may be unique in our capacity to reflect on our own origin, humans are certainly *not* separate from nature. Thus to posit a fundamental ontological divide, as Descartes (1989: 115) and others have done, is unecological (Keller 1997b).

Our vision of the origin of life has repercussions on our actions. The normative implication of an anthropocentric worldview is that humans' manipulation of natural systems can never be detrimental to us, because we are separate from and superior to nature. Unfortunately this belief is in direct contradiction to the lessons of ecology—one reason Paul Sears (1964) is justified in calling ecology a "subversive" science. A subversive science would interpret the deleterious effects of industrial chemicals on human health (Steingraber 1997) as calling into question the entire edifice of modern market civilization.

Human activity that degrades the biosphere also degrades human health and our ability to flourish. From an ecological standpoint, this is a tragic paradox: humans should *promote* human health and our ability to live well (we are, after all, the *wise* humans). The first step beyond this paradox is universal literacy about ecological systems (Golley 1998) and the myriad ways human beings are embedded in life's web.

Notes

1. Eldredge and Gould (1985) question the validity of gradualism, instead arguing for "punctuated equilibria," although Dennett (1996: 282–99) concludes that gradualism withstands their criticisms.

2. As Dennett (1996: 272) points out, technically, the curvilinear space created by the intersection of two arches and a dome is a *pendentive;* a *spandrel* is a wall with an arch punched through it.

CHAPTER 18

The Metaphysics of Evolution

David L. Hull

Extreme variation in the meaning of the term *species* throughout the history of biology has often frustrated attempts of historians, philosophers, and biologists to communicate with one another about the transition in biological thinking from the static species concept to the modern notion of evolving species. The most important change which has underlain all the other fluctuations in the meaning of the word *species* is the change from it denoting such metaphysical entities as essences, Forms, or Natures to denoting classes of individual organisms. Several authors have taken notice of the role of metaphysics in the work of particular biologists. An attempt will be made in this paper to present a systematic investigation of the role which metaphysics has played in the work of representative biologists throughout the history of biology, especially as it relates to their species concepts.

That biologists through the years have held radically different species concepts is evident from the different natural phenomena which each has thought consistent with this concept. For example, Aristotle's belief in the immutability of species was completely consonant with his belief in extensive interspecific hybridization, the inheritance of acquired characteristics, the failure of plants to breed true from seed, and spontaneous generation; whereas the occurrence of such phenomena was incompatible with the species concepts of most biologists after the seventeenth century, whether these biologists believed in static or evolving species.

The general purpose of this paper will be to explain the close connection which has always existed between metaphysics and the theory of evolution. To explain, for example, why evolution in the modern sense seemed inconceivable to essentialists and neoplatonists but perfectly plausible to nominalists. More specifically, the purpose of this paper will be to trace the transition from the static species concept of Plato and Aristotle to the static species concept of such seventeenth- and eighteenth-century biologists as Ray, Linnaeus, and Cuvier and from these notions to the disparate evolutionary concepts of species espoused by Buffon, Lamarck, and Darwin. Many of the ideas which modern

From *British Journal for the History of Science* 3 (1967): 309–37. Notes omitted.

readers find most incomprehensible in the works of early biologists appear not in the least bizarre when viewed against the background of the tacitly-held metaphysics of these men. To discuss these issues a few preliminary remarks must be made about the metaphysics of Plato and Aristotle.

Essences and Forms

The Greek correlate to the English word *species* was a term of very broad significance in the philosophies of Plato and Aristotle. Not only were there species of plant and animal but also there were species of rock, goodness, and triangularity. Everything was divided into species or natural kinds. *Species* was also a relative term when used in conjunction with *genus*. For example, the ruminants were a species of the genus hairy quadrupeds, and the hairy quadrupeds were a species of the genus blooded animals. But the most important aspect of the ancient species concept was that not all classes were species. For example, neither the class of things bigger than a breadbox nor the class of red-haired fifth-formers would have been considered species. In order for a species to be a species, it had to have an essence or a Form. In fact, for Plato and Aristotle and for centuries of philosophers to follow, species names referred primarily to these metaphysical entities and not simply to classes of things. It was these metaphysical entities which were *really* real. One of the most important transitions in Western philosophy was the shift from such things as essences and Forms being the basic constituents of the world to the objects of experience being the basic constituents. This shift was reflected in the word *species* in biology, but with a considerable time lag.

According to the traditional interpretation of Plato's philosophy, universals such as man, goodness, and triangularity existed in a realm separate from and independent of the realm of appearances. The name "Flicka" referred to an individual horse. The word *horse* referred to the Idea or Form horse in the separate, eternal, and unchanging realm of Ideas. Individual horses might come and go but the Idea of horse was immutable. Hence, science rightfully dealt with the Idea horse, not with individual horses or even classes of horses in the world of appearances. It was these immutable Forms, his *eide*, which Plato termed *species*.

The situation in the case of Aristotle is complicated by his more complex (and perhaps even inconsistent) terminology. Although he used the term *eidos*, his more fundamental term was *ousia*, translated into Latin as *substantia* and into English as *substance*. Aristotle distinguished between two senses of *ousia* or substance. Primary substance was the individual thing, the logical subject about which assertions could be made. Secondary substance was the essence of the thing. The situation is even further complicated by the fact that Aristotle also used half a dozen terms and phrases which can also be translated as *essence*. According to one of the oldest traditions, Aristotle's metaphysics is much like that of Plato, except that essences, the Aristotelian correlate of Forms, are not separate and independent of the realm of appearances but are somehow "in" individual things. Aristotelian essences like Platonic Forms are immutable.

Hence, whenever Plato or Aristotle claimed that species were immutable, it was primarily to the immutability of these Forms and essences that he was referring and not to individual organisms or even classes of organisms. In turn the immutability of essences and Forms was the metaphysical justification for claiming that species as classes of organisms were on a whole static. But the word *species* was also supposed to have some empirical import. There was also some empirical justification for the claim that species were static. It is to the empirical connotations of *species* in the biological works of Aristotle and his successor Theophrastus (370–285 B.C.E.) that we now turn. For the purpose of this paper Plato's philosophy can be ignored until the eighteenth and nineteenth centuries, when it emerges as neoplatonism in the works of such biologists as Buffon and Cuvier. . . .

Spontaneous Generation

Although Aristotle and Theophrastus believed that almost all higher animals, a few lower animals, and all plants were capable of reproducing their kind, they also believed that throughout the animal and plant kingdoms an individual of one species could be transformed directly into an individual of another species. In fact, for the lower animals such as the testaceans, acalephae, sponges, and many so-called insects, this was their *only* means of propagation. Aristotle believed that the highest forms of animal life reproduced viviparously, but by this he meant without the production of an egg of any kind. Only intermediate animals such as reptiles and fish produced eggs from which individuals like their parents developed. This is true even though Aristotle often refers to certain structures produced by lower forms of animal life as "eggs." He makes it perfectly clear that he does not take these structures to be true eggs and continues to use the word *egg* in reference to them for the same reason he continues to refer to whales and porpoises as *fish* after he has carefully explained why they are not fish but more like hairy quadrupeds. Similarly, Theophrastus believed that all plants could reproduce their kind, although not always by seed (*Enquiry into Plants*, III, i, 1 and 5 (1916); *Causes of Plants*, I, v (1927)).

Sexuality extended much further into the animal kingdom than did the production of eggs. Aristotle believed that with rare exception all higher and intermediate forms of animal life as well as some insects paired to reproduce themselves (*De Generatione Animalium* 721a3). With the exception of the date palm, sexuality among plants was unknown, although numerous pairs of plants were called "male" and "female" (*Enquiry into Plants*, II, i, 1 (1916)). The important consideration in all these various modes of generation is whether like produces like. Instead of like producing like directly (as Aristotle thought occurred among the higher animals), or like producing like indirectly via a true egg (as Aristotle thought occurred among intermediate animals) or a seed (as Theophrastus thought occurred among most plants), in most lower animals *there was a succession of species. Like consistently did not produce like.*

Among those organisms which were capable of breeding true to type, occasionally an individual would change spontaneously into an individual of another species. For example, Theophrastus mentions that when the marshes dry up the water snake changes into a viper (ibid., II, iv, 4). Such changes as these were termed "spontaneous" not because they were uncaused (in the sense of material or efficient cause); some material such as the body of another individual or a bit of mud was necessary as was some instigating cause like heat. Nor was the production of such individuals erratic. Whenever the conditions were right, the change would take place. Such changes were termed spontaneous because this was not the natural method of reproduction for the individuals concerned—there was no pairing, no egg or seed produced, but most importantly, no succession of generations, like producing like.

The interesting type of spontaneous generation, however, occurred among the lowest forms of animal life. In the typical case, a grub or scolex was generated initially from a bit of inorganic matter such as scum, slime, soil, or mud. Then in turn a chrysalis or pupa was generated from the scolex. Finally, from one of these structures came the perfect animal. In this last case, one is tempted to read our modern notion of metamorphosis back into Aristotle, but this would be historically inaccurate. Not until the seventeenth century would biologists appreciate the true nature of such changes. For Aristotle metamorphosis was literally just that, a change in form. He did not look upon it as the maturation of a single individual, a series of stages in a life cycle, but as a succession of species. Regardless of how we might view these phenomena today, for Aristotle the generation of a piece of earth into an earthworm, an earthworm into an eel, or the generation of a bit of slime at the bottom of a well into an ascarid and the ascarid into a gnat was exactly the same process as the generation of a piece of cabbage leaf into a grub, the grub into a caterpillar, the caterpillar into a chrysalis, and the chrysalis into a butterfly (*De Generatione Animalium* 721a3, 762b25; *Historia Animalium* 551b26, 551a13).

Aristotle's belief in spontaneous generation was, of course, in error. Most modern commentators, including those who emphasize the biological motivation of Aristotle's metaphysics, pass over it with little or no comment (Hantz 1930; Randall 1960; Grene 1963). But his belief in spontaneous generation, erroneous as it may be, poses serious problems for his metaphysics, almost as serious as those posed by organic evolution. When a metaphysics is confronted with a recalcitrant datum, there are two common ways to reconcile the two. One can deny the existence of the datum or one can redefine some of the basic terms in the metaphysical system. Aristotle chose the second alternative in the case of spontaneous generation; he chose the first in the case of organic evolution. Both conflict with his metaphysics. The history of Western thought would have been considerably different had he made just the opposite choice: if he had denied the existence of spontaneous generation and reworked his metaphysics to permit the acceptance of organic evolution. Such a decision, however, would have run counter to the superficial appearance of things. Gall insects and intestinal parasites gave every appearance of arising spontaneously, whereas most members of most species stayed well within the limits of their species. (On the basis of examples like these, some authors have perversely chastised Aristotle for being "too empirical.") Exactly what problems were posed by spontaneous generation and organic evolution will be discussed in the next section.

The Four Causes

Central to Aristotle's metaphysics was his doctrine of the four causes, the doctrine that in order to explain anything completely, the material, efficient, formal, and final causes had to be given. Aristotle claimed that his predecessors fell short in attempting to explain the regularities present in nature just in terms of material and efficient causation. Aristotle objected to the views of Empedocles, not because he had postulated material and efficient causes for the genesis of species, but because he had maintained that these were the *only* factors involved (*De Partibus Animalium* 640a17, 642a15). Empedocles claimed that species first arose by the spontaneous generation of various organs which combined by chance. Those combinations which were organized in a fitting way survived; those that were not perished. Aristotle argued that nature exhibited too much regularity for this to be the true view. "For teeth and all other natural things either invariably or normally come about in a given way; but of not one of the results of chance or spontaneity is this true" (*Physics* 198a22). Such regularity can be explained only in terms of formal and final causes. There are always horses because horses tend to beget horses. This happens so regularly because in these cases the efficient, formal, and final causes are one. The efficient cause of a horse is the essence of its male parent; its formal cause is this same essence embodied in itself; and its final cause is again its essence, since the individuals of a species naturally strive to realize as perfectly as they can the essence of their species (*Physics* 198a25; see also Ross 1937: 74; and Copleston 1960: 313).

It was this combination of factors which led Aristotle to argue against organic evolution. It should have led him to argue against spontaneous generation as well. If essences are static and if in general the goal toward which living organisms strive unconsciously is the realization of their essence, then the wholesale progressive change implied in any evolutionary theory is impossible. . . .

Species in the Seventeenth Century

If anything is characteristic of biology after the sixteenth century, it is the prevalence of what is still known as the species problem. Why after centuries of uncritical acceptance did the existence of species present such a problem for biologists? The answer can be found, I think, in the convergence of two trends. First, Andrea Cesalpino (1519–1603) had reintroduced Aristotelian metaphysics into biology but in a much less sophisticated form than Aristotle had proposed. To this was added a literal interpretation of Christian scripture. Aristotle had maintained both that essences were eternal and that they could exist only "in" individuals. Thus, it followed necessarily that species as classes of organisms were equally eternal. When Aristotle was Christianized, species retained all of their Aristotelian characteristics with one exception. As classes of organisms, they had a beginning in time. The first members of each species were created specially by God in the Garden of Eden.

But of even more importance, by the seventeenth century biologists were beginning to turn their attention away from essences and the like and to the organisms themselves as they discovered that species as classes of organisms possessed more of the properties of essences than Aristotle and medieval thinkers had supposed. Harvey (1518–1657) had voiced his famous dictum *ex ova omnia* (1651), which was at least verbally in opposition to Aristotle's views, and in a little more than a century DeGraaf (1641–1673) claimed to have actually discovered the mammalian ovum. The nature of reproduction in plants was beginning to be understood with the work of Nehemia Grew (1641–1712) and Rudolph Camerarius (1665–1721). Francesco Redi (1627–1693), Jan Swammerdam (1637–1683), and others were proving that insects and other metazoans, which supposedly generated spontaneously, produced their young in a manner similar to higher animals.

It was no accident that John Ray (1627–1705), one of the strongest advocates of the newly emerged Special Creationist static species concept, also argued most effectively against the existence of spontaneous generation and was one of the first systematists to consider caterpillars and perfect insects in the same species rather than in distinct species as Aristotle had done (Raven 1942: 375, 462, 469, 471). Under the joint influence of the biological discoveries of the period and the prevalent interpretation of Christian scripture, species as classes of organisms were taking on the properties of their metaphysical correlates. Not only were essences immutable, eternal, and static but also the individual organisms themselves, whether severally or in groups[,] were thought never to change their species (see Mayr 1959a; Zirkle 1959). Although Ray had a few reservations, he believed that plants as well as animals "preserve their distinct species permanently; one species never springs from the seed of another nor vice versa" (Ray 1686: 40). How difficult it is to maintain this position consistently *without* recourse to essences will be seen shortly.

Early in his career Linnaeus (1707–1778) agreed wholeheartedly with Ray and had no reservations on the subject. "No new species are produced nowadays" (1964). The *nowadays* will prove important. Aristotle needed essences to supply the stability and timelessness which he thought that the perceivable world lacked and the objects of science must have. Linnaeus no longer needed essences to fulfil this purpose. Species as classes of organisms were themselves stable and timeless. Coincidentally, the metaphysics had faded into the background. In its place Linnaeus substituted an extreme concern with the idiosyncrasies and technicalities of Aristotelian logic with little understanding of the metaphysics upon which it is based. It should occasion no surprise to discover that Aristotle's *Historia Animalium,* which Linnaeus prized so much, contains lengthy discussions of the logic of definition and classification but little in the way of metaphysics.

As is usually the case when a complex logical system is adopted with little understanding of its foundations, attempted modifications of the system result in confusion and inconsistency. Linnaeus's attempts to reconcile *Peloria* and other so-called "hybrid species" with his Special Creationist static species concept is a case in point. . . .

There are two possibilities as far as the status of these new productions is concerned. Either they are varieties or they too are species. On the first interpretation, two species hybridize to produce a variety, which is distinct from its parent species presumably because of its hybrid nature. Again, this cannot be the case for strictly logical reasons. According to Aristotelian logic, a variety must be a variety of some species, and at any one level it can be a variety of only one species. *Peloria* can embody the essence of only one of its parent species and will embody it completely. It cannot embody the essences of both of its parent species. Hence, the reference to hybrid origin is superfluous. On the second interpretation, two species hybridize to produce a third species, as Aristotle himself claimed happened in the cases of the Laconian hound and the Indian dog. But for Aristotle all the species concerned were separate species, all equally eternal and immutable. The reader need not be reminded that Aristotle is talking about essences; individuals can change their species and give rise on occasion to members of other species. But this is the exception. Linnaeus has made it the rule. If *Peloria* and its parent species are all species, then each must have its separate essence. All are equally mutable or immutable. The only edge that the parent species have over their hybrid species is that they were specially created and are older. Originally when Linnaeus said that no new species are produced nowadays, he meant that all species were created in the beginning and since then, including the present, no additional species had been produced. All species were immutable and equally old. On this interpretation of his claim for the hybrid origin of some species, no species as a class of organisms is truly immutable, although some were specially created and are older. The production of a new species is as damaging to the immutability claim for classes of organisms as is having been produced from an older species. Once again, only essences are immutable.

Eighteenth-Century Nominalism

Aristotle had maintained that species as essences were distinct and, thus, sharply distinguishable from each other by morphological characters. Perhaps an individual variant might exist, but essences did not overlap. With the emphasis now shifted from essences to individual organisms, the static species concept implied that individual organisms themselves should be sharply delineable into species by means of their morphological properties alone. Presented on the one hand with this implication of the static species concept and on the other hand with the continuous gradations present in nature, Buffon (1707–1788) became a nominalist. Medieval nominalism stemmed mainly from an overreaction to essentialism, to the real existence of substantial essences or Forms. Originally, nominalists asserted that such transcendent entities did not exist. The only thing that the members of a species had in common was a name. All that really existed were individuals. As A. C. Crombie (1950) has noted, the transition from the essentialist position to nominalism took place in philosophy and physics from the twelfth to the seventeenth centuries. It had little effect in biology until the eighteenth century, but by this time alternatives to the early strict nominalism were prevalent—conceptualism and neoplatonism.

Buffon was led to espouse nominalism, not because he explicitly rejected the existence of substantial essences, but because he simply did not consider them. In contemplating the gradual variation among contemporary organisms in the absence of any essentialist presuppositions, he came to the conclusion that only individuals really existed. The species, genera, families, and so on of the name-givers like Linnaeus were just figments of the imagination. Buffon reasoned that in order for species as classes of organisms to be real, they had to be separated by sharp morphological gaps. There were no such gaps. Hence, species were not real (Hull 1965a, 1965b). As he himself put it in volume 1 of his *Histoire Naturelle* (1749): "In general, the more one increases the number of one's divisions, in the case of natural products, the nearer one comes to the truth; since in reality individuals alone exist in nature, while genera, orders, classes, exist only in our imagination" (1785a: 38, 20). To talk otherwise, as Linnaeus and other systematists did, was to commit "an error in metaphysics in the very principles underlying their methods. This error is due to a failure to apprehend nature's processes, which take place always by gradation, and to the desire to judge a whole by one of its parts" (ibid.). The shift in metaphysics from Aristotle to Buffon is complete. For Aristotle, essences are really real; for Buffon, individual organisms.

Buffon's flirtation with nominalism was short-lived. By the second volume of his *Histoire Naturelle* (also published in 1749), Buffon had discovered a way to distinguish between species. Perhaps the members of two species cannot always be distinguished from each other by simple morphological properties, but the members of a species are also characterized by certain relational properties. These relational properties are what make species "organic wholes" rather than arbitrary collections of individuals. The relation which Buffon noticed was that of interbreeding with the production of fertile offspring:

> We should regard two animals as belonging to the same species if, by means of copulation, they can perpetuate themselves and preserve the likeness of the species; and we should regard them as belonging to different species if they are incapable of producing progeny by the same means. Thus the fox will be known to be a different species from the dog, if it proves to be the fact that from the mating of a male and female of these two kinds of animals no offspring is born; and even if there should result a hybrid offspring, a sort of mule, this would suffice to prove that fox and dog are not of the same species—inasmuch as the mule would not produce anything. (1785b: 10)

Buffon is not saying that species breed true as Aristotle, Ray, and Linnaeus have said before him. Rather he has formulated a physiological definition of *species* which comes close to the modern biological definition. Thus, Buffon had discovered a criterion of demarcation for the members of contemporary species. Never again will species be unreal for him—higher taxa maybe, but not species. Other biologists had been aware that species were generally intersterile, but no one until Buffon had made much of it. No biologist until Buffon needed to. Aristotle had his essences. Members of a species bred true. Although on occasion hybridization took place, the occurrence of such phenomena was unimportant in comparison to the immutability of species as essences. Linnaeus had his conviction that all species could be distinguished by certain invariable diagnostic properties, regardless of how the species might have originated. But Buffon had abandoned both essences and diagnostic characters. Failure to produce fertile offspring was another matter. It is also relevant that interbreeding is a relational property. One of the main bones of contention between nominalists and Aristotelians was the status of relations. Aristotelians tended to ignore them or to treat them as peculiar simple properties. Nominalists emphasized the importance of relations and argued that they were irreducible to simple properties (Crombie 1950: 261).

Throughout the history of biology, contemplation of the gradations present in nature had a tendency to lead biologists such as Buffon and later Lamarck and even Darwin to become (or at least to talk very much like) nominalists. In turn the acceptance of a nominalist metaphysics (or lack of one) made it easier to entertain evolutionary hypotheses. It was at this stage in his development that Buffon first suggested the idea that one species might "degenerate" into another: in fact that all species might have resulted by the degeneration of one single original species. Since the process was one of degeneration, the original species would have to be the least degenerate species—probably man. Immediately, Buffon rejects the idea, both for theological and empirical reasons (1785c: 383). Regardless of the explanation for his rejection of his evolutionary hypothesis, Buffon periodically returns to the question but without the advantage of a flexible metaphysics.

Buffon had said that genera, orders, and classes exist only in the imagination, which to a nominalist means they do not exist at all. But it is not a very big step from this position to neoplatonism. It was at this same time that Louis Daubenton (1716–1800) introduced him to the extensive homologies present among the members of the animal kingdom. Just as the study of the gradations present in nature tended to lead to nominalism and from there to ruminations over the possibility of evolution, the study of homologies tended to predispose biologists to certain metaphysical views. Today homologies are used as evidence for evolution and against any metaphysics which precludes it, and Buffon was not insensitive to this possible explanation of such similarities. But the facts of comparative anatomy also made the idea that there were basic "plans" inherent in nature . . . seem quite attractive. By successive abstractions of greater and greater generality, one could envisage the prototype perch, the prototype fish, the prototype vertebrate, and so on. . . .

The Rise of Neoplatonism

The conflict between Buffon and Linnaeus in the first half of the eighteenth century was to be reenacted in the second, but this time the line between the adversaries, Lamarck (1748–1829) and Cuvier (1769–1832), was more clearly drawn. Lamarck was committed unambiguously to "evolutionism" (Gillispie 1959). Cuvier was irrevocably a Special Creationist and antievolutionist. The conflict between these two great biologists was largely a matter of their holding different views of the philosophy of science based on radically different metaphysics. Lamarck was a thoroughgoing nominalist and viewed science as the advancement of bold hypotheses which were consistent with the known facts and which could be confirmed to a greater or a lesser degree by additional observations. Unfortunately, Lamarck's idea of confirmation was of a pretty low level. But one thing was certain. The inductive inference which led to the postulation of a scientific hypothesis need not itself be absolutely certain or even highly warranted. Cuvier was a thoroughgoing neoplatonist and saw science as the formulation of descriptive generalizations which never went beyond the evidence in the manner usually attributed to Francis Bacon (1561–1626). . . .

Cuvier's conception of species and higher taxa as types together with his conception of the individual organism combined to rule out the possibility of evolution. He says that "every organized creature forms whole, a unique and closed system, whose parts mutually correspond to one another and concur toward the same definite action by reciprocal reaction" (1830: 95). Every organism had a definite, finely balanced organization which imposed upon it very strict basic conditions of existence. It was these basic conditions of existence that were the experiential correlates of neoplatonic Forms and, as such, were necessarily static. An individual organism might vary slightly, but if it strayed too far from its original organization, it ceased to exist. Gaps existed between taxa at all levels because only certain combinations *could* exist. Cuvier looked upon this impossibility as akin to the logical impossibility to be found in geometry. Just as there could not be a plane closed figure halfway between a triangle and a quadrilateral, there could be no intermediate species of original beings. This was not speculation but a necessary consequence of the conditions of existence. Cuvier looked upon all of his generalizations as either as purely empirical, descriptive generalizations or as "rational laws" (1964: 189). Once again, such logical and metaphysical considerations as these lay behind the refusal to countenance evolution.

Cuvier's work with fossils made it impossible for him to dismiss them as nature's little tricks. They were the remains of individuals that had ceased to meet their basic conditions of existence. As much as Cuvier did not like to postulate causes for such remote occurrences, he did provide an explanation for the apparently sudden extinction of numerous species periodically throughout the history of the earth—his famous catastrophic inundations. He was unable to bring himself to do the same for vestigial organs or the serial appearance of new forms in the fossil record. In the first case, the existence of nonfunctional organs would have conflicted with his conception of the organism as a perfectly functioning closed system. Cuvier maintained lamely that what we now term vestigial organs had to be accepted as they were without explanation. He came to a similar conclusion with respect to the origin of species.

Although the periodic inundations were not total, permitting a few members of some but not all of the species existing at the time to repopulate the earth, many individuals of all species would have necessarily perished. Their remains should be in the fossil record. They were not. For example, in the case of the primates, Cuvier knew of no fossil remains of *any* primate including man. Given his Baconian principles, he should have concluded that man and other primates did not antedate the Great Flood. He does not. He concludes that the fossil remains of these forms must be hidden somewhere under the oceans or glaciers. It is hard not to detect a certain religious motivation in Cuvier's selective use of his Baconian principles. Perhaps Cuvier did not make use of any religious arguments. He certainly let religious motives sway his judgment in the acceptance of certain unverified hypotheses and his contemptuous rejection of others.

The Resurgence of Nominalism

Lamarck could not have been more different from Cuvier than he was. Like Buffon before him he was led from the observation of the continuous gradations present in nature to the conclusion that only individuals existed. He says, for example: "Meanwhile we should remember that nothing of all this exists in nature; that she knows neither classes, orders, genera, nor species, in spite of all the foundation which the portion of the natural series which our collections contain has seemed to afford them; and that of organisms or living bodies there are, in reality, only individuals and among different races which gradually pass into all degrees of organization" (1806: 14).

As it did in the case of Buffon, Lamarck's adopting a nominalist position enabled him to entertain the idea of organic evolution, but unlike Buffon, Lamarck [would] never waiver in his nominalism or "evolutionism." Species were for him merely collections of organisms sufficiently similar for the time being to be called species. The criterion of intersterility or the sterility of hybrids could be used to distinguish between the contemporary members of species, but it could not be used to divide a continuously evolving lineage into species. Lamarck's ideas on evolution also differed from those of Buffon. The process is not one of "degeneration" but one of progressive development from "lower" to "higher" forms. To use Gillispie's (1959: 271) metaphor, Lamarck envisaged a Great Escalator of Being, animals generated spontaneously at the bottom and moving through successive generations up the series. . . . Organisms begin at the bottom and gradually evolve upward, but all groups are equally contemporary. There are always individuals evolving through the various lineages. Lamarck had not developed an appreciation for the contingency which was to be at the heart of Darwin's theory. For Lamarck no major group had ever become extinct. In fact only very rarely had any species become extinct, and these species were terminal species which had been destroyed by the direct agency of man and which would reevolve in his absence (1963: 44).

This facet of Lamarck's theory is usually presented as a personal idiosyncrasy or as an overreaction to Cuvier's catastrophism. On the contrary, as is so often the case with such seemingly bizarre notions, it is a direct consequence of certain other beliefs which he held. Lamarck agreed with Cuvier that organisms were functionally integrated wholes which had to fulfil certain basic conditions of existence, but for Lamarck these conditions were not part of the makeup of the organism but a relation between its labile constitution and the variable environment. Lamarck believed that the integration present in organisms was sufficiently plastic to permit an organism to progress from one way of life to another. As long as these environmental niches were present, there would be organisms to fill them. For Lamarck evolution did not occur by random mutation in all possible directions but up several predetermined paths. Certain worms were always becoming annelids and others insects. Once an individual had started up the insect-arachnid-crustacean progression, it could not become a mollusc. It had progressed too far up its predetermined path. Whether a particular worm became an insect or an annelid was not predetermined but the paths were.

Lamarck's theory of evolution was little more than an expansion of Aristotle's vermiparous reproduction to include all species, just one possible variation on essentialism. Instead of individuals striving unconsciously to fulfil their own essences, Lamarck had individuals striving unconsciously to progress from one species to the next. There *must* be intermediaries between molluscs and fish, forms through which individuals could progress, just as there must be caterpillars and chrysalises between grubs and the perfect moth or butterfly.

Metaphysics and Evolution

From the earliest reviews of the *Origin of Species,* Darwin has been called unphilosophical. Given one rather unflattering view of philosophy, Darwin was the first to agree. "My power to follow a long and purely abstract train of thought is very limited. I should, moreover, never have succeeded with metaphysics or mathematics" (1958: 140). On this same view of philosophy as long, involved arguments about purely abstract matters, Darwin was not the least surprised that Herbert Spencer, the Philosopher of Evolution, "the greatest living philosopher in England, perhaps equal to any that have lived" (1888: 120), should say nothing relevant to evolutionary theory as a biological theory. If this is what philosophy is, then Darwin was no philosopher. One can go even further. It is doubtful whether knowledge of this sort of thing would have helped Darwin much in the formulation of his theory. In fact, it might well have been detrimental. As Ernst Mayr has said, "Instead of getting caught up in philosophical premises, as did Agassiz and the highly competent German and French zoologists and botanists, Darwin was able to pursue his biological speculations unfettered" (1959b: 215).

It is certainly true that when Darwin set out on his voyage, he had little to unlearn either in the history of his own field or in philosophy. While on the *Beagle* and afterward he was forced to compare his naive Christian beliefs with what he saw of species variation and distribution, viewed in the light of Lyell's geological theories. There were already in existence philosophical systems which purported to explain all such phenomena, but Darwin was unfamiliar with them. As it was, the only alternative to his own theory which Darwin saw was that of Special Creation and it was patently false. If Darwin had started out with a complete understanding of one of the philosophical worldviews, for instance neoplatonism, he might have been seduced by its spurious explanatory power. Only after he had his own theory firmly in mind and supported by large bodies of facts was he forced to confront some of the more sophisticated alternatives to his theory of evolution.

But neither Darwin's ineptness as a philosopher nor the advantages of philosophical naïveté should be exaggerated. As Darwin himself observed, "Some of my critics have said, 'Oh, he is a good observer, but he has no power of reasoning!' I do not think that this can be true, for the *Origin of Species* is one long argument from the beginning to the end, and it has convinced not a few able men. No one could have written it without some power of reasoning" (1958: 140). Darwin's critics were and still are confusing originality with ineptness. Darwin was breaking new ground both biologically and philosophically. Although there is, broadly speaking, a deductive core to the *Origin,* by and large it is one long, involved inductive argument conducted in the midst of a mass of very concrete facts. Darwin's argument as presented in the *Origin* is a genuinely inductive argument, not just a deduction set up on end. It is this aspect of Darwin's theory which leads Ernst Mayr to say:

> It is almost universally stated that Darwin was no philosopher, that he was totally unphilosophical. Even though he himself would probably have pleaded guilty, this accusation is quite misleading. To be sure, Darwin did not belong to any of the established schools of philosophy, nor did he publish an essay or volume explicitly devoted to an exposition of his philosophical ideas. Yet few writers in the last two hundred

years have had so profound an impact on our thinking. This holds for logic, metaphysics, and ethics. It has taken one hundred years to appreciate that Darwin's conceptual framework is, indeed, a new philosophical system. (1975: xviii)

It is this aspect of the *Origin* which Gertrude Himmelfarb misses completely when she complains: "What Darwin was doing, in effect, was creating a 'logic of possibilities.' Unlike conventional logic, where the compound of possibilities results not in a greater possibility, a probability, but in a lesser one, the logic of the *Origin* was one in which possibilities were assumed to add up to probability" (1959: 274). Darwin *was* creating a "logic of possibilities," a logic whose value logicians only now are beginning to appreciate (see, e.g., Kaplan 1964, chapters 6 and 7). Darwin received considerable support for his materialism and his opposition to teleology from many of the philosophers in his day, but he certainly received little support for the type of inductive logic which his book exhibited.

Although Darwin's early ignorance of certain basic metaphysical issues was undoubtedly an advantage in the formulation of his theory, such ignorance was a decided disadvantage when it came to defending his theory. Here his ineptness as a "philosopher" took its toll. It is one thing to recognize that evolutionary theory conflicts with some of the more simple-minded versions of teleology (Darwin 1975: 211). It is another to realize that it conflicts with some of the subtler tenets of teleology. Darwin was aware of some of the more obvious ramifications of his theory for philosophy, but he did not see just how extensive these ramifications actually were. For example, he did not appreciate the consequences which evolutionary theory had for the logic of definition (Hull 1965a, 1965b).

Darwin's notion of definition was the traditional one. A definition must provide a set of characters which are severally necessary and jointly sufficient. For instance, he declines any attempt to define "instinct" because none of the characteristics of instinct happens to be universal (1975: 207, 208). It was this notion of definition which led Darwin, like Buffon and Lamarck before him, to dismiss the divisions which systematists might make between species in evolving lineages as "imaginary" (ibid.: 119). Systematists need not "be incessantly haunted by the shadowy doubt whether this or that form be in essence a species," since after all species "are merely artificial combinations made for convenience" (ibid.: 484, 485). The only concern of the systematist is whether a form "be sufficiently constant and distinct from other forms, to be capable of definition; and if definable, whether the differences be sufficiently important to deserve a specific name" (ibid.: 484).

In an evolving lineage there will be no sharp break in the gradual changes among morphological characters, the characters traditionally used to establish systematic affinity. Hence, species delimitation will be arbitrary. For Darwin as for the philosophers in his day, the only acceptable distinctions were clear distinctions. A hazy border was no border at all (ibid.: 248; see Wittgenstein 1973 for the opposite view). Many contemporary biologists might agree and yet nevertheless argue that because of the biological definition of species, many of the divisions between contemporary species are real. But Darwin felt obliged to say that "neither sterility nor fertility offers any clear distinction between species and varieties" (1975: 248). He even accuses biologists who change their estimation of systematic affinity in the light of information concerning interbreeding of "arguing in a circle" (ibid.: 245; see also Hull 1967a).

Darwin was forced to reject anything like the biological definition of species because his primary mode of reasoning throughout the *Origin* was to argue that since gradual variation among contemporary forms was *actual,* gradual variation in time was *possible.* Varieties were incipient species. Since the chief criterion for distinguishing between species and varieties was sterility and fertility, he was forced to disparage it. Now that evolution has been universally accepted by the scientific community, biologists can afford to emphasize reproductive isolation. In fact, it has become one of the cornerstones of modern evolutionary theory.

What then, according to Darwin, is the metaphysical status of species? Are they real? Do they exist? These are metaphysical questions, and perhaps expecting a biologist to provide explicit answers to them is expecting too much. But a great deal of the controversy and dissatisfaction which has plagued evolutionary theory since the publication of the *Origin* might have been avoided had Darwin presented consciously formulated and explicitly stated answers to these questions. As it is, it is difficult to decide *what* Darwin thought about the reality of species. Evolving species do not sit comfortably in any of the traditional metaphysical niches. They are not as "real" as essentialists would have them. Nor are they as "unreal" as the nominalists would have them. After all, it is species that evolve.

Darwin's position seems to lie somewhere between these two extremes, but because of his reticence on the subject, it is impossible to decide where. In part because of the personality of its founder and in part because of the very nature of evolution by random mutation and natural selection, evolutionary theory has had to get along without a metaphysics. Only since the so-called new systematics have biologists begun to realize that such a thing as a metaphysics of evolution *as a biological phenomenon* is not only desirable but also necessary. In fact, the development of evolutionary theory since Darwin might well be described as a scientific theory in search of a metaphysics.

CHAPTER 19

The Spandrels of San Marco and the Panglossian Paradigm: A Critique of the Adaptationist Program

Stephen J. Gould and Richard C. Lewontin

Introduction

The great central dome of Saint Mark's Cathedral in Venice presents in its mosaic design a detailed iconography expressing the mainstays of Christian faith. Three circles of figures radiate out from a central image of Christ: angels, disciples, and virtues. Each circle is divided into quadrants, even though the dome itself is radially symmetrical in structure. Each quadrant meets one of the four spandrels in the arches below the dome. Spandrels—the tapering triangular spaces formed by the intersection of two rounded arches at right angles (figure 15)—are necessary architectural by-products of mounting a dome on rounded arches. Each spandrel contains a design admirably fitted into its tapering space. An evangelist sits in the upper part flanked by the heavenly cities. Below, a man representing one of the four biblical rivers (Tigris, Euphrates, Indus, and Nile) pours water from a pitcher into the narrowing space below his feet.

The design is so elaborate, harmonious, and purposeful that we are tempted to view it as the starting point of any analysis, as the cause in some sense of the surrounding architecture. But this would invert the proper path of analysis. The system begins with an architectural constraint: the necessary four spandrels and their tapering triangular form. They provide a space in which the mosaicists worked; they set the quadripartite symmetry of the dome above.

Such architectural constraints abound and we find them easy to understand because we do not impose our biological biases upon them. Every fan-vaulted ceiling must have a series of open spaces along the midline of the vault, where the sides of the fans intersect between the pillars. Since the spaces must exist,

From *Proceedings of the Royal Society of London: Biological Sciences* 205 (1979): 581–98.

Figure 15. Spandrels formed by intersecting arches

they are often used for ingenious ornamental effect. In King's College Chapel in Cambridge, for example, the spaces contain bosses alternately embellished with the Tudor rose and portcullis (figure 16). In a sense, this design represents an "adaptation," but the architectural constraint is clearly primary. The spaces arise as a necessary by-product of fan vaulting; their appropriate use is a sec-

Figure 16. Ceiling of King's College Chapel in Cambridge

ondary effect. Anyone who tried to argue that the structure exists because the alternation of rose and portcullis makes so much sense in a Tudor chapel would be inviting the same ridicule that Voltaire (1759) heaped on Dr. Pangloss: "Things cannot be other than they are. . . . Everything is made for the best purpose. Our noses were made to carry spectacles, so we have spectacles. Legs were clearly intended for breeches, and we wear them." Yet evolutionary biologists, in their tendency to focus exclusively on immediate adaptation to local conditions, do tend to ignore architectural constraints and perform just such an inversion of explanation.

As a closer example, recently featured in some important biological literature on adaptation, anthropologist Michael Harner (1977) has proposed that Aztec human sacrifice arose as a solution to chronic shortage of meat (limbs of victims were often consumed, but only by people of high status). Edward O. Wilson (1978) has used this explanation as a primary illustration of an adaptive, genetic predisposition for carnivory in humans. Harner and Wilson ask us to view an elaborate social system and a complex set of explicit justifications involving myth, symbol, and tradition as mere epiphenomena generated by the Aztecs as an unconscious rationalization masking the "real" reason for it all: need for protein. But Sahlins (1978) has argued that human sacrifice represented just one part of an elaborate cultural fabric that, in its entirety, not only represented the material expression of Aztec cosmology, but also performed such utilitarian functions as the maintenance of social ranks and systems of tribute among cities.

We strongly suspect that Aztec cannibalism was an "adaptation" much like evangelists and rivers in spandrels, or ornamented bosses in ceiling spaces: a secondary epiphenomenon representing a fruitful use of available parts, not a cause of the entire system. To put it crudely: a system developed for other reasons generated an increasing number of fresh bodies; use might as well be made of them. Why invert the whole system in such a curious fashion and view an entire culture as the epiphenomenon of an unusual way to beef up the meat supply? Spandrels do not exist to house the evangelists. Moreover, as Sahlins argues, it is not even clear that human sacrifice was an adaptation at all. Human cultural practices can be orthogenetic and drive toward extinction in ways that Darwinian processes, based on genetic selection, cannot. Since each new monarch had to outdo his predecessor in even more elaborate and copious sacrifice, the practice was beginning to stretch resources to the breaking point. It would not have been the first time that a human culture did itself in. And, finally, many experts doubt Harner's premise in the first place (Ortiz de Montellano 1978). They argue that other sources of protein were not in short supply, and that a practice awarding meat only to privileged people who had enough anyway, and who used bodies so inefficiently (only the limbs were consumed, and partially at that)[,] represents a mighty poor way to run a butchery.

We deliberately chose nonbiological examples in a sequence running from remote to more familiar: architecture to anthropology. We did this because the primacy of architectural constraint and the epiphenomenal nature of adaptation are not obscured by our biological prejudices in these examples. But we trust that the message for biologists will not go unheeded: if these had been biological systems, would we not, by force of habit, have regarded the epiphenomenal adaptation as primary and tried to build the whole structural system from it?

The Adaptationist Program

We wish to question a deeply engrained habit of thinking among students of evolution. We call it the adaptationist program, or the Panglossian paradigm. It is rooted in a notion popularized by A. R. Wallace (1899) and A. Weismann (but not, as we shall see, by Darwin) toward the end of the nineteenth century: the near omnipotence of natural selection in forging organic design and fashioning the best among possible worlds. This program regards natural selection as so powerful and the constraints upon it so few that direct production of adaptation through its operation becomes the primary cause of nearly all organic form, function, and behavior. Constraints upon the pervasive power of natural selection are recognized[,] of course (phyletic inertia primarily among them, although immediate architectural constraints, as discussed in the last section, are rarely acknowledged). But they are usually dismissed as unimportant or else, and more frustratingly, simply acknowledged and then not taken to heart and invoked.

Studies under the adaptationist program generally proceed in two steps:

1. An organism is atomized into "traits," and these traits are explained as structures optimally designed by natural selection for their functions. For lack of space, we must omit an extended discussion of the vital issue: "what is a trait?" Some evolutionists may regard this as a trivial, or merely a semantic problem. It is not. Organisms are integrated entities, not collections of discrete objects. Evolutionists have often been led astray by inappropriate atomization, as D'Arcy Thompson (1942) loved to point out. Our favorite example involves the human chin (Gould 1977: 381–82; Lewontin 1978). If we regard the chin as a "thing," rather than as a product of interaction between two growth fields (alveolar and mandibular), then we are led to an interpretation of its origin (recapitulatory) exactly opposite to the one now generally favored (neotenic).

2. After the failure of part-by-part optimization, interaction is acknowledged via the dictum that an organism cannot optimize each part without imposing expenses on others. The notion of "tradeoff" is introduced, and organisms are interpreted as best compromises among competing demands. Thus, interaction among parts is retained completely within the adaptationist program. Any suboptimality of a part is explained as its contribution to the best possible design for the whole. The notion that suboptimality might represent anything other than the immediate work of natural selection is usually not entertained. As Dr. Pangloss said in explaining to Candide why he suffered from venereal disease: "It is indispensable in this best of worlds. For if Colum-

bus, when visiting the West Indies, had not caught this disease, which poisons the source of generation, which frequently even hinders generation, and is clearly opposed to the great end of Nature, we should have neither chocolate nor cochineal." The adaptationist program is truly Panglossian. Our world may not be good in an abstract sense, but it is the very best we could have. Each trait plays its part and must be as it is.

At this point, some evolutionists will protest that we are caricaturing their view of adaptation. After all, do they not admit genetic drift, allometry, and a variety of reasons for nonadaptive evolution? They do, to be sure, but we make a different point. In natural history, all possible things happen sometimes; you generally do not support your favored phenomenon by declaring rivals impossible in theory. Rather, you acknowledge the rival, but circumscribe its domain of action so narrowly that it cannot have any importance in the affairs of nature. Then, you often congratulate yourself for being such an undogmatic and ecumenical chap. We maintain that alternatives to selection for best overall design have generally been relegated to unimportance by this mode of argument. Have we not all heard the catechism about genetic drift: it can only be important in populations so small that they are likely to become extinct before playing any sustained evolutionary role (but see Lande 1976).

The admission of alternatives in principle does not imply their serious consideration in daily practice. We all say that not everything is adaptive; yet, faced with an organism, we tend to break it into parts and tell adaptive stories as if tradeoffs among competing, well-designed parts were the only constraint upon perfection for each trait. It is an old habit. As Romanes complained about A. R. Wallace in 1900: "Mr. Wallace does not expressly maintain the abstract impossibility of laws and causes other than those of utility and natural selection. . . . Nevertheless, as he nowhere recognizes any other law or cause[,] he practically concludes that, on inductive or empirical grounds, there is no such other law or cause to be entertained."

The adaptationist program can be traced through common styles of argument. We illustrate just a few; we trust they will be recognized by all:

1. If one adaptive argument fails, try another. Zigzag commissures of clams and brachiopods, once widely regarded as devices for strengthening the shell, become sieves for restricting particles above a given size (Rudwick 1964). A suite viewed as weapons against predators become symbols of intraspecific competition among males (Davitashvili 1961). The Eskimo face, once depicted as "cold engineered" (Coon et al. 1950), becomes an adaptation to generate and withstand large masticatory forces (Shea 1977). We do not attack these newer interpretations; they may all be right. We do wonder, though, whether the failure of one adaptive explanation should always simply inspire a search

for another of the same general form, rather than a consideration of alternatives to the proposition that each part is "for" some specific purpose.

2. If one adaptive argument fails, assume that another must exist; a weaker version of the first argument. Costa and Bisol, for example, hoped to find a correlation between genetic polymorphism and stability of environment in the deep sea, but they failed. They conclude (1978: 132, 133): "The degree of genetic polymorphism found would seem to indicate absence of correlation with the particular environmental factors which characterize the sampled area. The results suggest that the adaptive strategies of organisms belonging to different phyla are different."

3. In the absence of a good adaptive argument in the first place, attribute failure to imperfect understanding of where an organism lives and what it does. This is again an old argument. Consider Wallace on why all details of color and form in land snails must be adaptive, even if different animals seem to inhabit the same environment (1899: 148): "The exact proportions of the various species of plants, the numbers of each kind of insect or of bird, the peculiarities of more or less exposure to sunshine or to wind at certain critical epochs, and other slight differences which to us are absolutely immaterial and unrecognizable, may be of the highest significance to these humble creatures, and be quite sufficient to require some slight adjustments of size, form, or colour, which natural selection will bring about."

4. Emphasize immediate utility and exclude other attributes of form. Fully half the explanatory information accompanying the full-scale fiberglass *Tyrannosaurus* at Boston's Museum of Science reads: "Front legs a puzzle: how *Tyrannosaurus* used its tiny front legs is a scientific puzzle; they were too short even to reach the mouth. They may have been used to help the animal rise from a lying position." (We purposely choose an example based on public impact of science to show how widely habits of the adaptationist program extend. We are not using glass beasts as straw men; similar arguments and relative emphases, framed in different words, appear regularly in the professional literature.) We don't doubt that *Tyrannosaurus* used its diminutive front legs for something. If they had arisen *de novo*, we would encourage the search for some immediate adaptive reason. But they are, after all, the reduced product of conventionally functional homologues in ancestors (longer limbs of allosaurs, for example). As such, we do not need an explicitly adaptive explanation for the reduction itself. It is likely to be a developmental correlate of allometric fields for relative increase in head and hindlimb size. This nonadaptive hypothesis can be tested by conventional allometric methods (Gould 1974 in general; Lande 1978 on limb reduction) and seems to us both more interesting and fruitful than untestable speculations based on secondary utility in the best of

possible worlds. One must not confuse the fact that a structure is used in some way (consider again the spandrels, ceiling spaces, and Aztec bodies) with the primary evolutionary reason for its existence and conformation.

Telling Stories

"All this is a manifestation of the rightness of things, since if there is a volcano at Lisbon it could not be anywhere else. For it is impossible for things not to be where they are, because everything is for the best" (Dr. Pangloss on the great Lisbon earthquake of 1755 in which up to fifty thousand people lost their lives).

We would not object so strenuously to the adaptationist program if its invocation, in any particular case, could lead in principle to its rejection for want of evidence. We might still view it as restrictive and object to its status as an argument of first choice. But if it could be dismissed after failing some explicit test, then alternatives would get their chance. Unfortunately, a common procedure among evolutionists does not allow such definable rejection for two reasons. First, the rejection of one adaptive story usually leads to its replacement by another, rather than to a suspicion that a different kind of explanation might be required. Since the range of adaptive stories is as wide as our minds are fertile, new stories can always be postulated. And if a story is not immediately available, one can always plead temporary ignorance and trust that it will be forthcoming, as did Costa and Bisol (1978), cited above. Secondly, the criteria for acceptance of a story are so loose that many pass without proper confirmation. Often, evolutionists use *consistency* with natural selection as the sole criterion and consider their work done when they concoct a plausible story. But plausible stories can always be told. The key to historical research lies in devising criteria to identify proper explanations among the substantial set of plausible pathways to any modern result.

We have, for example (Gould 1978), criticized Barash's (1976) work on aggression in mountain bluebirds for this reason. Barash mounted a stuffed male near the nests of two pairs of bluebirds while the male was out foraging. He did this at the same nests on three occasions at ten-day intervals: the first before eggs were laid, the last two afterward. He then counted aggressive approaches of the returning male toward both the model and the female. At time one, aggression was high toward the model and lower toward females but substantial in both nests. Aggression toward the model declined steadily for times two and three and plummeted to near zero toward females. Barash reasoned that this made evolutionary sense since males would be more sensitive to intruders before eggs were laid than afterward (when they can have some confidence that their genes are inside). Having devised this plausible story, he considered his work as completed (1976: 1099, 1100):

> The results are consistent with the expectations of evolutionary theory. Thus aggression toward an intruding male (the model) would clearly be especially advantageous early in the breeding season, when territories and nests are normally defended. . . . The initial aggressive response to the mated female is also adaptive in that, given a situation suggesting a high probability of adultery (i.e., the presence of the model near the female) and assuming that replacement females are available, obtaining a new mate would enhance the fitness of males. . . . The decline in male-female aggressiveness during incubation and fledgling stages could be attributed to the impossibility of being cuckolded after the eggs have been laid. . . . The results are consistent with an evolutionary interpretation.

They are indeed consistent, but what about an obvious alternative, dismissed without test by Barash? Male returns at times two and three, approaches the model, tests it a bit, recognizes it as the same phony he saw before, and doesn't bother his female. Why not at least perform the obvious test for this alternative to a conventional adaptive story: expose a male to the model for the *first* time after the eggs are laid.

Since we criticized Barash's work, Morton et al. (1978) repeated it, with some variations (including the introduction of a female model), in the closely related eastern bluebird *Sialia sialis*. "We hoped to confirm," they wrote, that Barash's conclusions represent "a widespread evolutionary reality, at least within the genus *Sialia*. Unfortunately, we were unable to do so." They found no "anti-cuckoldry" behavior at all: males never approached their females aggressively after testing the model at any nesting stage. Instead, females often approached the male model and, in any case, attacked female models more than males attacked male models. "This violent response resulted in the near destruction of the female model after presentations and its complete demise on the third, as a female flew off with the model's head early in the experiment to lose it for us in the brush" (ibid.: 969). Yet, instead of calling Barash's selected story into question, they merely devise one of their own to render both results in the adaptationist mode. Perhaps, they conjecture, replacement females are scarce in their species and abundant in Barash's. Since Barash's males can replace a potentially "unfaithful" female, they can afford to be choosy and possessive. Eastern bluebird males are stuck with uncommon mates and had best be respectful. They conclude: "If we did not support Barash's suggestion that male bluebirds show anticuckoldry adaptations, we suggest that both studies still had 'results that are consistent with the expectations of evolutionary theory' (Barash 1976: 1099), as we presume any careful study would." But what good is a theory that cannot fail in careful study (since by "evolutionary theory," they clearly mean the action of natural selection applied to particular cases, rather than the fact of transmutation itself)?

The Master's Voice Reexamined

Since Darwin has attained sainthood (if not divinity) among evolutionary biologists, and since all sides invoke God's allegiance, Darwin has often been depicted as a radical selectionist at heart who invoked other mechanisms only in retreat, and only as a result of his age's own lamented ignorance about the mechanisms of heredity. This view is false. Although Darwin regarded selection as the most important of evolutionary mechanisms (as do we), no argument from opponents angered him more than the common attempt to caricature and trivialize his theory by stating that it relied exclusively upon natural selection. In the last edition of the *Origin,* he wrote (1872: 395): "As my conclusions have lately been much misrepresented, and it has been stated that I attribute the modification of species exclusively to natural selection, I may be permitted to remark that in the first edition of this work, and subsequently, I placed in a most conspicuous position—namely at the close of the Introduction—the following words: 'I am convinced that natural selection has been the main, but not the exclusive means of modification.' This has been of no avail. Great is the power of steady misinterpretation." Romanes, whose once famous essay (1900) on Darwin's pluralism versus the panselectionism of Wallace and Weismann deserves a resurrection, noted of this passage (1900: 5): "In the whole range of Darwin's writings there cannot be found a passage so strongly worded as this: it presents the only note of bitterness in all the thousands of pages which he has published." Apparently, Romanes did not know the letter Darwin wrote to *Nature* in 1880, in which he castigated Sir Wyville Thomson for caricaturing his theory as panselectionist (1880: 32):

> I am sorry to find that Sir Wyville Thomson does not understand the principle of natural selection. . . . If he had done so, he could not have written the following sentence in the Introduction to the Voyage of the Challenger: "The character of the abyssal fauna refuses to give the least support to the theory which refers the evolution of species to extreme variation guided only by natural selection." This is a standard of criticism not uncommonly reached by theologians and metaphysicians when they write on scientific subjects, but is something new as coming from a naturalist. . . . Can Sir Wyville Thomson name any one who has said that the evolution of species depends only on natural selection? As far as concerns myself, I believe that no one has brought forward so many observations on the effects of the use and disuse of parts, as I have done in my "Variation of Animals and Plants under Domestication"; and these observations were made for this special object. I have likewise there ad-

duced a considerable body of facts, showing the direct action of external conditions on organisms.

We do not now regard all of Darwin's subsidiary mechanisms as significant or even valid, though many, including direct modification and correlation of growth, are very important. But we should cherish his consistent attitude of pluralism in attempting to explain Nature's complexity.

A Partial Typology of Alternatives to the Adaptationist Program

In Darwin's pluralistic spirit, we present an incomplete hierarchy of alternatives to immediate adaptation for the explanation of form, function, and behavior.

1. No adaptation and no selection at all. At present, population geneticists are sharply divided on the question of how much genetic polymorphism within populations and how much of the genetic differences between species is, in fact, the result of natural selection as opposed to purely random factors. Populations are finite in size and the isolated populations that form the first step in the speciation process are often founded by a very small number of individuals. As a result of this restriction in population size, frequencies of alleles change by *genetic drift,* a kind of random genetic sampling error. The stochastic process of change in gene frequency by random genetic drift, including the very strong sampling process that goes on when a new isolated population is formed from a few immigrants, has several important consequences. First, population and species will become genetically differentiated, and even fixed for different alleles at a locus in the complete absence of any selective force at all.

Secondly, alleles can become fixed *in a population in spite of natural selection.* Even if an allele is favored by natural selection, some proportion of [the] population, depending upon the product of population size N and selection intensity s, will become homozygous for the less fit allele because of genetic drift. If Ns is large this random fixation for unfavorable alleles is a rare phenomenon, but if selection coefficients are on the order of the reciprocal of population size ($Ns = 1$) or smaller, fixation for deleterious alleles is common. If many genes are involved in influencing a metric character like shape, metabolism, or behavior, then the intensity of selection on each locus will be small and Ns per locus may be small. As a result, many of the loci may be fixed for nonoptimal alleles.

Thirdly, new mutations have a small chance of being incorporated into a population, even when selectively favored. Genetic drift causes the immediate loss of most new mutations after their introduction. With a selection intensity *s*, a new favorable mutation has a probability of only 2*s* of ever being incorporated. Thus, one cannot claim that, eventually, a new mutation of just the right sort for some adaptive argument will occur and spread. "Eventually" becomes a very long time if only one in one thousand or one in ten thousand of the "right" mutations that do occur ever get incorporated in a population.

2. No adaptation and no selection on the part at issue; form of the part is a correlated consequence of selection directed elsewhere. Under this important category, Darwin ranked his "mysterious" laws of the "correlation of growth." Today, we speak of pleiotropy, allometry, "material compensation" (Rensch 1959: 179–87), and mechanically forced correlations in D'Arcy Thompson's sense (1942; Gould 1971). Here we come face to face with organisms as integrated wholes, fundamentally not decomposable into independent and separately optimized parts.

Although allometric patterns are as subject to selection as static morphology itself (Gould 1966), some regularities in relative growth are probably not under immediate adaptive control. For example, we do not doubt that the famous 0.66 interspecific allometry of brain size in all major vertebrate groups represents a selected "design criterion," though its significance remains elusive (Jerison 1973). It is too repeatable across too wide a taxonomic range to represent much else than a series of creatures similarly well designed for their different sizes. But another common allometry, the 0.2 to 0.4 intraspecific scaling among homeothermic adults differing in body size, or among races within a species, probably does not require a selectionist story though many, including one of us, have tried to provide one (Gould 1974). R. Lande (personal communication) has used the experiments of Falconer (1973) to show that selection upon *body size alone* yields a brain-body slope across generations of 0.35 in mice.

More compelling examples abound in the literature on selection for altering the timing of maturation (Gould 1977). At least three times in the evolution of arthropods (mites, flies, and beetles), the same complex adaptation has evolved, apparently for rapid turnover of generations in strongly *r*-selected feeders on superabundant but ephemeral fungal resources: females reproduce as larvae and grow the next generation within their bodies. Offspring eat their mother from inside and emerge from her hollow shell, only to be devoured a few days later by their own progeny. It would be foolish to seek adaptive significance in paedomorphic morphology per se; it is primarily a by-product of selection for rapid cycling of generations. In more interesting cases, selection for small size (as in animals of the interstitial fauna) or rapid maturation (dwarf males of many crustaceans) has occurred by progenesis (Gould 1977: 324–36), and descendant adults contain a mixture of ancestral juvenile and adult features. Many biologists have been tempted to find primary adaptive meaning for the mixture, but it probably arises as a by-product of truncated maturation, leaving some features "behind" in the larval state, while allowing others, more strongly correlated with sexual maturation, to retain the adult configuration of ancestors.

3. The decoupling of selection and adaptation.

i. Selection without adaptation. Lewontin (1979) has presented the following hypothetical example:

> A mutation which doubles the fecundity of individuals will sweep through a population rapidly. If there has been no change in efficiency of resource utilization, the individuals will leave no more offspring than before, but simply lay twice as many eggs, the excess dying because of resource limitation. In what sense are the individuals or the population as a whole better adapted than before? Indeed, if a predator on immature stages is led to switch to the species now that immatures are more plentiful, the population size may actually decrease as a consequence, yet natural selection at all times will favour individuals with higher fecundity.

ii. Adaptation without selection. Many sedentary marine organisms, sponges and corals in particular, are well adapted to the flow regimes in which they live. A wide spectrum of "good design" may be purely phenotypic in origin, largely induced by the current itself. (We may be sure of this in numerous cases, when genetically identical individuals of a colony assume different shapes in different microhabitats.) Larger patterns of geographic variation are often adaptive and purely phenotypic as well. Sweeney and Vannote (1978), for example, showed that many hemimetabolous aquatic insects reach smaller adult size with reduced fecundity when they grow at temperatures above and below their optima. Coherent, climatically correlated patterns in geographic distribution for these insects—so often taken as a priori signs of genetic adaptation—may simply reflect this phenotypic plasticity.

"Adaptation"—the good fit of organisms to their environment—can occur at three hierarchical levels with different causes. It is unfortunate that our language has focused on the common result and called all three phenomena "adaptation"; the differences in process have been obscured and evolutionists have often been misled to extend the Darwinian mode to the other two levels as well. First, we have what physiologists call "adaptation": the phenotypic plasticity that permits organisms to mold their form to prevailing circumstances during ontogeny. Human "adaptations" to high altitude fall into this category (while others, like resistance of sickling heterozygotes to malaria, are genetic and Darwinian). Physiological adaptations are not heritable, though the capacity to develop them presumably is. Secondly, we have a "heritable" form of non-Darwinian adaptation in humans (and, in rudimentary ways, in a few other advanced social species): cultural adaptation (with heritability imposed by learning). Much confused thinking in human sociobiology arises from a failure to distinguish this mode from Darwinian adaptation based on genetic variation. Finally, we have adaptation arising from the conventional Darwinian mechanism of selection upon genetic variation. The mere existence of a good fit between organism and environment is insufficient evidence for inferring the action of natural selection.

4. Adaptation and selection but no selective basis for differences among adaptations. Species of related organisms, or subpopulations within a species, often develop different adaptations as solutions to the same problem. When "multiple adaptive peaks" are occupied, we usually have no basis for asserting that one solution is better than another. The solution followed in any spot is a result of history; the first steps went in one direction, though others would have led to adequate prosperity as well. Every naturalist has his favorite illustration. In the West Indian land snail *Cerion*, for example, populations living on rocky and windy coasts almost always develop white, thick, and relatively squat shells for conventional adaptive reasons. We can identify at least two different developmental pathways to whiteness from the mottling of early whorls in all *Cerion*, two paths to thickened shells, and three styles of allometry leading to squat shells. All twelve combinations can be identified in Bahamian populations, but would it be fruitful to ask why—in the sense of optimal design rather than historical contingency—*Cerion* from eastern Long Island evolved one solution, and *Cerion* from Acklins Island another?

5. Adaptation and selection, but the adaptation is a secondary utilization of parts present for reasons of architecture, development, or history. We have already discussed this neglected subject in the first section on spandrels, spaces, and cannibalism. If blushing turns out to be an adaptation affected by sexual selection in humans, it will not help us to understand why blood is red. The immediate utility of an organic structure often says nothing at all about the reason for its being. . . .

We feel that the potential rewards of abandoning exclusive focus on the adaptationist program are very great indeed. We do not offer a council of despair, as adaptationists have charged; for nonadaptive does not mean nonintelligible. We welcome the richness that a pluralistic approach, so akin to Darwin's spirit, can provide. Under the adaptationist program, the great historic themes of developmental morphology . . . were largely abandoned; for if selection can break any correlation and optimize parts separately, then an organism's integration counts for little. Too often, the adaptationist program gave us an evolutionary biology of parts and genes, but not of organisms. It assumed that all transitions could occur step by step and underrated the importance of integrated developmental blocks and pervasive constraints of history and architecture. A pluralistic view could put organisms, with all their recalcitrant, yet intelligible, complexity, back into evolutionary theory.

CHAPTER 20

How to Carry Out
the Adaptationist Program?

Ernst Mayr

To have been able to provide a scientific explanation of adaptation was perhaps the greatest triumph of the Darwinian theory of natural selection. After 1859 it was no longer necessary to invoke design, a supernatural agency, to explain the adaptation of organisms to their environment. It was the daily, indeed hourly, scrutiny of natural selection, as Darwin had said, that inevitably led to ever greater perfection. Ever since then it has been considered one of the major tasks of the evolutionist to demonstrate that organisms are indeed reasonably well adapted, and that this adaptation could be caused by no other agency than natural selection. Nevertheless, beginning with Darwin himself (remember his comments on the evolution of the eye), evolutionists have continued to worry about how valid this explanation is. The more generally natural selection was accepted after the 1930s, and the more clearly the complexity of the genotype was recognized, particularly after the 1960s, the more often the question was raised as to the meaning of the word *adaptation*. The difficulty of the concept adaptation is best documented by the incessant efforts of authors to analyze it, describe it, and define it. Since I can do no better myself, I refer to a sample of such efforts (Muller et al. 1949; Wright 1949; Dobzhansky 1956, 1968; Bock and von Wahlert 1965; Williams 1966; Stern 1970; Munson 1971; Brandon 1978; Lewontin 1978, 1979; Bock 1980). The one thing about which modern authors are unanimous is that adaptation is not teleological, but refers to something produced in the past by natural selection. However, since various forms of selfish selection (e.g., meiotic drive, many aspects of sexual selection) may produce changes in the phenotype that could hardly be classified as "adaptations," the definition of adaptation must include some reference to the selection forces effected by the inanimate and living environment. It surely cannot have been anything but a lapse when Gould wanted to deny the designation "adaptation" to certain evolutionary innovations in clams, with this justification: "The first clam that fused its mantle margins or retained its byssus

From *American Naturalist* 121 (1983): 324–33.

to adulthood may have gained a conventional adaptive benefit in its local environment. But it surely didn't know that its invention would set the stage for future increases in diversity" (Gould and Calloway 1980: 395). Considering the strictly *a posteriori* nature of an adaptation, its potential for the future is completely irrelevant, as far as the definition of the term *adaptation* is concerned.

A program of research devoted to demonstrate the adaptedness of individuals and their characteristics is referred to by Gould and Lewontin (1979 [reproduced in this volume]) as an "adaptationist program." A far more extreme definition of this term was suggested by Lewontin (1979: 6) to whom the adaptationist program "assumes without further proof that all aspects of the morphology, physiology and behavior of organisms are adaptive optimal solutions to problems." Needless to say, in the ensuing discussion I am not defending such a sweeping ideological proposition.

When asking whether or not the adaptationist program is a legitimate scientific approach, one must realize that the method of evolutionary biology is in some ways quite different from that of the physical sciences. Although evolutionary phenomena are subject to universal laws, as are most phenomena in the physical sciences, the explanation of the history of a particular evolutionary phenomenon can be given only as a "historical narrative." Consequently, when one attempts to explain the features of something that is the product of evolution, one must attempt to reconstruct the evolutionary history of this feature. This can be done only by inference. The most helpful procedure in an analysis of historical narratives is to ask "why" questions; that is, questions (to translate this into modern evolutionary language) which ask what is or might have been the selective advantage that is responsible for the presence of a particular feature.

The adaptationist program has recently been vigorously attacked by Gould and Lewontin (1979 [reproduced in this volume]) in an analysis which in many ways greatly pleases me, not only because they attack the same things that I questioned in my "bean bag genetics" paper (Mayr 1959c), but also because they emphasize the holistic aspects of the genotype as I did repeatedly in discussions of the unity of the genotype (Mayr 1970, chap. 10; 1975). Yet I consider their analysis incomplete because they fail to make a clear distinction between the pitfalls of the adaptationist program as such and those resulting from a reductionist or atomistic approach in its implementation. I will try to show that basically there is nothing wrong with the adaptationist program, if properly executed, and that the weaknesses and deficiencies quite rightly pointed out by Gould and Lewontin are the result of atomistic and deterministic approaches.

In the period after 1859 only five major factors were seriously considered as the causes of evolutionary change, or, as they are sometimes called, the agents of evolution. By the time of the evolutionary synthesis (by the 1940s), three of these factors had been so thoroughly discredited and falsified that they are now no longer considered seriously by evolutionists. These three factors are: inheritance of acquired characters, intrinsic directive forces (orthogenesis, etc.), and saltational evolution (de Vriesian mutations, hopeful monsters, etc.). This left only two evolutionary mechanisms as possible causes of evolutionary change (including adaptation): chance, and selection forces. The identification of these two factors as the principal causes of evolutionary change by no means completed the task of the evolutionist. As is the case with most scientific problems, this initial solution represented only the first orientation. For completion it requires a second stage, a fine-grained analysis of these two factors: What are the respective roles of chance and of natural selection, and how can this be analyzed?

Let me begin with chance. Evolutionary change in every generation is a two-step process, the production of genetically unique new individuals and the selection of the progenitors of the next generation. The important role of chance at the first step, the production of variability, is universally admitted (Mayr 1962), but the second step, natural selection, is on the whole viewed rather deterministically: Selection is a nonchance process. What is usually forgotten is what an important role chance plays even during the process of selection. In a group of sibs it is by no means necessarily only those with the most superior genotypes that will reproduce. Predators mostly take weak or sick prey individuals but not exclusively so, nor do localized natural catastrophes (storms, avalanches, floods, etc.) kill only inferior individuals. Every founder population is largely a chance aggregate of individuals and the outcome of genetic revolutions, initiating new evolutionary departures, may depend on chance constellations of genetic factors. There is a large element of chance in every successful colonization. When multiple pathways toward the acquisition of a new adaptive trait are possible, it is often a matter of a momentary constellation of chance factors as to which one will be taken (Bock 1959).

When one attempts to determine for a given trait whether it is the result of natural selection or of chance (the incidental by-product of stochastic processes) one is faced by an epistemological dilemma. Almost any change in the course of evolution might have resulted by chance. Can one ever prove this? Probably never. By contrast, can one deduce the probability of causation by selection? Yes, by showing that possession of the respective feature would be favored by selection. It is this consideration which determines the approach of the evolutionist. He must first attempt to explain biological phenomena and processes as the product of natural selection. Only after all attempts to do so have failed, is he justified in designating the unexplained residue tentatively as a product of chance.

The evaluation of the impact of selection is a very difficult task. It has been demonstrated by numerous selection experiments that selection is not a phantom. That it also operates in nature is a conclusion that is usually based only on inference, but that is increasingly often experimentally confirmed. Very convincing was Bates's demonstration that the geographic variation of mimics parallels exactly that of their distasteful or poisonous models. The agreement of desert animals with the variously colored substrate also strongly supports the power of selection. In other cases the adaptive value of a trait is by no means immediately apparent.

As a consequence of the adaptationist dilemma, when one selectionist explanation of a feature has been discredited, the evolutionist must test other possible adaptationist solutions before he can resign and say: This phenomenon must be product of chance. Gould and Lewontin ridicule the research strategy: "If one adaptive argument fails, try another one." Yet the strategy to try another hypothesis when the first fails is a traditional methodology in all branches of science. It is the standard in physics, chemistry, physiology, and archaeology. Let me merely mention the field of avian orientation in which sun compass, sun map, star navigation, Coriolis force, magnetism, olfactory clues, and several other factors were investigated sequentially in order to explain as yet unexplained aspects of orientation and homing. What is wrong in using the same methodology in evolution research?

At this point it may be useful to look at the concept of adaptation from a historical point of view. When Darwin introduced natural selection as the agent of adaptation, he did so as a replacement for supernatural design. Design, as conceived by the natural theologians, had to be perfect, for it was unthinkable that God would make something that was less than perfect. It was on the basis of this tradition that the concept of natural selection originated. Darwin gave up this perfectionist concept of natural selection long before he wrote the *Origin*. Here he wrote, "Natural selection tends only to make each organic being as perfect as, or slightly more perfect than, the other inhabitants of the same country with which it has to struggle for existence. And we see that this is the degree of perfection attained under nature" (1975: 201). He illustrated this with the biota of New Zealand, the members of which "are perfect . . . compared with another" (ibid.), but "rapidly yielding" (ibid.) to recent colonists and invaders. After Darwin, some evolutionists forgot the modesty of Darwin's claims, but other evolutionists remained fully aware that selection cannot give perfection, by observing the ubiquity of extinction and of physiological and morphological insufficiencies. However, the existence of some perfectionists has served Gould and Lewontin as the reason for making the adaptationist program the butt of their ridicule and for calling it a Panglossian paradigm. Here I dissent vigorously. To imply that the adaptationist program is one and the same as the argument from design (satirized by Voltaire in *Can-*

dide) is highly misleading. When *Candide* was written (in 1759), a concept of evolution did not yet exist and those who believed in a benign creator had no choice but to believe that everything "had to be for the best." This is the Panglossian paradigm, the invalidity of which has been evident ever since the demise of natural theology. The adaptationist program, a direct consequence of the theory of natural selection, is something fundamentally different. Parenthetically one might add that Voltaire misrepresented Leibniz rather viciously. Leibniz had not claimed that this is the best possible world, but only that it is the best of the possible worlds. Curiously one can place an equivalent limitation on selection (see below). Selection does not produce perfect genotypes, but it favors the best which the numerous constraints upon it allow. That such constraints exist was ignored by those evolutionists who interpreted every trait of an organism as an ad hoc adaptation.

The attack directed by Gould and Lewontin against unsupported adaptationist explanations in the literature is fully justified. But the most absurd among these claims were made several generations ago, not by modern evolutionists. Gould and Lewontin rightly point out that some traits, for instance the gill arches of mammalian embryos, had been acquired as adaptations of remote ancestors but, even though they no longer serve their original function, they are not eliminated because they have become integral components of a developmental system. Most so-called vestigial organs are in this category. Finally, it would indeed be absurd to atomize an organism into smaller and smaller traits and to continue to search for the ad hoc adaptation of each smallest component. But I do not think that this is the research program of the majority of evolutionists. Dobzhansky well expressed the proper attitude when saying: "It cannot be stressed too often that natural selection does not operate with separate 'traits.' Selection favors genotypes. . . . The reproductive success of a genotype is determined by the totality of the traits and qualities which it produces in a given environment" (1956: 340). What Dobzhansky described reflects what I consider to be the concept of the adaptationist program accepted by most evolutionists, and I doubt that the characterization assigned to the adaptationist program by Gould and Lewontin, "An organism is atomized into traits and these traits are explained as structures optimally designed by natural selection for their functions" (1979 [reproduced in this volume]: 585), represents the thinking of the average evolutionist.

By choosing this atomistic definition of the adaptationist program and by their additional insistence that the adaptive control of every trait must be "immediate," Gould and Lewontin present a picture of the adaptationist program that is indeed easy to ridicule. The objections cited by them are all based on their reductionist definition. Of course, it is highly probable that not all secondary by-products of relative growth are "under immediate adaptive control." In the case of multiple pathways it is, of course, not necessary that every morphological detail in a convergently acquired adaptation be an ad hoc adaptation. This is true, for instance, in the case cited by them, of the adaptive complex for a rapid turnover of generations that evolved at least three times independently in the evolution of the arthropods. Evolution is opportunistic and natural selection makes use of whatever variation it encounters. As Jacob (1977) has said so rightly: "Natural selection does not work like an engineer. It works like a tinkerer."

Considering the evident dangers of applying the adaptationist program incorrectly, why are the Darwinians nevertheless so intent on applying it? The principal reason for this is its great heuristic value. The adaptationist question, "What is the function of a given structure or organ?" has been for centuries the basis for every advance in physiology. If it had not been for the adaptationist program, we probably would still not yet know the functions of thymus, spleen, pituitary, and pineal. Harvey's question "Why are there valves in the veins?" was a major stepping-stone in his discovery of the circulation of blood. If one answer turned out to be wrong, the adaptationist program demanded another answer until the true meaning of the structure was established or until it could be shown that this feature was merely an incidental by-product of the total genotype. It would seem to me that there is nothing wrong with the adaptationist program, provided it is properly applied.

Consistent with the modern theory of science, adaptationist hypotheses allow a falsification in most cases. For instance, there are numerous ways to test the thesis that the differences in beak dimensions of a pair of species of Darwin's finches on a given island in the Galápagos [are] the result of competition (Darwin's character divergence). One can correlate size of preferred seeds with bill size and study how competition among different assortments of sympatric species of finches affects bill size. Finally, one can correlate available food resources on different islands with population size (Boag and Grant 1981). As a result of such studies the adaptationist program leads in this case to a far better understanding of the ecosystem.

The case of the beak differences of competing species of finches is one of many examples in which it is possible, indeed necessary, to investigate the adaptive significance of individual traits. I emphasize this because someone might conclude from the preceding discussion that a dissection of the phenotype into individual characters is inappropriate in principle. To think so would be a mistake. A more holistic approach is appropriate only when the analysis of individual traits fails to reveal an adaptive significance.

What has been rather neglected in the existing literature is the elaboration of an appropriate methodology to establish adaptive significance. In this respect a recent analysis by Traub (1980) on adaptive modifications in fleas is exemplary. Fleas are adorned with a rich equipment of hairs, bristles, and spines, some of which are modified into highly specialized organs. What Traub (and various authors before him) found is that unrelated genera and species of fleas often acquire convergent specializations on the same mammalian or avian hosts. The stiffness, length, and other qualities of the mammalian hair are species specific and evidently require special adaptations that are independently acquired by unrelated lineages of fleas. "The overall association [between bristles and host hair] is so profound that it is now possible to merely glance at a new genus or species of flea and make correct statements about some characteristic attributes of its host" (ibid.: 64). Basically, the methodology consists in establishing a tentative correlation between a trait and a feature of the environment, and then to analyze in a comparative study, other organisms exposed to the same feature of the environment and see whether they have acquired the same specialization. There are two possible explanations for a failure of confirmation of the correlation. Either the studied feature is not the result of a selection force or there are multiple pathways for achieving adaptedness.

When the expanded comparative study results in a falsification of the tentative hypothesis, and when other hypotheses lead to ambiguous results, it is time to think of experimental tests. Such tests are not only often possible, but indeed are now being made increasingly often, as the current literature reveals (Clarke 1979). Only when all such specific analyses to determine the possible adaptive value of the respective trait have failed, is it time to adopt a more holistic approach and to start thinking about the possible adaptive significance of a larger portion of the phenotype, indeed possibly of the Bauplan[1] as a whole.

Thus, the student of adaptation has to sail a perilous course between a pseudoexplanatory reductionist atomism and stultifying nonexplanatory holism. When we study the literature, we find almost invariably that those who were opposed to nonexplanatory holism went too far in adopting atomism of the kind so rightly stigmatized by Gould and Lewontin, while those who were appalled by the simplistic and often glaringly invalid pseudoexplanations of the atomists usually took refuge in an agnostic holism and abandoned all further effort at explanation by invoking best possible compromise, or integral component of Bauplan, or incidental by-product of the genotype. Obviously neither approach, if exclusively adopted, is an appropriate solution. How do Gould and Lewontin propose to escape from this dilemma?

While castigating the adaptationist program as a Panglossian paradigm, Gould and Lewontin exhort the evolutionists to follow Darwin's example by adopting a pluralism of explanations. As much as I have favored pluralism all my life, I cannot follow Darwin in this case and, as a matter of fact, neither do Gould and Lewontin themselves. For Darwin's pluralism, as is well known to the historians of science, consisted of accepting several mechanisms of evolution as alternatives to natural selection, in particular the effects of use and disuse and the direct action of external conditions on organisms. Since both of these subsidiary mechanisms of Darwin's are now thoroughly refuted, we have no choice but to fall back on the selectionist explanation.

Indeed, when we look at Gould and Lewontin's "alternatives to immediate adaptation," we find that all of them are ultimately based on natural selection, properly conceived. It is thus evident that the target of their criticism should have been neither natural selection nor the adaptationist program as such, but rather a faulty interpretation of natural selection and an improperly conducted adaptationist program. Gould and Lewontin's proposals (1979 [reproduced in this volume]: 590–93) are not "alternatives to the adaptationist program," but simply legitimate forms of it. Such an improved adaptationist program has long been the favored methodology of most evolutionists. There is a middle course available between a pseudoexplanatory reductionist atomism and an agnostic nonexplanatory holism. Dobzhansky (1956) in his stress on the total developmental system and adjustment to a variable environment and my own emphasis on the holistic nature of the genotype (1963, 1970, 1975) have been attempts to steer such a middle course, to mention only two of numerous authors who adopted this approach. They all chose an adaptationist program, but not an extreme atomistic one.

Much of the recent work in evolutionary morphology is based on such a middle-course adaptationist program, for instance Bock's (1959) analysis of multiple pathways and my own work on the origin of evolutionary novelties (1960). A semiholistic adaptationist program often permits the explanation of seemingly counterintuitive results of selection. For instance, the large species of albatrosses (*Diomedea*) have only a single young every second year and do not start breeding until they are six to eight years old. How could natural selection have led to such an extraordinarily low fertility for a bird? However, it could be shown that in the stormy waters of the south temperate and subantarctic zones only the most experienced birds have reproductive success and this in turn affects all other aspects of the life cycle. Under the circumstances the extraordinary reduction of fertility is favored by selection forces and hence is an adaptation (Lack 1968).

The critique of Gould and Lewontin would be legitimate to its full extent if one were to adopt (1) their narrow reductionist definition of the adaptationist program as exclusively "breaking an organism into unitary traits and proposing an adaptive story for each considered separately" (1979 [reproduced in this volume]: 581) and (2) their characterization of natural selection, in the spirit of natural theology, as a mechanism that must produce perfection.

Since only a few of today's evolutionists subscribe to such a narrow concept of the adaptationist program, Gould and Lewontin are breaking in open doors.

To be sure, it is probable that many evolutionists have a far too simplistic concept of natural selection: they are neither fully aware of the numerous constraints to which natural selection is subjected, nor do they necessarily understand what the target of selection really is, nor, and this is perhaps the most important point, do they appreciate the importance of stochastic processes, as is rightly emphasized by Gould and Lewontin.

Darwin, as mentioned above, was aware of the fact that the perfecting of adaptations needs to be brought only to the point where an individual is "as perfect as, or slightly more perfect than" any of its competitors. And this point might be far from potentially possible perfection. What could not be seen as clearly in Darwin's day as it is by the modern evolutionist, is that there are numerous factors in the genetics, developmental physiology, demography, and ecology of an organism that make the achievement of a more perfect adaptation simply impossible. Gould and Lewontin (1979 [reproduced in this volume]) and Lewontin (1979) have enumerated such constraints and so have I (Mayr 1982a) based in part on independent analysis.

Among such constraints, the following seem most important.

1. *A capacity for nongenetic modification.* The greater the developmental flexibility of the phenotype, the better a species can cope with a selection pressure without genetic reconstruction. This is important for organisms that are exposed to highly unpredictable environmental conditions. When the phenotype can vary sufficiently to cope with varying environmental challenges, selection cannot improve the genotype.

2. *Multiple pathways.* Several alternative responses are usually possible for every environmental challenge. Which is chosen depends on a constellation of circumstances. The adoption of a particular solution may greatly restrict the possibilities of future evolution. When the ancestor of the arthropods acquired an external skeleton, his descendants henceforth had to cope with frequent molts and with a definite limitation on body size. Yet, to judge from the abundance and diversity of arthropods in the water and on land, it was apparently a fortunate choice in other respects.

3. *Stochastic processes.* An individual with a particular genotype has only a greater probability of reproductive success than other members of its population, but no certainty. There are far too many unpredictable chance factors in the environment to permit a deterministic outcome of the selection process. With the benefit of hindsight, one might come to the conclusion that selection has sometimes permitted a less perfect solution than would have seemed available. Virtually all evolutionists have underestimated the ubiquity and importance of stochastic processes. The kind of constraints to which natural selection is subjected becomes even more apparent when we look at the process of selection more closely.

4. *The target of selection* is always a whole individual, rather than a single

gene or an atomized trait, and an individual is a developmentally integrated whole, "fundamentally not decomposable into independent and separately optimized parts" (Gould and Lewontin 1979 [reproduced in this volume]: 591). For this reason, adaptation is by necessity always a compromise between the selective advantages of different organs, different sexes, different portions of the life cycle, and different environments. Even if the human chin is not the direct product of an ad hoc selection pressure, it is indirectly so as the compromise between two growth fields each of which is under the influence of selection forces. A pleiotropic gene or gene complex may be selected for a particularly advantageous contribution to the phenotype even if other effects of this gene complex are slightly deleterious. To uncouple the opposing effects is apparently not always easy. Since it is sufficient when an individual is competitively superior to most other individuals of its population, it may achieve this by particular features, indeed sometimes by a single trait. In that case natural selection "tolerates" the remainder of the genotype even when some of its components are more or less neutral or even slightly inferior.

5. *Cohesion of the genotype.* Development is controlled by a complex regulatory system, the components of which are often so tightly interconnected with each other, that any change of an individual part, a gene, could be deleterious. For instance, it is apparently less expensive in the development of a mammal to go through a gill arch stage than to eliminate this circuitous path and to approach the adult mammalian stage more directly. Allometry is another manifestation of regulatory systems. A selectively favored increase (or decrease) of body size may result in a slightly deleterious change in the proportions of certain appendages. Selection will determine the appropriate compromise between the advantages of a changed body size and the disadvantages of correlated changes in the proportion of appendages. The capacity of natural selection to achieve deviations from allometry has been established by numerous investigations. It was realized by students of morphology as far back as Étienne Geoffroy St. Hilaire, that there is competition among organs and structures. Geoffroy expressed this in his *loi de balancement.* The whole is a single interacting system. Organisms are compromises among competing demands. Wilhelm Roux (1881), almost one hundred years ago, referred to the competitive developmental interactions as the *struggle of parts* in organisms. The attributes of every organism show to what an extent it is the result of a compromise. Every shift of adaptive zones leaves a residue of morphological features that are actually an impediment. Reductionists have asked, Why has selection not been able to eliminate these weaknesses? The answer would seem to be that these are inseparable parts of a whole which, as a whole, is successful.

There are chance components in all these processes, but it must be stated emphatically that selection and chance are not two mutually exclusive alternatives, as was maintained by many authors from the days of Darwin to the earlier writings of Sewall Wright and to the arguments of some anti-Darwinians of today. Actually there are stochastic perturbations ("chance events") during every stage of the selection process.

The question whether or not the adaptationist program ought to be abandoned because of presumptive faults can now be answered. It would seem obvious that little is wrong with the adaptationist program as such, contrary to what is claimed by Gould and Lewontin, but that it should not be applied in an exclusively atomistic manner. There is no better evidence for this conclusion than that which Gould and Lewontin themselves have presented. Aristotelian "why" questions are quite legitimate in the study of adaptations, provided one has a realistic conception of natural selection and understands that the individual-as-a-whole is a complex genetic and developmental system and that it will lead to ludicrous answers if one smashes this system and analyzes the pieces of the wreckage one by one.

A partially holistic approach that asks appropriate questions about integrated components of the system needs to be neither stultifying nor agnostic. Such an approach may be able to avoid the Scylla and Charybdis of an extreme atomistic or an extreme holistic approach.

Note

1. [German architectural term usually translated as "floor plan," and meaning, roughly, "overall plan." See Gould and Lewontin (1979 [reproduced in this volume]: 593–95) and Dennett (1996: 277). Eds.]

CHAPTER 21

Evolutionary Ecology and the Use of Natural Selection in Ecological Theory

James P. Collins

Studies in ecology focus on two basic problems: the first is determining what controls the distribution and abundance of organisms, the second is describing the interrelationships of living organisms with their biotic and abiotic environment. These problems are complementary, because the size of a population depends on interactions between the individuals in that population and its biotic and abiotic environment. Similarly, describing the interrelationships of organisms helps to reveal why the relative abundance of a species varies from one area to another. Fundamentally, ecologists seek to explain where organisms occur by understanding the interactions affecting the relative abundance of plants and animals.

Studies in evolutionary biology, like those in ecology, focus on two basic problems: first, describing the historical record of life, the pattern of evolutionary change; second, discovering the mechanisms affecting the evolution of species.

Ecologists study processes such as competition, predation, energy flow, and nutrient availability to understand why species vary in abundance. In addition, some ecologists have argued for considering the mechanisms of organic evolution as factors that help explain ecological patterns. In an evaluation of six introductory textbooks in ecology, for example, one of the criteria Gordon Orians (1973) used was the degree of pervasiveness of the concept of natural selection. In this review he wrote, "Perhaps the greatest challenge for contemporary ecology is the development of theories about the properties of communities on the basis of selection for their component individuals."

As might be expected in a discipline encompassing the variety of research topics and methods found in ecology, there is a diversity of opinion concerning the need to incorporate evolutionary principles, especially the process of natural selection, into ecological theory (Smith 1975; McIntosh 1980; Simberloff 1980 [reproduced in this volume]). For example, in 1954, in their classic textbook on animal ecology, H. G. Andrewartha and L. C. Birch rarely used evo-

From *Journal of the History of Biology* 19 (1986): 257–88.

lutionary theory to develop answers to ecological questions such as why an organism is common in one place and rare in another. Orians argues that Andrewartha and Birch separated the concepts of ecology and evolution while they "claimed that a general and satisfying theory of ecology can and should be constructed without recourse to evolutionary thinking and concepts" (1962: 140). In his presidential address to the Ecological Society in 1967, John Harper reminded plant ecologists of the association between their discipline and Darwin's theory of evolution: "The theory of evolution by natural selection is an ecological theory—founded on ecological observation by perhaps the greatest of all ecologists. It has been adopted by and brought up by the science of genetics, and ecologists, being modest people, are apt to forget their distinguished parenthood" (1967: 247). He argued that in choosing research problems, the *Origin of Species* could be fruitfully utilized as a source for "questions which plant ecologists can usefully spend the next hundred years answering." Contemporary ecologists could especially benefit from emulating the manner in which Darwin summarized his ecological observations and posed his ecological questions. This attention would lead to more studies in plant population biology and microevolution, studies Harper judged to be "exciting" and relevant for understanding both theoretical and applied problems within ecology. J. B. C. Jackson observed that "the primary innovation of niche theory [during the late 1950s and early 1960s] was not simply a new appreciation of the importance of competition between animals, but the incorporation of an evolutionary perspective to distributional problems largely absent among ecologists since Darwin" (1981: 889). Orians's comments concerning Andrewartha and Birch and the observations by Harper suggest that ecologists vary in using evolutionary concepts to develop theory or choose research problems. Jackson reinforces this view through his conclusion that ecologists lacked an evolutionary perspective prior to development of niche theory in the late 1950s and early 1960s.

In this essay I address the following related questions: (1) How did ecologists use evolutionary theory before the late 1950s? and (2) In what sense was there a rapprochement between ecology and evolutionary biology during the 1950s and 1960s? My objective is not to develop complete answers, for there are none, but to sketch an outline that can serve as a starting point for further study.

The Association of Ecological and Evolutionary Theory before 1960

Adaptation and Selection for Traits of Individuals

Ecology emerged as a self-conscious discipline during the last decades of the nineteenth century and the early decades of the twentieth. The discipline was recognized at this time as "a logically consistent area of biology bringing together an extremely heterogeneous mix of natural history, physiology, hydrobiology, biogeography, and evolutionary concerns of nineteenth-century biology" (McIntosh 1976: 353). The close alliance between ecological and evolutionary concepts during the early stages of the development of ecology is illustrated by Robert McIntosh (1980), who shows that many ecologists specifically saw this new discipline as helping to explain adaptation and evolution. In 1866, for example, Haeckel claimed that the theory of evolution explained and formed the groundwork of ecology (Stauffer 1957). Stephen Forbes argued in 1895 that the whole Darwinian doctrine belonged to ecology. Just after 1900 H. C. Cowles wrote, "If ecology has a place at all in modern biology, certainly one of its great tasks is to unravel the mysteries of adaptation" (1904). At that same time, J. G. Needham described ecology as "the study of the phenomena of fitness" (1904).

The early close relationship between ecological and evolutionary concepts was also noted by W. C. Allee and his colleagues in their classic work on animal ecology (1949). In a review of ecological studies from 1900 to 1910 Thomas Park wrote:

> The reader may ask with justification: Why have there not been reviewed works on evolution as they contribute to ecological growth? The answer is that during this period the growth of ecology and evolution were so inextricably woven together that it seems artificial to separate the two. Many of the studies we have mentioned in the foregoing pages contain data, conclusions, or concepts that bear on evolution or speciation. In other words, certain ecologists of these times had a lively interest in such matters. This is as it should be, and it epitomizes the viewpoint of this book and its authors. (Allee et al. 1949: 48–49)

In a footnote to this paragraph he added the following caveat: "For the sake of accuracy, however, it should be mentioned that certain ecologists were veering away from an evolutionary viewpoint in the first decade [1900–10]. A good example, perhaps, was V. E. Shelford, who, during that period, was crystallizing his ideas on 'physiological animal geography' in contradistinction to historical or faunal animal geography." In fact, as ecology developed as an independent discipline during the early twentieth century, ecological studies separated into several subareas—for example "physiological animal geography," which was not primarily oriented toward evolution, in contrast to "historical or faunal animal geography," which was evolutionarily oriented (ibid.). . . .

Although adaptation was assumed, after the first decade of this century the emphasis on a physiological approach in plant and animal ecology led to the framing of questions in terms of a proximate or mechanistic, as opposed to an ultimate or evolutionary, view of causation (Mayr 1961). The questions were more the mechanistic "How does this structure function (and can therefore be regarded as an adaptation)?" than "Why does the organism have this trait (for this function), and not some other trait?" The latter question forces the investigator to consider the mechanistic aspect of function, as well as the processes by which the structure evolved; the former question does not necessarily lead to the historical, evolutionary part of an explanation. The union of ecology and evolution was conserved, however, in the associations of ecologists and taxonomists.

From 1920 to 1950 there were extensive interactions between ecologists, such as F. E. Clements, and taxonomists, such as M. V. Hall. These interactions gave rise first to the area of "experimental taxonomy" and then to "genecology" (Hagen 1984: 250). Hagen notes that "experimental taxonomy as originally envisioned by F. E. Clements was to be an ecological approach to the study of plant relationships" (ibid.: 257). In 1923 the Swedish botanist Gote Turesson wrote that genecology was characterized by the combined use of genetical, ecological, and taxonomical approaches to study "the patterns of intraspecific ecological adaptation, and elucidate the mechanisms whereby this adaptation was achieved" (Turesson 1923).

A principal research method used by experimental taxonomists and genecologists was the reciprocal transplant experiment. Commonly, this method involved transplanting individuals from one population to another habitat occupied by that species or another species, and vice versa. These experiments had "taxonomic as well as ecological significance" (Hagen 1984: 255). Further, "both Clements and Turesson approached experimental taxonomy primarily as ecologists interested in adaptation. The transplant experiments provided a method for quantitatively studying the responses of plants to environmental change" (ibid.). This line of research led ultimately to the classic transplant experiments of Jens Clausen, David Keck, and William Hiesey (1940).

This very brief survey illustrates that within the first several decades of this century there existed areas of research in which ecologists used both ecological and evolutionary concepts. These concepts complemented one another, especially in studies of adaptation at the individual level of organization. . . .

The Evolution of Population Regulation

From 1920 through the 1950s the population as an important unit for study attracted increasing numbers of researchers. Theoreticians such as R. A. Fisher (1930), Sewall Wright (1931), J. B. S. Haldane (1932), and S. Chetverikov (1926) focused primarily on integrating the Darwinian theory of natural selection with the rapidly developing concepts of Mendelian genetics. Empirical evolutionary geneticists like Dobzhansky (1937) described genetic changes within and between natural and laboratory populations. These theoretical and empirical studies contributed significantly to the demonstration of microevolution as a consequence of natural selection of individuals. Seminal theoretical and empirical works in population dynamics also were produced by investigators such as A. J. Lotka (1956), V. Volterra (1926), V. A. Kostitzin, Charles Elton (1927), and G. F. Gause (1964).

Among the many lines of inquiry in population dynamics, the question of what regulates changes in the size of a population drew much attention from 1920 through the 1950s. Research focused on biotic mechanisms (for example, competition) and abiotic mechanisms (such as temperature). Above all, there was significant debate concerning the role of density-dependent, density-independent, and to a smaller extent genetical factors in controlling population size. In hindsight, the question "How is population size regulated?" actually had two potential causal explanations. The first is seen by asking, "How, as a result of proximate or mechanistic causes, do populations fluctuate within apparent limits?" Possible causal mechanisms might be temperature, humidity, or density of competitors. Alternatively, one can ask, "Why, as a result of ultimate or evolutionary causes, do populations fluctuate within apparent limits?" By the late 1950s natural selection was the evolutionary causal mechanism frequently cited as capable of producing population regulation. Three views emerged to explain how natural selection might produce regulation; each differed in the level of biological organization at which natural selection was thought to act.

The first view interpreted change in population size as a compensatory adaptive response by the population, or even species, to ensure its persistence. Changes in population size were adaptive, because alterations were seen as responses that had evolved as a result of a selection pressure—for example, a regular shortage of food. The authors used such "species-advantage" or "population-advantage" explanations regardless of whether they ascribed change in size of the population to density-dependent, density-independent, or genetical factors. A good sense of the use of this argument may be obtained from Lamont Cole's (1954) classic work on life history characters. . . .

In this paper it will be regarded as axiomatic that the reproductive potentials of existing species are related to requirements for survival; that any life history features affecting reproductive potential are subject to natural selection; and such features observed in existing species should be considered adaptations, just as purely morphological or behavioral patterns are commonly so considered. . . . Some of the more striking life history phenomena have long been recognized as adaptations to special requirements. The great fecundity rather generally found in parasites and in many marine organisms is commonly regarded as an adaptation insuring the maintenance of a population under conditions where the probability is low that any particular individual will establish itself and reproduce successfully. (Cole 1954: 104)

A second view saw change in size of any single population as part of a homeostatic suprapopulation's response, to maintain the balance or equilibrium of the entire community. Here natural selection operated at the level of the entire community. This viewpoint can be illustrated using Allee and colleagues. In 1949 they developed in detail the argument that natural selection acts on populations as units and that consequently size of a population is regulated for the benefit of the group (684–85). There was no discussion of mechanisms for achieving such regulation, but they did speculate that natural selection acts on some genetic traits to influence the number of animals in the population. They also argued, like Clements and C. C. Adams previously, that the community is like a superorganism. Natural selection is the mechanism by which species evolve harmonious relationships between themselves and their environment. Manifestation of this adjustment over evolutionary time is a balance of nature in ecological time (see Simberloff 1980 [reproduced in this volume]). The strength of this viewpoint is worth illustrating with part of a chapter entitled "Evolution of Interspecies Integration and the Ecosystem," written by Alfred Emerson for a section on ecology and evolution.

> It may be concluded from these data that the community maintains a certain balance, establishes a biotic border, and has a certain unity paralleling the dynamic equilibrium and organization of other living systems. Natural selection operates upon the whole interspecies system, resulting in a slow evolution of adaptive integration and balance. Division of labor, integration, and homeostasis characterize the organism and the supraorganismic intraspecies population. The interspecies system has also evolved these characteristics of the organism and thus may be called an ecological supraorganism. . . . Homeostatic equilibrium within the ecosystem (balance of nature) is in large part the result of evolution. (Allee et al. 1949: 728–29)

In the third view, population size changed as a result of selection at the level of individual organisms. In response to the forces of natural selection, parents were favored who adjusted production of young to maximize the number who would survive to reproduce. Such parents would have a high Darwinian fitness. David Lack (1948) . . . show[s] that clutch size in starlings is not limited by the total number of eggs a bird can produce. Population size, therefore, is not limited by inability to lay additional eggs. Further, he explicitly rejects an argument such as Cole's: birds do not adjust clutch size in response to mortality at the species level. Rather, food is the ultimate factor that alters clutch size. Within the limits imposed by food resources, individual parents will raise as many young as possible who will have a high probability of surviving to adulthood. Elsewhere Lack makes it clear that natural selection adjusts this number of young through differential fitness of parents within a population. Finally, Lack notes that the "traditional ecologist" did not subscribe to this explanation based on selection of individuals; rather, it was a view acceptable to population geneticists.

Ecology and Evolution after 1960:
The Emergence of Evolutionary Ecology

By 1957 questions surrounding the mechanisms of population regulation supported an entire Cold Spring Harbor Symposium. A group of the world's leading animal ecologists debated the central issue of whether density-dependent or density-independent processes regulate population size. Genetical processes received some attention, but the poles of the argument were density dependence and density independence. This interesting theoretical problem of how and why populations are regulated persisted into the 1960s. To a significant extent the debate surrounding population regulation catalyzed several developments in ecology during this decade, including the emergence of evolutionary ecology.

In 1962 Orians first used the term *evolutionary ecology* as we understand it today. He argued that density-dependent and density-independent explanations for population regulation understood causation in biology differently, as Ernst Mayr summarized in 1961; the difference lay in a functional approach to causation as opposed to an evolutionary approach. Eventually this functional-evolutionary, or proximate-ultimate, perspective constituted one of three elements affecting the emergence of evolutionary ecology. The other two major elements were empirical results demonstrating the commonness of micro-evolution in natural communities and G. C. Williams's (1966) logical argument concerning the strength with which natural selection was expected to act at various levels of biological organization.

The Proximate-Ultimate Perspective
In his discussion of the differences between functional and evolutionary biology, it is evident that Mayr regards them as distinct fields of study. The functional biologist is "concerned with the operation and interaction of structural elements, from molecules up to organs and whole individuals. His ever-repeated question is 'How?' How does something operate, how does it function?" On the contrary, "the evolutionary biologist differs in his method and in the problems in which he is interested. His basic question is 'Why?' When we say 'why' we must always be aware of the ambiguity of this term. It may mean 'how come?' but it may also mean the finalistic 'what for?' It is obvious that the evolutionist has in mind the historical 'how come?' when he asks 'why?'" (1961: 1502).

Mayr illustrates each approach to causation, using the observation that some north temperate species of birds migrate south each winter. He asks the rhetorical question, "Why did the warbler on my summer place in New Hampshire start his southward migration on the 25th of August?" According to Mayr, four factors must be considered in explaining this behavior.

1. An ecological cause. Insects are not readily available during the winter in New Hampshire, so this insectivorous species must migrate south or starve.

2. A genetic cause. Over its evolutionary history this species has acquired a particular genotype. This genotype affects various aspects of the phenotype, causing the bird to respond to environmental stimuli that result in migration.

3. An intrinsic physiological cause. The physiology of the bird is slowly affected by decreasing day length throughout late summer. When the amount of daylight drops below a certain level, the bird is ready to migrate.

4. An extrinsic physiological cause. The bird, already physiologically prepared, migrated on August 25th because of a sudden drop in temperature that day.

In one sense, then, the immediate causes of migration are the physiological condition of the bird and the sudden decrease in temperature (for which the bird was physiologically prepared). These are the *proximate* causes of migration. Mayr contrasts them with lack of food during the winter in New Hampshire and the genotype of the bird, both of which are *ultimate* causes of migration. Ultimate causes are distinguished primarily by their historical component. Evolutionary processes, such as natural selection or genetic drift, have affected members of this species for many generations, and genotypes of contemporary warblers reflect the results. Thus a predictable lack of food in late summer serves as a selection pressure favoring migration at this time of year. Any genetic changes resulting in physiological alterations required for successful migration are favored. Similarly, any genetic changes in physiology would be favored if they led to improved detection of and response to external stimuli indicating a time to migrate with a high likelihood of success.

Concerning the question of population regulation, Orians argued in 1962 that functional ecologists such as Andrewartha and Birch explained changes in population size by using proximate factors as possible limiting mechanisms. Alternatively, evolutionary ecologists such as Lack explained variation in population size with ultimate factors as limiting mechanisms. In his discussion of proximate and ultimate explanations in ecology, Orians emphasized that they were complementary views of causality. In a particular situation neither explanation is necessarily wrong. The two answer different questions, and neither answer alone provides a complete explanation—say, of why different warblers migrate on different days. A proximate cause, perhaps individual variation in physiological factors, partially explains why each *individual* migrates when it does each year. Reference to an ultimate cause[,] however, could involve an attempt to explain why the *population* is composed of individuals that migrate at all, as opposed to remaining in New Hampshire for their entire lives. The proximate explanation concerns the life of individuals, whereas the ultimate explanation concerns the history of the population and the species.

In summary, Mayr's proximate versus ultimate "fields" are more easily distinguished if one considers the form of the question being asked, because this will reflect the level of causation, proximate or ultimate, at which the investigator is working. Functional biologists pursue causation largely by framing biological problems in a mechanistic fashion; consequently, they focus on "how" biologic processes operate. "Why" questions can generally be rephrased "how come." Evolutionary biologists pursue causation by asking why a particular trait has evolved, or alternatively why a trait is adaptive, thereby forcing the investigator to consider why this trait, as opposed to any other, would be favored under a particular set of environmental circumstances. In addition to the perspective brought to ecological problems by the proximate-ultimate argument, empirical results from at least three major areas of research contributed to the emerging evolutionary ecological view.

Genetics, Microevolution, and Ultimate Explanations

By the early 1960s an increasing amount of data showed that ecological characteristics of a population, such as density, and ecological characteristics of individuals, such as competitive ability, were influenced by the genetic constitution of the individual as well as by extrinsic factors such as temperature, humidity, and the like. In a complementary manner the probabilistic influence of population size on gene frequency of the population, especially in the case of genetic drift when there is small effective population size, was given an increasing role in explaining ecological processes such as the outcome of competition. . . .

As the evolutionary ecological approach developed during the 1960s, investigators used genetic and microevolutionary facts and principles within two somewhat different, but related, "traditions" or "approaches." Each analyst used evolutionary principles, but differed in the degree of emphasis given to genetics in the development of theory and design of research programs. One approach incorporated an evolutionary viewpoint into ecological studies by adopting the principles of population genetics and population ecology. Ecological traits were viewed as strongly affected by differences in the genetic background of the organisms under study. Thus, for example, black moths may suffer more predation on light trees than white moths in the same species, but ultimately this fact proves interesting only if there is a significant genetic basis for white and black body color.

Empirical results derived from the evolutionary synthesis and the general area of ecological genetics affected development of the genetical approach within evolutionary ecology. S. S. Shvarts (1977) argued that "ecological genetics," exemplified by Ford (1964) and Lerner (1963), could just as well be called evolutionary ecology or genetic ecology. . . . Roughgarden's (1979) introductory textbook provides a contemporary example of this genetical approach to evolutionary ecology. In the preface he writes: "Theoretical population biology is not a new field although its current visibility is unprecedented. Both theoretical population genetics and theoretical population ecology are easily traced into the 1920s and 1930s. . . . What is truly recent is the beginnings of a union of population genetic theory with the theory of population ecology. This unified theory is evolutionary population ecology" (vii).

A second approach, which also constitutes evolutionary ecology, is the adaptationist approach. Not all ecologists who used evolutionary principles in the 1960s did so via population genetics. Some ecologists, as Shvarts (1977: 7) noted, focused primarily on studying the origin and development of ecological adaptation. Ecological traits such as clutch size, body size, and body color were assumed to have a significant genetic basis. That basis was usually not evaluated, however, nor was genetics explicitly incorporated into the analysis or discussion of the evolution of these traits. Among ecologists Lack (1944) and G. E. Hutchinson (1951) used this adaptationist approach. Hutchinson's 1965 book, *The Ecological Theater and the Evolutionary Play*, summarizes several facets of the adaptationist viewpoint. There Hutchinson argues that biotic and abiotic characteristics of the environment (= the ecological theater) provide conditions (= niches) into which organisms with appropriate (= adapted) phenotypes can potentially evolve (= the evolutionary play). An ecologist, therefore, needs to understand the environmental circumstances favoring success of one phenotype over another. In theory, this level of understanding would allow predictions about population or community structure. For example, an environment with certain specified conditions should favor evolution of a feeding specialist phenotype, or perhaps a large-bodied phenotype. On a shorter time scale, predictions about changes in community structure can be made, because environments with the specified conditions will favor one phenotype over any others. Thus, the optimum phenotype should displace less optimal phenotypes from any community it invades, or alternatively resist displacement by less optimal phenotypes in communities where it already exists.

A significant advantage of the adaptation approach is that it allows the investigator to conceive of the evolution of important ecological traits, such as competitive ability or feeding pattern, that undoubtedly have a complex genetic basis. An adaptation approach assumes this genetic complexity, and analysis of the ecology and evolution of the system proceeds through the study of phenotypes. This assumption hastens analysis of complex ecological problems, but the danger is that one may be analyzing a trait that does not show significant evolutionary change, or perhaps any evolutionary change, because of the nature of its genetic bases.[1]

Population Regulation and Levels of Natural Selection

Prior to 1960 it was not uncommon to explain changes in population size as a population-level adaptation. Alterations in numbers were regarded as evolved responses to a selection pressure—perhaps a regular shortage of food. Change in population size was interpreted as a compensatory response to ensure the population's persistence. In 1962 V. C. Wynne-Edwards claimed that the number of individuals in a population was regulated to benefit the group or population. He argued that within many species individuals employed behavioral patterns which he called epideictic displays, as a mechanism to monitor density of the population. Using information gleaned from these displays, individuals should regulate their reproduction, or perhaps emigrate, to prevent overexploitation of their resources. Wynne-Edwards reasoned that such restraint was expected, since ultimately it would assure survival of the population.

Publication of this argument seems to have focused the divergent elements of the debate surrounding the question of population regulation. Wynne-Edwards's hypothesis was strongly criticized by Maynard Smith (1964), Perrins (1964), and especially by Lack (1966) and Williams (1966). They raised two principal objections. First, an individual decreasing its number of offspring, and thereby its fitness relative to its conspecifics, solely for the benefit of the group was acting in a genotypically altruistic manner; that is, the individual was deliberately decreasing its own genetic contribution to future generations solely to benefit the group. Natural selection, however, could probably lead to the evolution of such genotypically altruistic traits only in very specialized circumstances, the most likely being situations where the altruist conferred benefits not upon a group of unrelated individuals, but rather upon individuals to whom the altruist was closely related genetically. Second, evolution occurs via natural selection of alternative alleles in Mendelian populations. Strictly speaking, therefore, if genic selection is sufficient to explain the evolution of a trait, then it is the preferred explanation (Occam's razor). Any other is less parsimonious, and therefore less satisfactory (Wimsatt 1980a). Considerable effort was directed to showing not only that selection at the level of the gene, or at most at the level of individuals, was sufficient, but also that it explained the facts better.[2]

Prior to the stimulus of Wynne-Edwards's 1962 book, population-advantage or species-advantage arguments to explain changes in the number of individuals in a population were commonplace. These arguments represented the balance-of-nature concept drawn from community ecology and expressed at the population level, seen especially clearly in the discussion by Alfred Emerson of how natural selection operates at various organismic levels (Allee et al. 1949: 683–95). Because the homeostatic multispecies association had become a dominant paradigm in community ecology after 1900, populations could be regarded as functioning to minimize the likelihood of their own extinction, or even the likelihood of drastic changes in the number of individuals (see, e.g., Nicholson 1933). The consequence of such processes at the population level could be homeostasis or balance at the level of the community. Thus the concept of a homeostatic community provided a context supporting the concept of control over the number of individuals within a population for the benefit of the species or population itself.

With his hypothesis of group selection Wynne-Edwards (1977) explicitly stated that the number of individuals in an area was regulated to benefit the population. In doing so he made it clear that group selection was different from strict Darwinian natural selection of individual organisms. . . . This view focused discussion on group selection as an alternative to the Darwinian model of natural selection of individuals. Williams's defense of the Darwinian model was clear and unequivocal. . . .

In summary[,] during the first half of this century population-[adaptive] arguments were used within the context of a balance-of-nature viewpoint to explain changes in density of populations. Such explanations were viewed as a logical use of Darwin's model of natural selection at higher organismic levels. Wynne-Edwards, however, argued that a mechanism other than Darwinian natural selection of individuals was responsible for the patterns of change in the size of a population. Group selection and natural selection were juxtaposed as contradictory alternatives and drew attention to the general problem of levels of selection, specifically to the question of how effective natural selection could be at any level other than that of the gene, or at most the individual organism. Williams's very effective argument was that genes were the primary focus of natural selection. By the mid-1970s selection of genes—or, for some investigators, of individuals—was viewed almost exclusively as the ultimate cause of changes in population size. Within many areas of ecological study these levels of explanation virtually replaced explanations requiring the operation of natural selection at the population or community levels.

Discussion

The question of how ecologists have used evolutionary theory (natural selection in particular) in developing their research programs and interpreting their results has no simple answer. The relationship of developments within ecological and evolutionary theory during this century is complex, but there are several salient points.

Ecologists have a long history of conducting research within the general framework of evolutionary theory. At the beginning of this century ecology and evolutionary theory had close associations in some areas, especially with respect to the mechanism of natural selection. Relationships between organisms and their environment, for example, were frequently structured around the concept adaptation. Experimental taxonomy and genecology were areas in which investigators used concepts from both ecology and evolution. By the 1950s arguments had appeared that natural selection at the community, population, or individual levels was the process that ultimately produced population regulation. Finally, during the late 1950s and early 1960s, evolutionary ecology began to develop. The approach relied heavily on evolutionary theory for explanation of patterns of distribution and abundance of organisms, and developed along two lines. Some ecologists used an "adaptation approach" to investigate the evolution of traits such as clutch size, body size, and competitive ability. Understanding the adaptive significance of a trait or traits helped to explain why an organism occurred where it did at a particular abundance. This adaptation view was complemented by a "genetical approach" closely allied with ecological genetics.[3] During the 1960s investigators drew on elements from both approaches to address the theoretical question of how populations are regulated. Specifically, this question was studied within G. C. Williams's logical framework of levels of selection, using the perspective provided by the distinction between proximate and ultimate explanations, with an awareness of empirical results concerning microevolution of ecologically important characters.

Historically, ecologists have varied in the extent to which they have explicitly incorporated evolutionary theory into developing and studying research problems. Again, reasons for such differences are predictably complex, but two contributing factors stand out.

First is the degree to which a research program was designed to evaluate proximate or ultimate causes. By 1960, for example, animal ecologists studied the mechanistic question of what regulates population size; that is, how factors such as temperature, humidity, and the like affect the number of organisms in a community and thus cause changes in population size. Ecologists also considered the evolutionary question of why population size might be regulated; that is, whether populations were limited because of community-level or population-level adjustments to ensure persistence of these levels of organization, or because of adjustments by individual organisms to ensure their own fitness. Ecologists addressed this interesting theoretical problem of population regulation well into the 1960s. Some did so, however, with an increasing implicit or explicit reliance on the distinction between proximate and ultimate causes. Awareness of this distinction, coupled with the ever-increasing influence of genetics on ecology and the debate over the effectiveness of selection at various levels of organization, initiated significant developments within ecology.

Second is the way in which a line of questioning within ecology directly responded to empirical results or theoretical arguments within evolutionary biology. For example, evidence that emerged from the evolutionary synthesis (1936–1947) relative to microevolution and research in ecological genetics directly affected several lines of research within ecology during the 1960s. The important empirical findings related to (1) evidence for a genetic basis for ecologically significant traits, (2) the demonstration that some kinds of evolutionary change progress within the range of time required for many ecological processes to reach completion, and (3) the demonstration that natural selection commonly operates over spatial scales sufficiently small that microevolution partially explains why populations differ in distribution and abundance over relatively short distances. Empirical results in these areas suggested that within some research programs ultimate causes needed to be considered as part of the explanation for the distribution and abundance of organisms.

Evolutionary ecology is probably best described as an approach to the study of ecology. In this respect it is not recognized principally by the topics investigated, but rather by the way in which ecological problems are conceived and analyzed using ecological, genetic, and evolutionary principles. In some historical respects the approach is not new, but simply reflects the current way in which evolutionary concepts are incorporated into ecological theory. There is, however, a novel conceptual focus distinguishing ecological explanations developed within this framework from those developed prior to the early 1960s, which also combined genetics and evolutionary theory in addressing ecological questions.

Historically, many ecologists explained the evolution of traits such as competitive ability of individuals, regulation of population size, and integration of a community as resulting from natural selection operating at one or more levels of organization. Since the 1960s, however, special attention has been directed to addressing two significant elements needed for making an evolutionary argument. First, there is an increasing sensitivity to the fact that phenotypic variation in a trait is attributable to variation of environmental as well as genetic factors. Failure to consider the pattern of inheritance, therefore, significantly reduces the power of any argument concerning the evolution of a trait. Second, in attributing a possible role to natural selection, many ecologists have become conscious of the need to indicate clearly the level at which selection is presumed most effective in shaping a trait, because the effectiveness of selection varies at different levels. Gene frequencies can theoretically change as a result of natural selection at one or more levels of biological organization.

But it is not a simple matter to decide which level of selection is having the most significant effect, or even any effect, on final expression of a trait. A gene frequency will change as a function of genetic variation and selection intensity within a level, as well as genetic variation and selection intensity between levels; in other words, change or maintenance of a gene frequency is the sum of the effects of selection within and between levels of organization (Price 1970, 1972). Consider, for example, an argument about the relative contribution of individual and population-level selection to expression of a trait. Deciding which level of selection is more significant requires understanding how total genetic variation within the entire population is divided. Total variation can be partitioned into an amount attributable to selection intensity and genetic differences among individuals within each group constituting the population, as opposed to selection intensity and genetic differences among the various groups of which the population is composed. The magnitude and direction of selection at each level can then help determine how selection within groups versus selection among groups contributes to change or maintenance of gene frequencies (Wade 1980). An awareness of such partitioning of variation as an element of an evolutionary argument is a novel characteristic of some recent studies that use evolutionary theory in developing explanations for the distribution and abundance of organisms.

This review of the history of the association between ecology and evolutionary theory suggests two concerns for contemporary ecologists. The first is that the current emphasis on the operation of natural selection at the level of the gene or individual organism reflects the present state of an evolving view of how natural selection operates to effect the distribution and abundance of plants and animals. Second, when an argument concerning the evolution or maintenance of a trait is needed, it should be framed carefully with an awareness of the relative effectiveness with which selection at different levels of organization can act in order to influence the final expression of a trait.

Notes

1. Two things should be noted concerning this dichotomy. (1) The term *genetical* in opposition to the term *adaptation* does not imply that investigators using the genetical approach are not using the concept of adaptation. I use these words for the convenience of distinguishing an approach based primarily on the study of phenotypes from one based on genotypes. (2) Some investigators commonly use aspects of each approach. I use these terms, therefore, not so much for purposes of categorization as for summarizing the general emphasis, phenotype versus genotype, in a particular study. M. Slatkin and J. Maynard Smith (1979) have reviewed models of coevolution in which ecological

and genetic changes were taken into account. They too recognized one class of models in which genetics was explicitly included and a second group in which this was not the case. They viewed the former approach as deriving from the theoretical population genetics of Fisher, Wright, and Haldane, whereas they saw the latter as originating with the mathematical models of Volterra and Lotka.

2. The premise that an understanding of the effects of selection at one level of organization is "in principle" extensible to understanding phenomena at a higher level of organization (for instance, selection of genes of individuals can explain the evolution of group properties) is not a trivial assumption. Arguments surrounding this hypothesis, especially as used by Williams (1966) relative to group selection, may be found in Wimsatt (1980a) and Sober and Lewontin (1982).

3. In summarizing his understanding of the content of ecological genetics, Lerner (1963: 492) wrote: "To an ecologist . . . a population is mainly characterized by its density, natality and mortality rates, age distribution, dispersion and growth rate. A geneticist would, of course, want to know about the genetical structure of the population, its variability, its mating system, its fitness. . . . And an ecological geneticist would wish to know the genetic basis of adaptation, the mechanisms of niche selection or adjustment to limited niches, the homeostasis of population size not only in relation to the nonliving environment but in relation to the whole system." Further, he argued: "The interface of the two branches of evolutionary biology, genetics and ecology, would to my mind not be a mere combination of two disciplines, but a synthesis which would possess an emergent methodology and address itself to questions, admittedly not newly stated, but which have newly acquired the property of being meaningful, that is to say, solvable" (ibid.: 489).

CHAPTER 22

Evolution: The Missing Ingredient in Systems Ecology

Craig Loehle and Joseph H. K. Pechmann

The theory of evolution forms one of the foundations of modern ecology. Research on the process of natural selection has enhanced the understanding of many subjects. For example, ecological genetics, biogeography, and population ecology are intimately tied to evolutionary processes and theory. Behavioral ecology and physiological ecology are fundamentally based on an adaptationist approach; that is, they attempt to discover the significance of, evolution of, or mechanisms behind various traits.

At present, however, evolutionary theory is rarely an integral part of systems ecology or ecosystem modeling. By evolution, we mean only organic evolution resulting from natural selection, not simply directional change in biological or physical systems over time. Systems ecology comprises those subdisciplines of ecology that are concerned with the dynamics, structure, and function of whole ecosystems, particularly using quantitative methods. Included in this definition are ecosystem models, food-web analyses, stability analyses of multispecies models, general systems-theory applications, nutrient-cycling studies, and some management models (e.g., whole-crop or timber-stand growth models).

In spite of some attempts to address evolution in the systems literature (Dunbar 1972; May 1973a; Cody 1974; Walters 1975; Hall and Day 1977; Innis and O'Neill 1979: 104; H. Odum 1983; O'Neill et al. 1986: 123), evolution is not well represented or adequately incorporated. When organic evolution is mentioned in the systems literature, it is often treated as a "black box" process that organizes ecosystem functioning in some optimal way. Many systems ecologists believe that, given sufficient time, evolution will provide mutual adaptation of species to form organized, functionally integrated ecosystems (see Patten and Odum 1981; E. P. Odum and Biever 1984), but there has been little critical examination of how evolutionary mechanisms might achieve this result. System-level constraints and causal feedback loops are often treated as higher-order phenomena that can be studied and modeled without consider-

From *American Naturalist* 132 (1988): 884–99.

ing the evolution of individual species. This holistic approach is seen as a practical way of dealing with the overwhelming complexity of ecosystems and the lack of natural-history data on their components. The goal of systems ecology is an ecosystem phenomenology that does not necessarily require detailed information about individual species (Ulanowicz 1986). Comparisons of leaf-area index among forest types or of trophic structure between terrestrial and aquatic habitats can be made without detailed examination of species characteristics. Similarly, the global carbon balance and rates of watershed nutrient loss are generally appropriately addressed at the whole-ecosystem level.

This holistic approach contrasts with the orientation of population ecologists and evolutionary biologists, who focus primarily on adaptation and selection at the level of individual organisms (by which we do not imply extreme Panglossian adaptationism; Gould and Lewontin 1979 [reproduced in this volume]). Most of these investigators believe that selection at levels of organization above the individual occurs rarely, if ever (Williams 1966; Maynard Smith 1976). Some do not believe that ecosystems are actual entities. They hold that ecosystems are no more than a collection of species in a particular environment, with all interesting questions centering on species' adaptations. For evolutionarily oriented scientists willing to consider higher-level systems, ecosystems and their properties are generally viewed as results of the dynamics of competition and natural selection among individual organisms.

The dichotomy between these two schools of thought results partially from different underlying philosophies concerning the nature of scientific inquiry and explanation. Scientific theories that explain in terms of composition (structure and function) are fundamentally different from those that explain in terms of time development (Shapere 1974). The basic question here is whether ecosystems can be satisfactorily understood and modeled by theories that address only current relationships and phenomena and not their long-term historical antecedents, namely, evolution (Orians 1962).

Incorporating evolutionary theories and concepts into systems ecology is an underexplored but potentially fruitful avenue. We believe that evolutionary concepts can contribute significantly to systems theory and models, which in turn can contribute to evolutionary biology by helping define more completely the arenas within which natural selection occurs. Reiners (1986) has proposed that unifying ecosystem ecology requires at least three separate but complementary theoretical frameworks: energetics, matter (chemical stoichiometry), and some aspect of population interactions or ecosystem "connectedness." We suggest that evolutionary theory might make up his third framework (or perhaps a fourth one) for studying the ecosystem. According to Reiners, some predictions can be made about ecosystems based largely on only one of the three frameworks, but in other cases, interactions among frameworks may need to be considered. We are arguing for the relevance of evolutionary theory in this kind of context. We do not believe that it is a panacea or that evolutionary theory should or even can be applied to all ecosystem problems. Although we also believe the converse, that systems ecology can contribute to evolutionary theory, we deal with this issue only tangentially here.

Our goal is to demonstrate the benefits of a closer integration. We address potential applications of evolutionary theory to systems theory in the areas of ecosystem organization and optimality principles. Further, we discuss how evolutionary theory might assist systems modelers in defining hierarchies, analyzing stability, deriving process functions, and estimating parameters. Our coverage is not exhaustive; nor do we pretend to resolve all issues that are raised. Complete integration of the two fields cannot be accomplished in one paper.

Applying Evolutionary Theory to Systems Theory

Ecosystem Organization

Ecological genetics (Istock 1984), coevolution theory (Wilson 1980; Futuyma and Slatkin 1983), evolutionary aspects of spatial heterogeneity (Whitham et al. 1984), and studies of evolutionary strategies (Maynard Smith 1982) are the focus of much current research. But is this information applicable to the structure and function of whole ecosystems or simply to the species populations found in the ecosystem? Because ecosystem function depends in part on species adaptations and because adaptations are evolutionary constructs, we must look to evolution to determine the constraints on ecosystem organization and function. Only traits that are biologically feasible and allowed by phylogenetic constraints (i.e., considering evolutionary history) are available as raw materials for constructing an ecosystem. This is because organisms are not infinitely adaptable and all selective pressures are not translated into adaptations, especially if selective pressures conflict (Whitham et al. 1984).

Some systems ecologists have proposed that ecosystems evolve to maximize measured and hypothetical quantities such as "exergy," persistent biomass, power, "[ascendency]," ecological buffering capacity, specific growth rate, the ratio of live biomass to maintenance metabolism, stability, resilience, and elasticity (table 4). Debate concerning these principles has been based on logical plausibility, general thermodynamics (H. Odum 1983), mathematical analysis (May 1973a; Fontaine 1981; O'Neill et al. 1986), and empirical observations of ecosystem dynamics (e.g., successional trends; Margalef 1968; Jørgensen and Mejer 1977, 1981). In an evolutionary sense, we should ask to what degree and in what ways natural selection acts on individuals to affect ecosystem-level properties. Various system-level properties may thereby be shown to be ca-

TABLE 4.
Hypothesized ecosystem properties optimized by evolution.

Property	Source
Hierarchical organization	Miller 1978; O'Neill et al. 1986
Maximum exergy	Jørgensen and Mejer 1981
Ecological buffering capacity	Jørgensen and Mejer 1977
Maximum power or efficiency	H. Odum 1983; H. Odum and Pinkerton 1955
Thermodynamic organization	Glansdorff and Prigogine 1971; Schneider 1987
Stability, resilience, elasticity, or persistence	Whittaker and Woodwell 1971; Dunbar 1972; May 1973
Maximum ascendency, average mutual information	Ulanowicz 1980, 1986
Maximum persistent biomass	O'Neill et al. 1986
Maximum ratio of live biomass to maintenance metabolism	Margalef 1968

Source: Adapted and expanded from Fontaine 1981.

pable of being optimized by natural selection, to be in conflict with selection acting on individuals, or even to be epiphenomena of individual selection.

A first step in testing the hypothesized attributes in the table is to check whether they are formulated using extensive or intensive variables. Slobodkin (1972) pointed out that natural selection operates at the level of individuals that are governed by intensive variables (e.g., food availability, temperature) rather than extensive variables (e.g., carrying capacity, diversity, entropy, total population). Information on extensive variables is available to humans but is generally not available to organisms in the system. Natural selection is thus incapable of responding to extensive properties per se. Theories formulated in terms of extensive variables are not necessarily incorrect, because an intensive variable may mediate the effect at the individual level. For example, the extensive variable "population size" can be mediated through intensive variables such as food availability at the individual level. But when a theory depends on evolutionary optimization of extensive variables, then there is cause for close scrutiny. An example is the hypothesis that evolution promotes ecosystem stability (an extensive property). Examination of relevant intensive variables and how natural selection is affected by or affects them can provide insight into mechanisms accounting for proposed extensive properties.

For examining the role of evolution in relation to ecosystem properties, a second key question is whether the properties are emergent or collective (Salt 1979; E. P. Odum 1983). A collective property of a system is one that is simply the sum of properties of its components (e.g., total site biomass is additive). A true system or emergent property is one that results from interactions among system components and is qualitatively different from the properties of those components. Alternatively, an emergent property can be defined as a property of a system that is not predictable from studies of isolated system components (but see Edson et al. 1981).

An explanation of collective ecosystem properties in terms of classical individual natural selection is often straightforward. For example, there is no need to invoke system integration to explain efficient use of water by desert ecosystems; natural selection acting on heritable traits of individuals seems a sufficient explanation. The system uses water efficiently because only organisms that use water efficiently can survive there, not necessarily because of any emergent causality or higher-level structuring. When ecosystem properties are shown to be collective, traditional evolutionary arguments can sometimes be extended directly to the ecosystem level. This provides a means for generating whole-system predictions. For truly emergent properties, however, the origin of these properties and whether they are somehow optimized or evolving toward optimality (as often assumed) is a far more complex question. If system-level organizational principles are important, then it becomes necessary to ask both how evolution might affect emergent system properties and how ecosystem dynamics might affect natural selection of individual species.

Consider the assertion that terrestrial communities maximize persistent organic matter (table 4). The accumulation of biomass (much of it dead) over time does not demonstrate that evolution promotes maximal biomass per se; organic matter may persist for other reasons. One of a plant's main defenses against herbivory is to create tissues that are inedible because of toughness, high concentrations of toxins, digestion inhibitors, or silica, or a lack of protein and nutrients (Zucker 1983; Whitham et al. 1984; Rhoades 1985; Loehle 1988b). In addition, competition among plants can result in selection for increased height, which results in woody, lignified structures. A consequence of these selection pressures is that most plant tissues decompose more slowly than animal tissue. Dryness of standing dead material also inhibits decomposition. Thus, litter and standing biomass may accumulate simply as a side effect of plant defensive measures, competition, and climate. Moreover, litter accumulation is not necessarily optimal for some ecosystem functions because seedling establishment may be inhibited, and net primary production (NPP) may be reduced by shading, precipitation interception, and nutrient immobilization (e.g., in tallgrass prairie; Knapp and Seastedt 1986). This same criticism can be applied to the principle of ecological buffering capacity (Jørgensen and Mejer 1977). Nutrients that are held in organic matter may provide a buffer against disturbance (nutrients are released after fire, for example) and help prevent nutrient loss from the site; however, this may be merely a fortuitous consequence of slow decomposition resulting from other selective pressures. If nutrient storage in organic matter were selected for per se, we would expect plant litter and wood to have a high mineral-nutrient content as a storage device, which is generally not the case. Thus, evolutionary arguments based on individual selection resulting in collective properties can provide additional insights concerning proposed system-level properties. If, however, we reject these alternative models, it is necessary to ask how selection might act to maximize persistent biomass or buffering capacity as emergent properties.

Another maximization theory for ecosystem organization is the maximum-power principle, which states that evolution favors systems that maximize the flow rate of useful energy (H. Odum 1983). Using nonequilibrium thermodynamics, Odum argued that systems capturing more power and channeling it into adaptive mechanisms will be favored. Phylogenetic constraints may, however, limit the achievement of optimal energy capture. For example, many remote oceanic islands have been invaded by human-dispersed foreign plants that have a growth rate much greater than that of the native vegetation. In such instances, the native vegetation apparently had not evolved to take full advantage of the resources available, either because of time limitation since arrival or because of phylogenetic or other constraints. Thus, predictions of energetic (thermodynamic) principles are not necessarily an adequate guide to what we actually find because evolution does not necessarily produce an optimal result.

This examination of system theories illustrates how evolutionary theory can contribute to hypotheses about ecosystem structure and function and complement the energetic and chemical approaches. It also suggests that additional insight about whether observed ecosystem characteristics have evolved or represent an optimum can be obtained by considering if and how natural selection might operate on those characteristics and by considering what constraints there might be on the process. We are not arguing that the hypotheses in table 4 are necessarily false, but we hope we have demonstrated the usefulness of an evolutionary approach and pointed out directions for further evaluations of these theories.

Group Selection

Classical individual natural selection can account for each species' having a functional role in the ecosystem (e.g., decomposition) because the ability to exploit an empty or uncrowded niche would be likely to increase an individual's fitness. But, can organisms derive individual fitness benefits by contributing to the benefit of the overall ecosystem via indirect pathways (Wilson 1980)? In other words, is ecosystem structure and function any more optimal than what we would expect from the collective result of each organism's attempting to maximize its own fitness? Any such extraoptimal functioning suggests emergent properties. To explain these (but not necessarily all) emergent properties, one must assume that those genotypes that are more "beneficial" to the system will increase relative to others that are less beneficial. This is not likely to occur under individual selection because any system-level benefits generated by one genotype would increase the survival and/or reproductive success of all genotypes in the ecosystem equally. Group-selection theories suggest possible mechanisms by which this might occur. Many evolutionary biologists and geneticists, however, do not believe that group selection is an important or widespread phenomenon in nature (Williams 1966; Maynard Smith 1976).

What is the likely scope for group selection as a mechanism contributing to ecosystem organization? Theoretical and experimental work by Wade (1978), Wilson (1980), and others suggests that group selection might be a more significant force for evolutionary change than previously thought. The prevailing skepticism toward group selection is based on genetic models predicting that individual selection within groups will override selection between groups except under excessively restrictive conditions. However, the predictions of these anti–group selection models depend greatly on their assumptions about extinction, dispersion, colonization, and other factors (Wade 1978). Different and perhaps more biologically meaningful assumptions lead to alternative models (see Wilson 1980), in which group selection plays a much more important role. In these models, traits that enhance ecosystem structure and function can be selected for because groups containing the individuals with the preferred traits profit differentially from any enhancement and thus contribute more genes to the next generation. Traits advantageous to the system therefore increase in frequency, sometimes even when individual selection acts simultaneously to decrease their frequency. Wilson's (1980) models, however, are limited to small spatial scales and do not apply at the whole-landscape level.

In order to understand the mechanisms by which natural selection might operate at levels of organization above the individual, the precise meanings of "best" and "benefit" at these higher levels need to be clarified. For example, is ecosystem stability analogous to individual survival and reproductive success as suggested by Dunbar (1972)? The tradeoffs and balances between individual and group selection must also be considered. Even at an organismal level, many conflicting selective pressures must be reconciled. Because of these constraints and other factors, such as phylogenetic history and chance, evolutionary "designs" are necessarily imperfect (Gould and Lewontin 1979 [reproduced in this volume]). This may be especially true for higher levels of organization, such as communities and ecosystems. Phenomena are likely to be more and more constrained to nonoptimal conditions as we proceed up the hierarchical scale because additional conflicting demands (group as well as individual) must be balanced (Cody 1974).

Models that address such questions have the potential to build a critical bridge between systems ecology and evolutionary biology. We are not trying to promote group selection. But because group-selection models seem to be required by certain theories of ecosystem organization, their limits and results should be kept in mind by systems ecologists.

Mutualism's Role

Examination of mutualisms may be another avenue by which evolutionary ecology can increase our understanding of ecosystem organization (E. P. Odum and Biever 1984). In mutualism, the reciprocal benefits gained from any seemingly altruistic activity are focused directly and immediately back onto the individuals involved. This can be explained in terms of traditional individual selection, although group selection may also be a factor. Recent work (see, e.g., Futuyma and Slatkin 1983; Addicott 1984) clarifies mutualism in terms of energetic costs and benefits, survival, and reproduction.

Although mutualism helps us understand some aspects of ecosystem linkages, models of the evolution of mutualisms indicate that the extent of tight linkage formation is limited. For example, most species are facultative rather than obligate mutualists because of spatial and temporal heterogeneity (Howe 1984), a generally less efficient form of mutualism.

A problem with invoking either group selection or mutualistic coevolution as mechanisms for generating integrated properties of ecosystems is that many species are part of more than one community, even on a local scale. Can we reasonably expect a species to evolve toward optimal integration into a multitude of communities? In addition, different species experience selection pressures and segregation into demes on vastly different geographical scales. Furthermore, most communities are composed of species that previously belonged to different communities because of migrations of species across the landscape over time, in many cases as recently as the last glaciation (Whittaker and Woodwell 1971; Davis 1986).

Community Assembly

Another mechanism for explaining system properties is provided by the theory of community assembly (H. Odum 1983; Taylor 1985), which states that an optimal community is assembled from the available species pool by a self-organizing, competitive process. The "best" primary producers for the particular physical environment are assembled because they outcompete the others. Other components are similarly assembled. Succession is the typical example of this process. If community assembly is a valid concept, then research on assembly mechanisms may help explain ecosystem organization and processes and the relative importance of evolutionary mechanisms at this level. For example, by understanding the assembly process of gap replacement in forests (Shugart 1984), we can better understand the ecosystem property of diversity. Community assembly may be collective (e.g., each piece of ground acquires the plant species that are most productive at that site), or it might be integrative and involve emergent properties (e.g., a stable assemblage of species might be formed by a trial-and-elimination process; Taylor 1985). It is probably better to refer to the assembly process as ecosystem development (Ulanowicz 1986) rather than community evolution, as it sometimes is. It is thus important to understand the assembly process better and assess its importance and how it relates to the evolution of ecosystems per se.

System-Level Constraints on Evolution

Research on system-level organization and functioning has under-utilized implications for evolutionary studies. There are ecosystem-level constraints on the evolution of individual species; not just any random combination of organisms can form a workable, persistent ecosystem (Smith 1975; Patten and Odum 1981). Biota must fit into some appropriate "niche" in their ecosystem in order to grow and reproduce successfully (Whittaker and Woodwell 1971). An understanding of basic ecosystem design could therefore help to clarify the selective pressures impinging on individual organisms. Research and modeling in mineral cycling, ecosystem energetics, and so on can thus make a useful contribution to studies of evolutionary processes and adaptations.

For example, analysis of carbon and nutrient flows in wetland systems under different hydrologic regimes may advance our understanding of the frequency and extent to which food resources limit anuran larvae developing at these sites. Food levels (determined by whole-system trophic structure) can affect such life-history traits in anurans as growth rate and body size at metamorphosis (Wilbur 1977). It would be useful to embed a genetic-selection analysis within an ecosystem model, particularly concerning the effect of system-level constraints or forcing functions on population genetics.

More detailed treatment of this subject would take us too far afield, but it is clear that the interaction of systems ecology and evolutionary theory cuts both ways.

Applying Evolutionary Theory to Ecosystem Modeling

A major thrust of systems ecology is constructing theoretical and applied ecosystem models, a process that involves conceptualization, construction, and application or testing. We believe that these models can be improved by applying evolutionary theory to the modeling process.

Hierarchy Theory

Studies of ecosystems inevitably must deal with processes occurring on various spatial and temporal scales. From ant-aphid dynamics on individual leaves to nutrient loss from entire watersheds, numerous processes simultaneously interact on different spatial scales. When a model is being constructed, it is usually necessary to aggregate components (e.g., into guilds or trophic levels) to make the model tractable. Hierarchy theory has been proposed as a means of dealing with these modeling problems and as a framework for understanding ecosystem organization (Rowe 1961; Pattee 1973; Miller 1978; Conrad 1979; Webster 1979; Totafurno et al. 1980; Allen and Starr 1982 [reproduced in this volume]; Salthe 1985; O'Neill et al. 1986). How do we test whether a given hierarchical scheme actually represents ecosystem structure or function? To begin by examining the entire ecosystem at once may not be feasible, although Miller (1978) made a preliminary effort. It would, perhaps, be best to begin with smaller subsystems. However, Rowe (1961) argued that there are no intermediate subsystems between individual organisms and ecosystems and that populations and communities cannot be considered subsystems because they are not "objects" that can be physically separated from other subsystems. Webster (1979) stated that populations and communities do not necessarily have stronger internal interactions than external interactions and thus violate Simon's (1962, 1973) criteria for hierarchical systems. Webster added that trophic levels, functional groups, and guilds cannot be considered subsystems of ecosystems for this same reason. However, distinct physical subdivisions of the ecosystem such as a forest's canopy, floor, and soil can perhaps be viewed as subsystems, according to Webster.

We propose, as a partial solution, using evolutionary criteria for determining groups of species with stronger internal than external interactions and hence for defining hierarchies and aggregations. Groups of highly interactive, coevolved (or genetically mutually interacting) species that form mutualistic, commensal, competitive, predator-prey, or parasitic associations should therefore be examined as possible hierarchical subsystems within ecosystems.

In a plant-mycorrhizae system, for example (figure 17), the plant supplies food and in return receives mineral nutrients (particularly phosphorus) from the fungal symbionts located in or on its roots. The benefit to both parties and the mechanism of exchange are both clear. The mycorrhizae populations act as subsystems of the plant and can be excluded by the plant when not needed, though perhaps not completely (P. Kormanik, personal communication). In

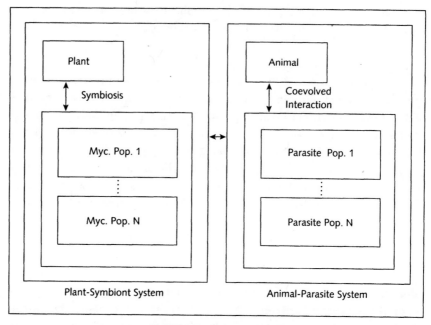

Figure 17. Hierarchical aggregation based on the intensity of mutual genetic interaction. There is a strong mutual genetic adaptation between an animal and its parasitic load (including pathogenic load), in many cases leading to an endemic load of parasites. The animal and its retinue can be aggregated together in a model, as can a plant and its symbionts. Finally, if the animal is an obligate pollinator of the plant, they may be aggregated together. This provides evolutionary criteria for aggregation in ecosystem models. Myc. Pop., mycorrhizal population.

practice and for modeling purposes, we often aggregate the linked system and ignore that it consists of more than one species. We can add pollinators to the system for greater complexity (figure 17). The pollinators in turn have parasites with which they are tightly functionally linked and coevolved. Here, instead of aggregating various animal species as a compartment and animal parasites as a compartment, we aggregate an animal species with its characteristic retinue (e.g., parasites), to form a unit. Finally, we can view the plant-symbiont system and its connected pollinator-parasite system together as a single plant-pollinator system.

This example suggests that our criteria for defining hierarchical systems in ecology or aggregating species in our models should include both functional and evolutionary interactions (i.e., mutual genetic interactions) among components. By studying the evolved species interactions in such a system, we can form a clearer picture of appropriate subsystem boundaries.

Evaluating Community Equilibria

Mathematical treatments of community dynamics have repeatedly addressed the question of stability properties of the population equations (see, e.g., May 1973a). It is useful to know, after a disturbance, if species will become extinct, if there will be a new set of dominant species, or if the system will return to its original state. If there is a unique stable point or attracting region for the population equations, then we can say that succession will have a unique end point, even if it never arrives there because of disturbance. Systems with several attractors will have several final states, depending on initial conditions and perturbations. Establishing the uniqueness of the attractor is a major step toward finding it when nonlinear equations are being analyzed. One approach to evaluating whether there is a unique attractor is to define sufficient conditions for uniqueness (DeAngelis and Goldstein 1978). Evaluation of these uniqueness conditions, however, is not straightforward.

Evolutionary theory can help solve this problem. If stability of the ecosystem is analogous to survival of the organism, or if natural selection tends to reduce fluctuations in individual populations, then unique attractors should be favored. Some systems ecologists (e.g., Patten and Odum 1981) assume that this is the case, and in fact Patten's work with linear models reflects this assumption and depends on the existence of a unique system equilibrium. However, as discussed previously, it is not clear whether ecosystems evolve as units (Whittaker and Woodwell 1971). The efficacy of selection in reducing wide population fluctuations is still under debate (Maynard Smith 1976). Selection for group attack in insects, for example, may actually promote massive instability (Berryman et al. 1985; Rhoades 1985).

As discussed previously, Howe (1984) has shown that mutualists tend to be facultative rather than obligate. This suggests that several species are capable of performing the same role in the system, depending on chance and community composition. If so, then existence of a single attractor seems inherently unlikely, though an attractor region might be possible in a model, with facultative mutualists replacing each other stochastically or as conditions change. If competitors or groups of facultative mutualists are aggregated in a model, then a unique attractor is much more likely.

In certain cases, it may be possible to build up from life histories of individual species to ecosystem dynamics (Walters 1975; Noble and Slatyer 1980). In such cases, evolutionary studies of adaptations and strategies at the species level can contribute to dynamic analysis and questions of stability.

Deriving Model Equations

During model construction, it is necessary to obtain or develop equations for such processes as photosynthesis and forage consumption. An engineering approach is to develop empirical curves or regression equations from available data, yielding a predictive or calculation-tool model (Loehle 1983). With such an approach, however, the function that fits the data may be biologically arbitrary (e.g., as in models tested in Loehle and Rittenhouse 1982). Moreover, given several empirical curves that each fit the data, there may be no criterion for choosing a best model. What is needed is a procedure that yields results that are more meaningful.

Evolutionary considerations can help in building biologically meaningful models. For example, we often ask, given physical and phylogenetic constraints on the system, what combination of traits is optimal for an organism's growth, survival, and reproduction. This adaptationist point of view (Mayr 1983 [reproduced in this volume]) forms a strong heuristic device for investigating biological processes in relation to their function. This approach yields models that would probably never be obtained by an empirical or regression approach, particularly those that are algorithmic, such as those for feeding-territory size (Schoener 1983b) or for foraging strategies (Pyke 1984). It also helps to explain tradeoffs in features, to predict relations of features to environment, and to explain anomalies. Examples include models for photosynthesis (Orians and Solbrig 1977) and for leaf shape (Givnish and Vermeij 1976).

Developing functions for models based on an evolutionary paradigm thus has several clear advantages. Models can be derived that could not be obtained otherwise. There is a more defensible rationale for choosing a particular model. And, finally, the context of derivation of the model defines when and where it is applicable.

Parameter Derivation

In order to build an ecosystem model, it is necessary to have specific parameter values for model equations. Methods discussed in the preceding section may yield a functional form, but further work may be required to obtain parameter estimates for particular cases. Required specific values can be obtained by (1) direct measurement of the system of interest, (2) use of literature values from similar systems or species, or (3) a combination of educated guessing, sensitivity analysis, and "fine-tuning." Direct measurement is usually considered the most desirable method, but in most cases the available measurements cover only a limited time span and range of conditions. This can lead to major problems in predicting long-term community dynamics or the effect of perturbations on the system, especially if species with different life-

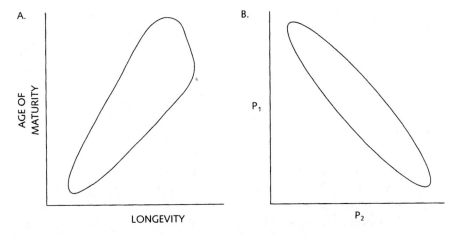

Figure 18. Evolutionary considerations can constrain possible values of parameters. Enclosed region shows biologically possible combinations of parameters. A, Evolution of life-history characteristics constrains age at maturity by longevity (from data in Loehle 1988b). B, If P_1 and P_2 are parameters for partitioning energy during growth, they must be inversely correlated.

history traits become dominant (Walters 1975). Even if the species composition remains the same, phenotypic plasticity and rapid genetic changes (Istock 1984) can be important. These potential changes in model parameters must somehow be taken into account.

The information provided by evolutionary studies of variation in life-history traits, intra and interspecific competition, and adaptive strategies under different conditions may be precisely what is needed in ecosystem models. Limits on and tradeoffs among strategic parameters, such as age at maturity, mode of reproduction (sexual vs. asexual), and longevity can sometimes be obtained from the literature. Models can be constructed on the basis of these strategic parameters (see, e.g., Noble and Slatyer 1980). Walters (1975) suggested that the possible values of these parameters be organized into a mathematical "strategy space." This can be used to supplement the values actually measured in the system and define changes that might be expected to occur, even if species dominance changes.

Covariance of traits or restriction of traits to certain regions of strategy space can result from evolutionary processes: some traits may genetically or physiologically correlate or conflict; and other combinations of traits may not be stable over evolutionary time. For example, there is a basic correlation between age at sexual maturity and longevity in trees (figure 18A). Parameters can also be constrained by physical bounds, as Walters (1975) illustrated for the nutrient uptake of algal cells of different sizes. There can also be constraints caused by competition between sinks during growth (figure 18B). If P_1 is the fraction of total plant biomass going to seeds and P_2 is the fraction going to vegetative reproduction, then there is a tradeoff between P_1 and P_2 (figure 18B). Walters showed how the combinations of constraints on the parameter axes (some physical, some evolutionary) form a bounded strategy space, the interior of which is feasible. The combination of feasible parameters that yields a maximal growth rate under a given set of conditions can then be used as the best *a priori* estimate of parameter values. Walters tested this approach and found good agreement between predicted and actual parameters of species living in the given environment. The same method was useful for predicting species that became dominant under new conditions.

The existence of bounds on feasible parameters has consequences for parameter tuning and sensitivity analysis. If natural selection acts to push a parameter (trait) toward a physically imposed upper bound, species values cluster near that bound but trail off below it. The confidence limits around the parameter include part of the infeasible zone when normality is assumed, making parameter tuning or sensitivity analysis potentially incorrect unless this skewness is considered.

Unfortunately, Walters's concept of strategy space and tradeoffs between and constraints on parameter values has rarely been applied to parameter-estimation problems (Walters, personal communication). It is typically assumed that parameters are completely free or are free within some percentage of the nominal values, as they are in many engineering and statistical applications, perhaps because constraints and tradeoffs lead to less tractable problems. In any event, evolutionary studies yield a great deal of information on tradeoffs and constraints that should be used to advantage.

A possible extension of the strategy-space concept has been provided by MacFarlane (1981). He showed that the relationships defining the structure of a given system (e.g., the food web) can exist only if critical parameters are confined to a certain range of values. Beyond these bounds the structure changes, producing a new set of relationships (e.g., linkage of the food web is altered). This set of feasible parameters is the support for the structure. MacFarlane developed a general mathematical framework for dealing with this situation that is intriguing in its potential application to the development of models of systems undergoing perturbation or environmental change and to the determination of evolutionarily stable combinations of traits.

Summary

Although applying evolutionary theory to questions about ecosystem structure and function is not simple, it is essential. For example, certain hypotheses of ecosystem optimality seem to require the operation of group selection or coevolution at levels above the individual. Such assumptions should not go untested, and an examination of evolutionary mechanisms in an ecosystem context could lead to new insights. Explaining ecosystem properties that are clearly additive or collective in terms of classical individual selection may often be straightforward. Practical aspects of constructing ecosystem models, including hierarchy applications, stability analysis, and parameter estimation, are also amenable to input from evolutionary theory. The concept of an evolutionarily based strategy space for model parameters should prove particularly fruitful. Applying evolutionary theory should help establish greater credibility for systems ecology and facilitate unification of what should be a holistic science.

Acknowledgments

This work was supported by contract DE-AC09-76SR00819 between the U.S. Department of Energy and the University of Georgia Research Foundation.

Afterword

Ecological science creates for itself a paradox, as its subject matter reaches beyond the boundaries of normal science. Ecology moves easily from the investigation of organisms to geology, hydrology, meteorology, even astronomy—the totality of nature. The subject matter cannot be rigorously delimited, so neither can its methods.

But perhaps a more serious obstacle to making ecology a conventional scientific discipline is the fact that the study of biotic communities leads to issues of human behavior. The inclusion of humans as a subject matter for ecology nudges us away from natural science proper (*Naturwissenschaften*) to the humanities and social sciences (*Geisteswissenschaften*). Thus, the attempt to establish ecology as a mechanistic science like physics, chemistry, and molecular biology, full of mathematical rigor and free of moral overtones, is self-defeating.

From ecology to ethics: the step is inevitable. When the issue of human behavior arises, it is difficult—and may be impossible—not to ask: Is there any difference between how humans are acting, and how humans *should* act? Now, at the end of the second millennium, we live on a planet where the activities of one species have an impact on all processes of the biosphere. The hegemony of *Homo sapiens* constricts the freedom of all organisms. Whether the cause of these changes is primarily industrialization or overpopulation is not the point. The phenomena are global; the solution (if there is one) must also be global. And any global solution will be political. The old injunction against scientists uttering moral assertions, based on the notion that nature is devoid of intrinsic value or purpose, is misguided. Ecologists cannot, and ought not, refrain from making moral judgments. Yes, ecology is political.

In conclusion, all aspects of natural process and pattern are part of any ecological investigation. To say that ecology should be a mechanistic science free of moral overtones, and should not become involved in making judgments about human impact on biotic systems, is to draw boundaries on its subject matter proper. To draw boundaries by separating the human from the non-human, by separating ethics from science, is itself unecological. To think ecologically is to think comprehensively, to think across the boundaries of natural science and the humanities. Thinking ecologically means *synthesizing* the many fields of human knowledge into a coherent worldview. Ultimately, the scientific ecologist includes in his or her purview ethics, values, and politics. As a consequence, there will never be overall consensus on the form and objectives of ecological science. This lack of complete consensus is not a sign of the science's sickness or weakness; rather, it is a sign of its inclusive scope and normative significance. This is why ecology is so captivating, and timely.

(N.B. Dates in parentheses are reprint dates.)

Abbey, Edward. 1969 (1988). *Desert Solitaire*. Tucson: University of Arizona Press.

Abrams, P. A. 1980. "Resource Partitioning and Interspecific Competition in a Tropical Hermit Crab Community." *Oecologia* 46: 365–79.

Abrams, P. A. 1981. "Alternative Methods of Measuring Competition Applied to Two Australian Hermit Crabs." *Oecologia* 51: 233–39.

Abrams, P. A. 1996. "Dynamics and Interactions in Food Webs with Adaptive Foragers." In G. Polis and K. Winemiller (eds.), *Food Webs: Integration of Patterns and Dynamics*. New York: Chapman Hall, pp. 113–21.

Ackermann, R. 1988. *Wittgenstein's City*. Amherst: University of Massachusetts Press.

Addicott, J. F. 1984. "Mutualistic Interactions in Population and Community Processes." In P. W. Price, C. N. Slobodchikoff, and W. S. Gaud (eds.), *A New Ecology*. New York: John Wiley, pp. 437–56.

Adelman, H. 1974. "Rational Explanation Reconsidered: Case Studies and the Hempel-Dray Model." *History and Theory* 13: 208–24.

Ahl, Valerie, and T. F. H. Allen. 1996. *Hierarchy Theory: A Vision, Vocabulary, and Epistemology*. New York: Columbia University Press.

Alexander, Samuel. 1927. *Space, Time and Deity: The Gifford Lectures at Glasgow, 1916–1918*. Volume 2. London: Macmillan.

Allee, Warder C., Alfred E. Emerson, Orlando Park, Thomas Park, and Karl P. Schmidt. 1949. *Principles of Animal Ecology*. Philadelphia: W. B. Saunders.

Allen, T. F. H., and Thomas W. Hoekstra. 1992. *Toward a Unified Ecology*. New York: Columbia University Press.

Allen, T. F. H., and Thomas B. Starr. 1982. *Hierarchy: Perspectives for Ecological Complexity*. Chicago: University of Chicago Press.

Amsterdamski, S. 1981. "Riduzione." In *Enciclopedia Einaudi*. Volume 12. Torino: Einaudi, pp. 62–75.

Anderson, Stanford (ed.). 1966. *Planning for Diversity and Choice: Possible Futures and Their Relations to the Man-Controlled Environment*. Cambridge: MIT Press.

Andrewartha, H. G., and L. C. Birch. 1954. *The Distribution and Abundance of Animals*. Chicago: University of Chicago Press.

Aristotle. (1979). *Metaphysics*. Trans. Hippocrates G. Apostle. Grinnell, Iowa: Peripatetic Press.

Aristotle. (1985). *Nicomachean Ethics*. Trans. Terence Irwin. Indianapolis: Hackett.

Auerbach, S. I., R. L. Burgess, and R. V. O'Neill. 1977. "The Biome Programs: Evaluating an Experiment." *Science* 195: 902–4.

Ayala, F. J., and T. Dobzhansky (eds.). 1974. *Studies in the Philosophy of Biology: Reduction and Related Problems*. Berkeley: University of California Press.

Ayer, Alfred J. 1952. *Language, Truth, and Logic*. New York: Dover.

Baker, G. 1986. "Following Wittgenstein." In J. Canfield (ed.), *The Philosophy of Wittgenstein*. Volume 10. New York: Garland, pp. 223–63.

Baker, G., and P. Hacker. 1985. *Wittgenstein*. Oxford: Blackwell.

Baker, G., and P. Hacker. 1986. "Critical Study: On Misunderstanding Wittgenstein: Kripke's Private Language Argument." In J. Canfield (ed.), *The Philosophy of Wittgenstein*. Volume 10. New York: Garland, pp. 321–64.

Bambrough, J. R. 1979. *Moral Scepticism and Moral Knowledge*. London: Routledge and Kegan Paul.

Barash, D. P. 1976. "Male Response to Apparent Female Adultery in the Mountain Bluebird: An Evolutionary Interpretation." *American Naturalist* 110: 1097–1101.

Barbour, M. G. 1996. "American Ecology and American Culture in the 1950s: Who Led Whom?" *Bulletin of the Ecological Society of America* 77: 44–51.

Bartlett, M. S. 1957. "On Theoretical Models for Competitive and Predatory Biological Systems." *Biometrika* 44: 27–42.

Bartlett, M. S. 1960. *Stochastic Population Models in Ecology and Epidemiology*. London: Methuen.

Bartlett, M. S., J. C. Gower, and P. H. Leslie. 1960. "A Comparison of Theoretical and Empirical Results for Some Stochastic Population Models." *Biometrika* 47: 1–11.

Bayes, Thomas. (1908). *Versuch zur Losung eines Problems der Wahrscheinlichkeitsrechnung*. Leipzig: W. Engelmann.

Beckman, T. A. 1971. "On the Use of Historical Examples in Agassiz's 'Sensationalism.'" *Studies in History and the Philosophy of Science* 1: 293–96.

Beckner, M. 1974. "Reduction, Hierarchies and Organicism." In F. J. Ayala and T. Dobzhansky (eds.), *Studies in the Philosophy of Biology: Reduction and Related Problems*. Berkeley: University of California Press, pp. 163–77.

Beddington, J. R., C. A. Free, and J. H. Lawton. 1978. "Modelling Biological Control: On the Characteristics of Successful Natural Enemies." *Nature* 273: 513–19.

Beddington, J. R., and Robert M. May. 1977. "Harvesting Natural Populations in a Randomly Fluctuating Environment." *Science* 197: 463–65.

Belovsky, G. E. 1984. "Moose and Snowshoe Hare Competition and a Mechanistic Explanation from Foraging Theory." *Oecologia* 61: 150–59.

Belovsky, G. E. 1986. "Generalist Herbivore Foraging and Its Role in Competitive Interactions." *American Zoologist* 26: 51–69.

Bergandi, Donato. 1995. "'Reductionist Holism': An Oxymoron or a Philosophical Chimera of E. P. Odum's Systems Ecology?" *Ludus Vitalis: Journal of Philosophy and Life Sciences* 3: 145–78.

Berkowitz, A. R., J. Kolsds, R. H. Peters, and S. T. Pickett. 1989. "How Far in Space and Time Can the Results from a Single Long-Term Study Be Extrapolated?" In G. E. Likens (ed.), *Long Term Studies in Ecology: Approaches and Alternatives*. New York: Springer-Verlag, pp. 192–198.

Bernstein, C., and B. Woodward. 1974. *All the President's Men*. New York: Simon and Schuster.

Berryman, A. A., B. Dennis, K. F. Raffa, and N. C. Stenseth. 1985. "Evolution of Optimal Group Attack, with Particular Reference to Bark Beetles (Coleoptera: Scolytidae)." *Ecology* 66: 898–903.

Bertalanffy, L. von. 1952. *Problems of Life.* New York: John Wiley.

Bertalanffy, L. von. 1969. *General System Theory.* New York: George Braziller.

Bird, Elizabeth Ann R. 1987. "The Social Construction of Nature: Theoretical Approaches to the History of Environmental Problems." Paper presented at "Forests, Habitats, and Resources: A Conference in World Environmental History," Duke University, Durham, North Carolina, April 30–May 2.

Blitz, David. 1992. *Emergent Evolution: Qualitative Novelty and the Levels of Reality.* Boston: Kluwer.

Bloor, D. 1983. *Wittgenstein.* New York: Columbia University Press.

Boag, P. T., and P. R. Grant. 1981. "Intense Natural Selection in a Population of Darwin's Finches (Geospizinae) in the Galápagos." *Science* 214: 82–85.

Bock, W. J. 1959. "Preadaptation and Multiple Evolutionary Pathways." *Evolution* 13: 194–211.

Bock, W. J. 1980. "The Definition and Recognition of Biological Adaptation." *American Zoologist* 20: 217–27.

Bock, W., and G. von Wahlert. 1965. "Adaptation and the Form-Function Complex." *Evolution* 19: 269–99.

Boltzmann, L. 1872. "Weitere Studien über das Wärmegleichtgewicht unter Gasmolekülen." *Wien. Ber.* 66: 275–370.

Bookchin, Murray. 1991. *The Ecology of Freedom: The Emergence and Dissolution of Hierarchy.* New York: Black Rose Books.

Born, M. 1949. *Natural Philosophy of Cause and Chance.* London: Oxford University Press.

Botkin, D. B., and M. J. Sobel. 1975. "Stability and Time Varying Ecosystems." *American Naturalist* 109: 625–46.

Boyd, R. 1972. "Determinism, Laws and Predictability in Principle." *Philos. Sci.* 39: 431–50.

Bramwell, Anna. 1989. *Ecology in the Twentieth Century: A History.* New Haven: Yale University Press.

Brandon, R. 1990. *Adaptation and Environment.* Princeton: Princeton University Press.

Brandon, R. N. 1978. "Adaptation and Evolutionary Theory." *Studies in the History and Philosophy of Science* 9: 191–206.

Branham, J. M. 1973. "The Crown-of-Thorns on Coral Reefs." *BioScience* 73: 219–26.

Brennan, Andrew. 1988. *Thinking about Nature: An Investigation of Nature, Value and Ecology.* Athens: University of Georgia Press.

Brooks, D. R., and E. O. Wiley. 1986. *Evolution as Entropy: Toward a Unified Theory of Biology.* Chicago: University of Chicago Press.

Brown, W. L., and Edward O. Wilson. 1956. "Character Displacement." *Systematic Zoology* 5: 49–64.

Buffon, George Louis Leclerc, Comte de. 1749 (1785a). *Natural History, General and Particular.* Trans. William Smellie. Volume 1. London.

Buffon, George Louis Leclerc, Comte de. 1749 (1785b). *Natural History, General and Particular.* Trans. William Smellie. Volume 2. London.

Buffon, George Louis Leclerc, Comte de. 1749 (1785c). *Natural History, General and Particular.* Trans. William Smellie. Volume 4. London.

Bunge, M. 1983. *Epistémologie.* Paris: Maloine S. A. Editeur.

Burden-Sanderson, John S. 1893. Inaugural Address. *Nature* 48: 464–72.

Caccamise, D. F. 1974. "Competitive Relationships of the Common and Lesser Nighthawk." *Condor* 76: 1–20.

Cain, S. A. 1944. *Foundations of Plant Geography.* New York: Harper.

Cain, Stanley A. 1947. "Characteristics of Natural Areas and Factors in Their Development." *Ecological Monographs* 17: 185–200.

Callahan, D., and S. Bok. 1980. *The Teaching of Ethics in Higher Education: A Report.* New York: The Hastings Center.

Callicott, J. Baird. 1994. *Earth's Insights.* Berkeley: University of California Press.

Campbell, D. T. 1969. "Reforms as Experiments." *American Psychologist* 24: 409–29.

Campbell, D. T. 1975. "Degrees of Freedom and the Case Study." *Comparative Political Studies* 8: 178–93.

Campbell, D. T. 1984. Foreword to R. K. Yin, *Case Study Research.* Beverly Hills: Sage, pp. 7–8.

Campbell, John H. 1985. "An Organizational Interpretation of Evolution." In David Depew and Bruce Weber (eds.), *Evolution at a Crossroads: The New Biology and the New Philosophy of Science.* Cambridge: MIT Press, pp. 133–67.

Capra, Fritjof. 1988. "Systems Theory and the New Paradigm." In Carolyn Merchant (ed.), *Ecology: Key Concepts in Critical Theory.* New Jersey: Humanities Press, pp. 334–41.

Carnap, Rudolp. 1938. "Logical Foundation of the Unity of Science." In *International Encyclopedia of Unified Science* 1: 42–62.

Carson, R. A. 1986. "Case Method." *Journal of Medical Ethics* 12: 36–37.

Carson, Rachel. 1962 (1994). *Silent Spring.* Boston: Houghton Mifflin.

Cartwright, N. 1989. "Capacities and Abstractions." In P. Kitcher and W. C. Salmon (eds.), *Scientific Explanation.* Minnesota Studies in the Philosophy of Science 13. Minneapolis: University of Minnesota Press, pp. 349–56.

Chamberlin, Thomas C. 1890 (1965). "The Method of Multiple Working Hypotheses." *Science* 148: 754–59.

Chase, Alston. 1995. *In Dark Wood: The Fight over Forests and the Rising Tyranny of Ecology.* Boston: Houghton Mifflin.

Chetverikov, S. S. 1926. "On Certain Aspects of the Evolutionary Process from the Standpoint of Modern Genetics." Trans. M. Barker. *Proceedings of the American Philosophical Society* 105: 167–95.

Chiang, C. L. 1954. "Competition and Other Interactions between Species." In O. Kempthorne, T. A. Bancroft, J. W. Gowen, and J. L. Lusk (eds.), *Statistics and Mathematics in Biology.* Ames: Iowa State College Press, pp. 197–215.

Chiang, C. L. 1968. *Introduction to Stochastic Processes in Biostatistics.* New York: John Wiley.

Chomsky, Noam. 1996. *Powers and Prospects.* Boston: South End Press.

Clarke, B. C. 1979. "The Evolution of Genetic Diversity." *Proceedings of the Royal Society of London: Biological Sciences* 205: 453–74.

Clarke, G. L. 1954. *Elements of Ecology.* New York: John Wiley.

Clausen, J., D. D. Keck, and W. M. Hiesey. 1940. *Experimental Studies on the Nature of Species, Part 1: Effect of Varied Environments on Western North American Plants.* Carnegie Institution of Washington Publication 520. Washington, D.C.: Carnegie Institution.

Clements, Frederic E. 1904. "Development and Structure of Vegetation." *Rep. Bot. Surv. Nebr. 7.*

Clements, Frederic E. 1905. *Research Methods in Ecology.* Lincoln: University of Nebraska Press.

Clements, Frederic E. 1916. *Plant Succession: An Analysis of the Development of Vegetation.* Carnegie Institution of Washington Publication 242. Washington, D.C.: Carnegie Institution.

Clements, Frederic E. 1934. "The Relict Method in Dynamic Ecology." *Journal of Ecology* 22: 39–68.

Clements, Frederic E. 1936. "Nature and Structure of the Climax." *Journal of Ecology* 24: 252–84.

Clements, Frederic E., and Victor E. Shelford. 1939. *Bioecology.* New York: John Wiley.

Cody, M. L. 1974. "Optimization in Ecology." *Science* 183: 1156–64.

Cohen, J. E. 1970. "Review of E. C. Pielou's *An Introduction to Mathematical Ecology.*" *American Scientist* 58: 699.

Cohen, J. E. 1971. "Mathematics as Metaphor." *Science* 172: 674–75.

Cohen, Michael P. 1984. *The Pathless Way: John Muir and American Wilderness.* Madison: University of Wisconsin Press.

Cole, Lamont C. 1954. "The Population Consequences of Life History Phenomena." *Quarterly Review of Biology* 29: 103–7.

Colinvaux, P. A. 1973. *Introduction to Ecology.* New York: John Wiley.

Collier, B. D., G. W. Cox, A. W. Johnson, and P. C. Miller. 1973. *Dynamic Ecology.* Englewood Cliffs, N.J.: Prentice-Hall.

Collins, James P. 1986. "Evolutionary Ecology and the Use of Natural Selection in Ecological Theory." *Journal of the History of Biology* 19: 257–88.

Colwell, T. B. 1970. "Some Implications of the Ecological Revolution for the Construction of Value." In E. Lazlo and J. B. Wilbur (eds.), *Human Values and Natural Science.* New York: Gordon and Breach, pp. 245–58.

Commoner, B. 1971. *The Closing Circle.* New York: Knopf.

Connell, J. H. 1983. "On the Prevalence and Relative Importance of Interspecific Competition: Evidence from Field Experiments." *American Naturalist* 129: 661–96.

Connell, J. H., and R. O. Slatyer. 1977. "Mechanisms of Succession in Natural Communities and Their Role in Community Stability and Organization." *American Naturalist* 111: 1119–44.

Conrad, M. 1979. "Hierarchical Adaptability Theory and Its Cross-Correlation with Dynamical Ecological Models." In E. Halfron (ed.), *Theoretical Systems Ecology.* New York: Academic Press, pp. 132–50.

Conrad, M. 1983. *Adaptability: The Significance of Variability from Molecule to Ecosystem.* New York: Plenum Press.

Cook, T. D., and D. T. Campbell. 1979. *Quasi-Experimentation: Design and Analysis Issues for Field Settings.* Chicago: Rand-McNally.

Coon, C. S., S. M. Garn, and J. B. Birdsell. 1950. *Races.* Springfield, Ohio: C. Thomas.

Cooper, W. S. 1926. "Fundamentals of Vegetational Change." *Ecology* 7: 391–413.

Copleston, Frederick. 1960. *A History of Philosophy.* Volume 1. London: Westminster.

Costa, R., and P. M. Bisol. 1978. "Genetic Variability in Deep-Sea Organisms." *Biological Bulletin* 155: 125–33.

Cousins, S. 1987. "Can We Count Natural Ecosystems?" *British Ecological Society Bulletin* 18: 156–58.

Cowles, Henry C. 1904. "The Work of the Year 1903 in Ecology." *Science* 19: 879–85.

Cowles, Henry C. 1911. "The Cause of Vegetative Cycles." *Botanical Gazette* 51: 161.

Cracraft, J. 1983. "Species Concepts and Speciation Analysis." In R. F. Johnson (ed.), *Current Ornithology.* Volume 1. New York: Plenum Press, pp. 159–187.

Crombie, A. C. 1950. "The Notion of Species in Medieval Philosophy and Science." *VIe Congrès international d'histoire des sciences, Actes:* 261–69.

Cronon, William. 1995 (1996). *Uncommon Ground: Rethinking the Human Place in Nature.* New York: W. W. Norton.

Crumley, Carole. 1987. "A Dialectical Critique of Hierarchy." In T. C. Patterson and C. W. Gailey (eds.), *Power Relations and State Formulation.* Philadelphia: American Anthropological Association.

Culver, D. C. 1975. "The Relationship between Theory and Experiment in Community Ecology." In S. A. Levin (ed.), *Ecosystem Analysis and Prediction.* Philadelphia: SIAM-SIMS, pp. 103–10.

Curtis, John T. 1955. "A Prairie Continuum in Wisconsin." *Ecology* 36: 558–66.

Curtis, John T. 1959. *The Vegetation of Wisconsin.* Madison: University of Wisconsin Press.

Curtis, John T., and R. P. McIntosh. 1951. "An Upland Continuum in the Prairie-Forest Border Region of Wisconsin." *Ecology* 32: 476–96.

Cushing, D. H. 1973. "The Dependence of Recruitment on Parent Stock." *J. Fish. Res. Bd. Canada* 30: 1965–76.

Cushing, D. H. 1977. *Science and the Fisheries.* London: Edward Arnold.

Cuvier, Georges. 1830. *Discours sur les révolutions de la surface du globe et sur les changements qu'elles ont produits dans le règne animal.* Third edition. Paris.

Cuvier, Georges. (1964). "Essay on Zoological Analogies." In *Georges Cuvier: Zoologist.* Cambridge, Mass.

Dalton, E. 1979. "The Case as Artifact." *Man and Medicine* 4: 15–17.

Danto, Arthur C. 1972. "Naturalism." In *The Encyclopedia of Philosophy.* Volume 5. New York: Macmillan and Free Press, pp. 448–50.

Darden, L., and N. Maull. 1977. "Interfield Theories." *Philosophy of Science* 44: 43–64.

Darling, Frank Frasier. 1937. *A Herd of Red Deer: A Study in Animal Behavior.* Oxford: Oxford University Press.

Darwin, Charles. 1859 (1975). *On the Origin of Species by Means of Natural Selection, or The Preservation of Favored Races in the Struggle for Life.* First edition. Cambridge: Harvard University Press.

Darwin, Charles. 1859 (1872). *On the Origin of Species.* Sixth edition. London: John Murray.

Darwin, Charles. 1880. "Sir Wyville Thomson and Natural Selection." *Nature* 23: 32.

Darwin, Charles. 1888. *The Life and Letters of Charles Darwin.* Volume 3. London.

Darwin, Charles. 1958. *The Autobiography of Charles Darwin.* Ed. Nora Barlow. New York: W. W. Norton.

Darwin, Charles. 1988. *The Correspondence of Charles Darwin.* Volume 4. New York: Cambridge University Press.

Darwin, Charles. 1991. *The Correspondence of Charles Darwin.* Volume 7. New York: Cambridge University Press.

Darwin, Charles. 1993. *The Correspondence of Charles Darwin.* Volume 8. New York: Cambridge University Press.

Davis, M. B. 1986. "Climatic Instability, Time Lags, and Community Disequilibrium." In J. Diamond and T. J. Case (eds.), *Community Ecology.* New York: Harper and Row, pp. 269–84.

Davitashvili, L. S. 1961. *Teoriya polovogo otbora* [Theory of sexual selection]. Moscow: Akademii Nauk.

Dawkins, Richard. 1986 (1996). *The Blind Watchmaker: Why the Evidence of Evolution Reveals a Universe without Design.* New York: W. W. Norton.

Dawson, W. R., J. D. Ligon, J. R. Murphy, J. P. Myers, D. Simberloff, and J. Verner. 1987. "Report of the Scientific Advisory Panel on the Spotted Owl." *Condor* 89: 205–29.

Dayton, Paul K. 1973. "Two Cases of Resource Partitioning in an Intertidal Community: Making the Right Prediction for the Wrong Reason." *American Naturalist* 107: 662–70.

Dayton, Paul K. 1979. "Ecology: A Science and a Religion." In R. J. Livingston (ed.), *Ecological Processes in Coastal and Marine Ecosystems.* New York: Plenum Press, pp. 3–18.

DeAngelis, D. L. 1975. "Stability and Connectance in Food Web Models." *Ecology* 56: 238–43.

DeAngelis, D. L., and R. A. Goldstein. 1978. "Criteria that Forbid a Large, Nonlinear Food-Web Model from Having More Than One Equilibrium Point." *Mathematical Biosciences* 41: 81–90.

DeAngelis, D. L., W. M. Post, and C. C. Travis. 1986. *Positive Feedback in Natural Systems.* New York: Springer-Verlag.

DeBach, P. 1974. *Biological Control by Natural Enemies.* London: Cambridge University Press.

Decker, C. A. 1975. *Journal of the International Society of Technology Assessment* 1: 5.

Dennett, Daniel C. 1996. *Darwin's Dangerous Idea: Evolution and the Meanings of Life.* New York: Simon and Schuster.

Depew, D. J., and B. H. Weber. 1994. *Darwinism Evolving: Systems Dynamics and the Genealogy of Natural Selection.* Cambridge: MIT Press.

Descartes, René. 1641 (1989). *Meditations on First Philosophy.* In *Descartes: Selected Philosophical Writings.* Trans. and ed. John Cottingham, Robert Stoothoff, and Dugald Murdoch. New York: Cambridge University Press.

de Vries, P. 1986. "The Discovery of Excellence." *Journal of Business Ethics* 5: 193–201.

Diamond, J. M. 1975. "Assembly of Species Communities." In *Ecology and Evolution of Communities.* Cambridge: Belknap Press of Harvard University Press, pp. 342–445.

Diamond, J. M., and T. J. Case. 1985. *Ecological Communities.* New York: Harper and Row.

Diamond, J. M., and T. J. Case (eds.). 1986. *Community Ecology.* New York: Harper and Row.

Dilman, I. 1973. *Induction and Deduction: A Study in Wittgenstein.* Oxford: Blackwell.

Dobzhansky, T. 1937. *Genetics and the Origin of Species.* New York: Columbia University Press.

Dobzhansky, T. 1956. "What Is an Adaptive Trait?" *American Naturalist* 90: 337–47.

Dobzhansky, T. 1968. "Adaptedness and Fitness." In R. Lewontin (ed.), *Population Biology and Evolution.* Syracuse: Syracuse University Press, pp. 109–21.

Drury, William H., and Ian C. T. Nisbet. 1973. "Succession." *Journal of the Arnold Arboretum* 54: 331–68.

Dunbar, M. J. 1972. "The Ecosystem as Unit of Natural Selection." *Transactions of the Connecticut Academy of Arts and Sciences* 44: 113–30.

Edelman, Gerald M. 1976. "Scientific Quests and Governmental Principles." *Science* 192: 99.

Edelson, M. 1988. *Psychoanalysis: A Theory in Crisis.* Chicago: University of Chicago Press.

Edson, Michael M., Theodore C. Foin, and Charles M. Knapp. 1981. "'Emergent Properties' and Ecological Research." *American Naturalist* 118: 593–96.

Egerton, Frank N. 1973. "Changing Concepts of the Balance of Nature." *Quarterly Review of Biology* 48: 322–50.

Egerton, Frank N. 1977. "A Bibliographic Guide to the History of General Ecology and Population Ecology." *History of Science* 15: 189–215.

Egerton, Frank N. 1983. "History of Ecology: Achievements and Opportunities, Part 1." *Journal of the History of Biology* 16: 259–310.

Ehrenfeld, David. 1993. *Beginning Again: People and Nature in the New Millennium.* New York: Oxford University Press.

Ehrenfels, C. von. 1890. "Über Gestaltqualitaten." *Vjschr. Wiss Phil.* 14: 149–92.

Eigen, M. 1971. "Self-Organization of Matter and the Evolution of Biological Macromolecules." *Naturwissenschaften* 58: 465–523.

Eldredge, Niles. 1985. *Unfinished Synthesis: Biological Hierarchies and Modern Evolutionary Thought.* Oxford: Oxford University Press.

Eldredge, Niles, and Stephen Gould. 1972 (1985). "Punctuated Equilibria: An Alternative to Phyletic Gradualism." In N. Eldredge, *Time Frames: The Rethinking of Darwinian Evolution and the Theory of Punctuated Equilibria.* New York: Simon and Schuster, pp. 193–223.

Eldridge, R. 1987. "Hypotheses, Criterial Claims, and Perspicuous Representations: Wittgenstein 'Remarks on Frazer's *The Golden Bough.*'" *Philosophical Investigations* 10: 226–45.

Elger, Frank E. 1947. "Arid Southeast Oahu Vegetation, Hawaii." *Ecological Monographs* 17: 383–435.

Elton, Charles S. 1927. *Animal Ecology.* New York: Macmillan.

Elton, Charles S. 1930. *Animal Ecology and Evolution.* Oxford: Oxford University Press.

Elton, Charles S. 1958. *The Ecology of Invasions by Animals and Plants.* London: Chapman and Hall.

Emerson, A. E. 1939. "Social Coordinations of the Superorganism." *American Midland Naturalist* 21: 182–209.

Engleberg, J., and L. L. Boyarsky. 1979. "The Noncybernetic Nature of Ecosystems." *American Naturalist* 114: 317–24.

Ervin, K. 1989. *Fragile Majesty.* Seattle: The Mountaineers.

Evernden, Neil. 1992. *The Social Creation of Nature.* Baltimore: Johns Hopkins University Press.

Falconer, D. S. 1973. "Replicated Selection for Body Weight in Mice." *Genetical Research* 22: 291–321.

Feibleman, J. K. 1954. "Theory of Integrative Levels." *British Journal for the Philosophy of Science* 5: 59–66.

Fenchel, Tom. 1989. "Comment on Carney's Article." *Functional Ecology* 3: 641.

Ferré, Frederick. 1988. *Philosophy of Technology.* Englewood Cliffs, N.J.: Prentice-Hall.

Ferré, Frederick. 1996. *Being and Value: Toward a Constructive Postmodern Metaphysics.* Albany: State University of New York Press.

Fetzer, J. 1974a. "A Single Case Propensity Theory of Explanation." *Synthese* 28: 171–98.

Fetzer, J. 1974b. "Statistical Probabilities: Single-Case Propensities versus Long-Run Frequencies." In W. Leinfellner and E. Kohler (eds.), *Development in the Methodology of Social Science.* Dordrecht: Reidel, pp. 387–97.

Fetzer, J. 1975. "On the Historical Explanation of Unique Events." *Theory and Decision* 6: 87–97.

Feyerabend, Paul. 1975. *Against Method.* London: Verso.

Findlay, Alexander. 1965. *A Hundred Years of Chemistry.* London: Duckworth.

Fisher, R. A. 1930. *The Genetical Theory of Natural Selection.* Oxford: Clarendon Press.

Fontaine, T. D. 1981. "A Self-Designing Model for Testing Hypotheses of Ecosystem Development." In D. M. Dubois (ed.), *Progress in Ecological Engineering and Management by Mathematical Modelling.* Proceedings of the Second International Conference on the State-of-the-Art in Ecological Modelling, Liège, Belgium, April 18–24, 1980. New York: Elsevier.

Forbes, Stephen A. 1880. "On Some Interactions of Organisms." *Bulletin of the Illinois State Laboratory of Natural History,* pp. 3–18.

Forbes, Stephen A. 1887 (1925). "The Lake as a Microcosm." *Illinois Natural History Survey Bulletin* 15: 537–550.

Ford, E. B. 1964. *Ecological Genetics.* London: Methuen.

Fortenbaugh, William W., Pamela M. Huby, Robert W. Sharples, and Dimitri Gutas (trans. and eds.). 1992. *Theophrastus of Eresus: Sources for His Life, Writings, Thought and Influence.* Part 2. New York: E. J. Brill.

Francoeur, R. T. 1984. "A Structured Approach to Teaching Decision-Making Skills in Biomedical Ethics." *Journal of Bioethics* 5: 145–54.

Franklin, A. 1986. *The Neglect of Experiment.* New York: Cambridge University Press.

Fraser, D. F. 1976. "Coexistence of Salamanders in the Genus *Plethodon:* A Variation of the Santa Rosalia Theme." *Ecology* 57: 238–51.

Freeland, W. J., and D. H. Jansen. 1974. "Strategies in Herbivory by Mammals: The Role of Plant Secondary Compounds." *American Naturalist* 108: 269–90.

Friederichs, K. 1930. *Die Grundfragen und Gesetzmässigkeiten der land und forstwirtschaftlichen Zoologie.* 2 volumes. Berlin: Verschlag Paul Parey.

Futuyma, D. J., and M. Slatkin. 1983. *Coevolution.* Sunderland, Mass.: Sinauer.

Gafiychuk, V. V., and R. E. Ulanowicz. 1996. "Self-Development and Distributed Self-Regulation in Dissipative Networks." Ref. No. CBL 96-010. Solomons, Md.: Chesapeake Biological Laboratory.

Gardner, M. R., and W. R. Ashby. 1970. "Connectance of Large, Dynamical (Cybernetic) Systems: Critical Values for Stability." *Nature* 228: 784.

Gause, George F. 1934 (1964). *The Struggle for Existence.* New York: Macmillan.

Georgescu-Roegen, N. 1971. *The Entropy Law and the Economic Process.* Cambridge: Harvard University Press.

Ghiselin, M. T. 1969. *The Triumph of the Darwinian Method.* Berkeley: University of California Press.

Ghiselin, M. T. 1987. "Species Concepts, Individuality, and Objectivity." *Biology and Philosophy* 2: 127–43.

Gibson, J. J. 1979. *The Ecological Approach to Visual Perception.* Boston: Houghton Mifflin.

Gillispie, C. C. 1959. "Lamarck and Darwin in the History of Science." In B. Glass, O. Temkin, and W. L. Straus (eds.), *Forerunners of Darwin: 1745–1859.* Baltimore: Johns Hopkins University Press, pp. 265–91.

Gilpin, M. E., and K. E. Justice. 1972. "Reinterpretation of the Invalidation of the Principle of Competitive Exclusion." *Nature* 36: 273–74.

Gini, A. R. 1985. "The Case Method." *Journal of Business Ethics* 4: 351–52.

Givnish, T. J., and G. J. Vermeij. 1976. "Sizes and Shapes of Liane Leaves." *American Naturalist* 110: 743–78.

Glacken, Clarence J. 1967. *Traces of the Rhodian Shore: Nature and Culture in Western Thought from Ancient Times to the End of the Eighteenth Century.* Berkeley: University of California Press.

Glansdorff, P., and I. Prigogine. 1971. *Thermodynamic Theory of Structure, Stability and Fluctuations.* London: John Wiley.

Gleason, Henry A. 1917. "The Structure and Development of the Plant Association." *Bulletin of the Torrey Botanical Club* 43: 463–81.

Gleason, Henry A. 1926. "The Individualistic Concept of the Plant Association." *Bulletin of the Torrey Botanical Club* 53: 7–26.

Gleason, Henry A. 1939. "The Individualistic Concept of the Plant Association." *American Midland Naturalist* 21: 92–110.

Gleick, James. 1987. *Chaos: Making a New Science.* New York: Viking Penguin.

Glymour, C. 1980. *Theory and Evidence.* Princeton: Princeton University Press.

Glynn, P. W. 1974. "The Impact of Acanthaster on Corals and Coral Reefs in the Eastern Pacific." *Environment and Conservation* 1: 245–304.

Goldstein, K. 1963. *The Organism.* New York: American Book.

Golley, Frank B. 1972. "Energy Flux in Ecosystems." In J. A. Wiens (ed.), *Ecosystem Structure and Function.* Corvallis: Oregon State University Press.

Golley, Frank B. 1993. *A History of the Ecosystem Concept in Ecology: More Than the Sum of the Parts.* New Haven: Yale University Press.

Golley, Frank B. 1998. *A Primer for Environmental Literacy.* New Haven: Yale University Press.

Gorovitz, S., and A. MacIntyre. 1976. "Toward a Theory of Medical Fallibility." *Journal of Medicine and Philosophy* 1: 51–71.

Gosselink, J. G., E. P. Odum, and R. M. Pope. 1974. *The Value of the Tidal Marsh.* Baton Rouge: Louisiana State University Center for Wetland Resources.

Goudge, T. A. 1972. "Pierre Teilhard de Chardin." In *The Encyclopedia of Philosophy.* Volume 8. New York: Macmillan and Free Press.

Gould, Stephen J. 1966. "Allometry and Size in Ontogeny and Phylogeny." *Biological Review* 41: 587–640.

Gould, Stephen J. 1971. "D'Arcy Thompson and the Science of Form." *New Literary History* 2: 229–58.

Gould, Stephen J. 1974. "Allometry in Primates, with Emphasis on Scaling and the Evolution of the Brain." *Approaches to Primate Paleobiology. Contrib. Primatol.* 5: 244–92.

Gould, Stephen J. 1977. *Ontogeny and Phylogeny.* Cambridge: Belknap Press of Harvard University Press.

Gould, Stephen J. 1978. "Sociobiology: The Art of Storytelling." *New Scientist* 80: 530–33.

Gould, Stephen J. 1981. *The Mismeasure of Man.* New York: W. W. Norton.

Gould, Stephen J. 1997. "Darwinian Fundamentalism." *New York Review of Books* 44 (10): 34–37.

Gould, Stephen J., and C. B. Calloway. 1980. "Clams and Brachiopods: Ships that Pass in the Night." *Paleobiology* 6: 383–96.

Gould, Stephen J., and Richard C. Lewontin. 1979. "The Spandrels of San Marco and the Panglossian Paradigm: A Critique of the Adaptationist Programme." *Proceedings of the Royal Society of London: Biological Sciences* 205: 581–98.

Graham, M. 1952. "Overfishing and Optimum Fishing." *Cons. Int. Explor. Mer Proc. Verb.* 132: 72–78.

Grene, Marjorie. 1963. *A Portrait of Aristotle.* Chicago: University of Chicago Press.

Grene, Marjorie. 1980. "A Note on Simberloff's 'Succession of Paradigms in Ecology.'" *Synthese* 43: 41–45.

Griffin, David R. 1993. Introduction to *Being and Value: Toward a Constructive Postmodern Metaphysics.* Albany: State University of New York Press, pp. xv–xviii.

Griffiths, P. E., and R. D. Knight. 1998. "What Is the Developmental Challenge?" *Philosophy of Science* 65: 253–58.

Grinnell, J. 1904. "The Origin and Distribution of the Chestnut-Backed Chickadee." *Auk* 2: 364–82.

Gruber, H. E., and P. H. Barrett (eds.). 1974. *Darwin on Man: A Psychological Study of Scientific Creativity, Together with Darwin's Early and Unpublished Notebooks.* New York: E. P. Dutton.

Grünbaum, A. 1984. *The Foundations of Psychoanalysis: A Philosophical Critique.* Berkeley: University of California Press.

Grünbaum, A. 1988. "The Role of the Case Study: Method in the Foundation of Psychoanalysis." *Canadian Journal of Philosophy* 18: 623–58.

Guba, E. G., and Y. S. Lincoln. 1981. *Effective Evaluation*. San Francisco: Jossey-Bass.

Gulland, J. A. 1979. "Dynamics of Populations and Fishery Management." *Inv. Pesq.* 43: 223–39.

Guthrie, W. K. C. 1962. *A History of Greek Philosophy*. Volume 1. Cambridge: Cambridge University Press.

Gutierrez, J. R., and A. B. Carey (eds.). 1985. "Ecology and Management of the Spotted Owl in the Pacific Northwest." U.S. Forest Service, General Technical Report 185.

Gutting, G. 1982. "Can Philosophical Beliefs Be Rationally Justified?" *American Philosophical Quarterly* 19: 315–30.

Haeckel, Ernst H. 1866a. *Allgemeine Entwickelungsgeschichte der Organismen oder Wissenschaft von den enstehenden organischen Formen*. Berlin: Ed. G. Reiner.

Haeckel, Ernst H. 1866b. *Generelle Morphologie der Organismen*. 2 volumes. Berlin: G. Reimer.

Haeckel, Ernst H. 1869 (1879). "Ueber Entwickelungsgang und Aufgabe der Zoologie." In *Gesammelte populäre Vorträge aus dem Gebiete der Entwickelungslehre*. Heft 2. Bonn, Germany: Strauss.

Haeckel, Ernst H. 1891. "Planktonic Studies: A Comparative Investigation of the Importance and Constitution of the Pelagic Fauna and Flora." Trans. G. W. Field. In *Report of United States Commissioner of Fish and Fisheries, 1889–1891*, pp. 565–641.

Hagen, Joel B. 1984. "Experimentalists and Naturalists in Twentieth-Century Botany: Experimental Taxonomy, 1920–1950." *Journal of the History of Biology* 17: 249–70.

Hagen, Joel B. 1989. "Research Perspectives and the Anomalous Status of Modern Ecology." *Biology and Philosophy* 4: 433–55.

Hagen, Joel B. 1992. *An Entangled Bank: The Origins of Ecosystem Ecology*. New Brunswick: Rutgers University Press.

Hairston, N. G., F. E. Smith, and L. B. Slobodkin. 1960. "Community Structure, Population Control and Competition." *American Naturalist* 94: 421–25.

Haken, H. 1988. *Information and Self-Organization*. Berlin: Springer-Verlag.

Hakim, Albert B. (ed.). 1997. *Historical Introduction to Philosophy*. Third edition. Upper Saddle River, N.J.: Prentice-Hall.

Haldane, J. B. S. 1932. *The Causes of Evolution*. London: Longmans, Green.

Haldane, J. S. 1884. "Life and Mechanism." *Mind* 9: 27–47.

Hall, C. A. S., and J. W. Day Jr. 1977. *Ecosystem Modeling in Theory and Practice*. New York: John Wiley.

Hantz, H. D. 1930. *The Biological Motivation in Aristotle*. New York.

Harner, Michael. 1977. "The Ecological Basis for Aztec Sacrifice." *American Ethnologist* 4: 117–35.

Harper, J. L. 1967. "A Darwinian Approach to Plant Ecology." *Journal of Ecology* 55: 247–70.

Harré, R. 1967. "History of the Philosophy of Science." In Paul Edwards (ed.), *The Encyclopedia of Philosophy*. Volume 6. New York: Macmillan and Free Press, pp. 289–96.

Harvey, W. 1651. *Exercitationes de Generatione Animalium.* London.

Haskill, Edward F. 1940. "Mathematical Systematization of 'Environment,' 'Organisms,' and 'Habitat.'" *Ecology* 21: 1–16.

Hassell, M. P. 1978. *The Dynamics of Arthropod Predator-Prey Associations.* Princeton: Princeton University Press.

Hawking, S. W. 1988. *A Brief History of Time: From the Big Bang to Black Holes.* New York: Bantam Books.

Heck, K. L. 1976. "Some Critical Considerations of the Theory of Species Packing." *Evolutionary Theory* 1: 247–58.

Heisenberg, Werner. 1927. "Über den anschaulichen Inhalt der quantentheoretischen Kinematik und Mechanik." *Zeitschrift für Physik* 43: 172–98.

Hempel, Carl, and Paul Oppenheim. 1948. "Studies in the Logic of Explanation." *Philosophy of Science* 15: 135–75.

Herodotus. (1862). *History.* Trans. G. Rawlinson. London: New Editions.

Himmelfarb, Gertrude. 1959. *Darwin and the Darwinian Revolution.* London: Chatto and Windus.

Hobbes, Thomas. 1651 (1985). *Leviathan.* C. B. MacPherson (ed.). New York: Penguin Books.

Hoering, W. 1980. "On Judging Rationality." *Studies in History and the Philosophy of Science* 11: 123–36.

Holling, C. S. 1966. "The Strategy of Building Models of Complex Ecological Systems." In K. E. F. Watt (ed.), *Systems Analysis in Ecology.* New York: Academic Press, pp. 195–214.

Holling, C. S. 1973. "Resilience and Stability of Ecological Systems." *Annual Review of Ecology and Systematics* 4: 1–23.

Horn, H. S. 1974. "The Ecology of Secondary Succession." *Annual Review of Ecology and Systematics* 5: 25–37.

Howe, H. F. 1984. "Constraints on the Evolution of Mutualisms." *American Naturalist* 123: 764–77.

Hoyningen-Huene, P., and F. M. Wuketits. 1989. "Epistemological Reductionism in Biology: Intuitions, Explications, and Objections." In P. Hoyningen-Huene and F. M. Wuketits (eds.), *Reductionism and Systems Theory in the Life Sciences.* Dordrecht: Kluwer, pp. 29–44.

Huberman, B. A. (ed.). 1988. *The Ecology of Computation.* Amsterdam: North-Holland.

Hull, David L. 1965a. "The Effect of Essentialism on Taxonomy: Two Thousand Years of Stasis. Part 1." *British Journal for the Philosophy of Science* 15: 314–26.

Hull, David L. 1965b. "The Effect of Essentialism on Taxonomy: Two Thousand Years of Stasis. Part 2." *British Journal for the Philosophy of Science* 16: 1–18.

Hull, David L. 1967a. "Certainty and Circularity in Evolutionary Taxonomy." *Evolution* 21: 174–89.

Hull, David L. 1967b. "The Metaphysics of Evolution." *British Journal for the History of Science* 3: 309–37.

Hull, David L. 1974. *The Philosophy of Biological Science.* Englewood Cliffs, N.J.: Prentice-Hall.

Hull, David L. 1976a." Are Species Really Individuals?" *Systematic Zoology* 25: 174–91.

Hull, David L. 1976b. "Informal Aspects of Theory Reduction." In R. S. Cohen, C. A. Hooker, A. C. Michalos, and J. W. Van Evra (eds.), *PSA 1975*. Dordrecht: Reidel pp. 653–70.

Hull, David L. 1978. "A Matter of Individuality." *Philosophy of Science* 45: 335–60.

Hull, David L. 1988. *Science as a Process: An Evolutionary Account of the Social and Conceptual Development of Science*. Chicago: University of Chicago Press.

Hull, David L. 1993. "Testing Claims about Science." In David Hull, Mickey Forbes, and Kathleen Okruhlik (eds.), *Proceedings of the 1992 Biennial Meeting of the Philosophy of Science Association*. Volume 2. East Lansing, Mich.: Philosophy of Science Association.

Hume, David. 1740 (1992). *A Treatise of Human Nature*. Ed. P. H. Nidditch. New York: Oxford University Press.

Hume, David. (1990). *Dialogues Concerning Natural Religion*. Martin Bell (ed.). London: Penguin Books.

Humphreys, P. 1989. "Scientific Explanation: The Causes, Some of the Causes, and Nothing but the Causes." In P. Kitcher and W. C. Salmon (eds.), *Scientific Explanation*. Minnesota Studies in the Philosophy of Science 13. Minneapolis: University of Minnesota Press, pp. 283–306.

Humphreys, P. 1991. *The Chances of Explanation: Causal Explanation in the Social, Medical, and Physical Sciences*. Princeton: Princeton University Press.

Hutchinson, G. Evelyn. 1951. "Copepodology for the Ornithologist." *Ecology* 32: 571–77.

Hutchinson, G. Evelyn. 1958a. "Homage to Santa Rosalia, or Why Are There So Many Kinds of Animals?" *American Naturalist* 93: 145–59.

Hutchinson, G. Evelyn. 1958b. Concluding Remarks. *Cold Spring Harbor Symposium on Quantitative Biology* 22: 415–27.

Hutchinson, G. Evelyn. 1965. *The Ecological Theater and the Evolutionary Play*. New Haven: Yale University Press.

Hutchinson, G. Evelyn. 1967. "A Treatise on Limnology." In *Introduction to Lake Biology and the Limno-Plankton*. New York: John Wiley.

Hutchinson, G. Evelyn. 1978. *An Introduction to Population Ecology*. New Haven: Yale University Press.

Innis, G. S., and R. V. O'Neill. 1979. *Systems Analysis of Ecosystems*. Fairland, Md.: International Cooperative Publishing House.

International Whaling Commission. 1978. *Report of the International Whaling Commission 28*. Cambridge: International Whaling Commission.

International Whaling Commission. 1979. *Report of the International Whaling Commission 29*. Cambridge: International Whaling Commission.

International Whaling Commission. 1980. *Report of the International Whaling Commission 30*. Cambridge: International Whaling Commission.

Istock, C. A. 1984. "Boundaries to Life History Variation and Evolution." In P. W. Price, C. N. Slobodchikoff, and W. S. Gaud (eds.), *A New Ecology*. New York: John Wiley, pp. 143–68.

Jackson, J. B. C. 1981. "Interspecific Competition and Species' Distributions: The Ghosts of Theories and Data Past." *American Zoologist* 21: 889–901.

Jacob, F. 1977. "Evolution and Tinkering." *Science* 196: 1161–66.

Jeffers, Robinson. 1938. *The Selected Poetry of Robinson Jeffers.* New York: Random House.

Jerison, H. J. 1973. *Evolution of the Brain and Intelligence.* New York: Academic Press.

Johannes, R. E., et al. 1972. "The Metabolism of Some Coral Reef Communities: A Team Study of Nutrient and Energy Flux at Eniwetok." *BioScience* 22: 541.

Johnson, P. L. (ed.). 1977. *An Ecosystem Paradigm for Ecology.* Oak Ridge, Tenn.: Oak Ridge Associated Universities.

Jones, Clive G., and J. M. Lawton (eds.). 1994. *Linking Species and Ecosystems.* New York: Chapman and Hall.

Jørgensen, S. E., and H. F. Mejer. 1977. "Ecological Buffer Capacity." *Ecological Modelling* 3: 39–61.

Jørgensen, S. E., and H. F. Mejer 1981. "Exergy as Key Function in Ecological Models." In W. J. Mitsch, R. W. Bosserman, and J. M. Klopatek (eds.), *Energy and Ecological Modeling.* Proceedings of a Symposium, Louisville, Kentucky, April 20–23. New York: Elsevier, pp. 587–90.

Kampis, G. 1991. *Self-Modifying Systems in Biology and Cognitive Science: A New Framework for Dynamics, Information, and Complexity.* Oxford: Pergamon Press.

Kant, Immanuel. 1781 (1965). *Critique of Pure Reason.* Unabridged edition. Trans. Norman Kemp Smith. New York: St. Martin's Press.

Kaplan, A. 1964. *The Conduct of Inquiry: Methodology for Behavioral Science.* San Francisco: Chandler.

Kauffman, Stuart. 1995. *At Home in the Universe: The Search for the Laws of Self-Organization and Complexity.* New York: Oxford University Press.

Keller, David R. 1995. "Self in Nature: Centers of Value for Environmental Ethics." Ph.D. diss., University of Georgia, Athens.

Keller, David R. 1997a. "Gleaning Lessons from Deep Ecology." *Ethics and the Environment* 2: 139–48.

Keller, David R. 1997b. "Deleuze's Ecological Philosophy of Self." *Encyclia* 74: 71–85.

Kerner von Marilaun, Anton. 1863 (1951). *The Plant Life of the Danube Basin.* Trans. H. S. Conrad. Ames: Iowa State University Press.

Kidder, T. 1981. *The Soul of a New Machine.* Boston: Little, Brown.

Kiester, A. 1982. "Natural Kinds, Natural History, and Ecology." In E. Saarinen (ed.), *Conceptual Issues in Ecology.* Boston: Reidel, pp. 345–56.

King, Roger J. H. 1990. "How to Construe Nature: Environmental Ethics and the Interpretation of Nature." *Between the Species* 6: 101–8.

Kingsland, Sharon E. 1985. *Modeling Nature: Episodes in the History of Population Ecology.* Chicago: University of Chicago Press.

Kingsland, Sharon E. 1991. "Defining Ecology as a Science." In Leslie A. Real and James H. Brown (eds.), *Foundations of Ecology: Classic Papers with Commentaries.* Chicago: University of Chicago Press.

Kitcher, P. 1985a. *Species.* Cambridge: MIT Press.

Kitcher, P. 1985b. "Two Approaches to Explanation." *Journal of Philosophy* 82: 632–39.

Knapp, A. K., and T. R. Seastedt. 1986. "Detritus Accumulation Limits Productivity of Tallgrass Prairie." *BioScience* 36: 662–68.

Kodric-Brown, Astrid, and James H. Brown. 1984. "Truth in Advertising: The Kinds of Traits Favored by Sexual Selection." *American Naturalist* 124: 309–23.

Koehler, W. 1929. *The Gestalt Psychology.* New York: Liveright.

Koestler, A. 1967. *The Ghost in the Machine.* New York: Macmillan.

Koestler, A., and J. R. Smythies. 1969. *Beyond Reductionism: New Perspectives in the Life Sciences.* London: Macmillan.

Koffka, K. 1935. *Principles of Gestalt Psychology.* New York: Harcourt Brace.

Kormondy, Edward J. 1974. "Natural and Human Ecosystems." In Frederick Sargent II (ed.), *Human Ecology.* Amsterdam: North-Holland, pp. 27–43.

Krebs, C. J. 1972 (1978). *Ecology: The Experimental Analysis of Distribution and Abundance.* New York: Harper and Row.

Krebs, J. R., D. W. Stephens, and W. J. Sutherland. 1983. "Perspectives in Optimal Foraging." In A. B. Brush and G. A. Clark Jr. (eds.), *Perspectives in Ornithology: Essays Presented for the Centennial of the American Ornithologists' Union.* Boston: Cambridge University Press, pp. 165–216.

Kripke, S. 1982. "Wittgenstein on Rules and Private Language." In I. Black (ed.), *Perspective on the Philosophy of Wittgenstein.* Oxford: Blackwell, pp. 239–96.

Kuhn, Thomas S. 1962 (1970). *The Structure of Scientific Revolutions.* Chicago: University of Chicago Press.

Kuhn, Thomas S. 1977. *Essential Tension: Selected Studies in Scientific Tradition and Change.* Chicago: University of Chicago Press.

Kulesza, George. 1975. "Comment on 'Niche, Habitat, and Ecotope.'" *American Naturalist* 109: 476–79.

Lack, David. 1944. "Ecological Aspects of Species-Formation in Passerine Birds." *Ibis* 86: 260–86.

Lack, David. 1948. "Natural Selection and Family Size in the Starling." *Evolution* 2: 95–110.

Lack, David. 1966. *Population Studies of Birds.* Oxford: Clarendon Press.

Lack, David. 1968. *Ecological Adaptations for Breeding in Birds.* London: Methuen.

Lamarck, Jean-Baptiste. 1806. *Système des Animaux sans vertèbres.* Paris.

Lamarck, Jean-Baptiste. 1809 (1963). *Zoological Philosophy: An Exposition with Regard to the Natural History of Animals.* Trans. Hugh Elliot. New York: Hafner.

Lande, R. 1976. "Natural Selection and Random Genetic Drift in Phenotypic Evolution." *Evolution* 30: 314–34.

Lande, R. 1978. "Evolutionary Mechanisms of Limb Loss in Tetrapods." *Evolution* 32: 73–92.

Lane, P. A., G. H. Lauff, and R. Levins. 1975. "The Feasibility of Using a Holistic Approach in Ecosystem Analysis." In S. A. Levin (ed.), *Ecosystem Analysis and Prediction.* Philadelphia: SIAM-SIMS, pp. 111–28.

Langton, Christopher G. 1989. *Artificial Life.* Santa Fe Institute Studies in the Sciences of Complexity 16. Redwood City, Calif.: Addison-Wesley.

Laszlo, Ervin. 1971. "Systems Philosophy." *Main Currents in Modern Thought* 28: 55–60.

Leibniz, Gottfried W. 1714 (1965). *Monadology.* Trans. Paul Schrecker and Anne Martin Schrecker. Indianapolis: Bobbs-Merrill.

Leigh, E. 1971. "The Energy Ethic." *Science* 172: 644.

Leopold, Aldo. 1949 (1987). *A Sand County Almanac and Sketches Here and There.* New York: Oxford University Press.

Lerner, I. M. 1963. "Synthesis." In S. J. Geerts (ed.), *Genetics Today.* Volume 2. New York: Macmillan, pp. 489–94.

Leslie, P. H. 1958. "A Stochastic Model for Studying the Properties of Certain Biological Systems by Numerical Methods." *Biometrika* 45: 16–31.

Leslie, P. H. 1962. "A Stochastic Model for Two Competing Species of *Triobolium* and Its Application to Some Experimental Data." *Biometrika* 49: 1–25.

Leslie, P. H., and J. C. Gower. 1958. "The Properties of a Stochastic Model for Two Competing Species." *Biometrika* 45: 316–30.

Levandowsky, M. 1977. "A White Queen Speculation." *Quarterly Review of Biology* 52: 383–86.

Levin, S. A. 1970. "Community Equilibria and Stability, an Extension of the Competitive Exclusion Principle." *American Naturalist* 104: 413–23.

Levin, S. A. 1974. "Dispersion and Population Interactions." *American Naturalist* 108: 207–28.

Levin, S. A., and R. T. Paine. 1974. "Disturbance, Patch Formation, and Community Structure." *Proceedings of the National Academy of Sciences* 71: 2744–47.

Levins, Richard. 1966. "The Strategy of Model Building in Population Biology." *American Science* 54: 421–31.

Levins, Richard. 1968a. "Ecological Engineering: Theory and Technology." *Quarterly Review of Biology* 43: 301–5.

Levins, Richard. 1968b. *Evolution in Changing Environments: Some Theoretical Explorations.* Princeton: Princeton University Press.

Levins, Richard. 1974. "The Qualitative Analysis of Partially Specified Systems." *Annals of the New York Academy of Sciences* 571: 2744–47.

Levins, Richard. 1975. "Evolution in Communities near Equilibrium." In M. L. Cody and J. M. Diamond (eds.), *Ecology and Evolution of Communities.* Cambridge: Harvard University Press, pp. 16–50.

Levins, Richard, and Richard C. Lewontin. 1980. "Dialectics and Reductionism in Ecology." *Synthese* 43: 47–78.

Levy, H. 1932. *The Universe of Science.* London.

Lewin, R. 1984. "Why Is Development So Illogical?" *Science* 224: 1327–29.

Lewontin, Richard C. 1969. "The Meaning of Stability." In G. M. Woodwell and H. H. Smith (eds.), *Diversity and Stability in Ecological Systems.* Brookhaven, N.Y.: Brookhaven National Laboratory, pp. 13–24.

Lewontin, Richard C. 1970. "The Units of Selection." *Review of Ecology and Systematics* 54: 421–31.

Lewontin, Richard C. 1974. "Darwin and Mendel: The Materialist Revolution." In J. Neyman (ed.), *The Heritage of Copernicus: Theories "More Pleasing to the Mind."* Cambridge: MIT Press.

Lewontin, Richard C. 1978. "Adaptation." *Scientific American* 239: 156–69.

Lewontin, Richard C. 1979. "Sociobiology as an Adaptationist Program." *Behavioral Science* 24: 5–14.

Lewontin, Richard C. 1983. "The Corpse in the Elevator." *New York Review of Books* 29 (January 20): 34–37.

Lewontin, Richard C., and R. Levins. 1976. "The Problem of Lysenkoism." In H. Rose and S. Rose (eds.), *The Radicalisation of Science*. London: Macmillan, pp. 32–64.

Liddell, Henry G., and Robert Scott. 1968. *A Greek-English Lexicon*. Oxford: Clarendon Press.

Lincoln, Y. S., and G. Guba. 1985. *Naturalistic Inquiry*. Newbury Park, Calif.: Sage.

Lindeman, Raymond L. 1942. "The Trophic-Dynamic Aspect of Ecology." *Ecology* 23: 399–418.

Linnaeus, Carolus. 1744. *Oratio de telluris habitabilis incremento*. Leiden: Cornelium Haak.

Linnaeus, Carolus. 1749. *Specimen academicum de oeconomia naturae*. Upsala, I. J. Biberg, respondent.

Linnaeus, Carolus. (1964). *Systema Naturae*. Trans. M. S. J. Engel-Lederboer and E. Nieuwkoop. Leyden.

Locke, John. 1690 (1992). "Of Property." *Second Treatise of Government*. In Michael Morgan (ed.), *Classics of Moral and Political Theory*. Indianapolis: Hackett.

Locy, William A. 1925. *The Growth of Biology*. New York: Henry Holt.

Loehle, Craig. 1983. "Evaluation of Theories and Calculation Tools in Ecology." *Ecological Modelling* 19: 239–47.

Loehle, Craig. 1988a. "Philosophical Tools: Potential Contribution to Ecology." *Oikos* 51: 97–104.

Loehle, Craig. 1988b. "Tree Life History Strategies: The Role of Defenses." *Can. J. For. Res.* 18: 209–22.

Loehle, Craig, and Joseph H. K. Pechmann. 1988. "Evolution: The Missing Ingredient in Systems Ecology." *American Naturalist* 132: 884–99.

Loehle, Craig, and L. R. Rittenhouse. 1982. "An Analysis of Forage Preference Indices." *Journal of Range Management* 35: 316–19.

Lopez, Barry. 1990. *The Rediscovery of North America*. Lexington: University Press of Kentucky.

Lotka, Alfred J. 1925 (1956). *Elements of Mathematical Biology*. New York: Dover.

Lovejoy, Arthur O. 1927. "The Meanings of 'Emergence' and Its Modes." In Edgar S. Brightman (ed.), *Proceedings of the Sixth International Congress of Philosophy*. New York: Longmans, Green, pp. 20–33.

Lovejoy, Arthur O. 1936 (1964). *The Great Chain of Being: A Study of the History of an Idea*. Cambridge: Harvard University Press.

Lovelock, James E., and Lynn Margulis 1974. "Atmospheric Homeostasis by and for the Biosphere: The Gaia Hypothesis." *Tellus* 26: 1–10.

Lubashevskii, I. A., and V. V. Gafiychuk. 1995. "A Simple Model of Self-Regulation in Large Natural Hierarchical Systems." *J. Env. Syst.* 23(3):281–89.

MacArthur, Robert H. 1955. "Fluctuations of Animal Populations and a Measure of Community Stability. *Ecology* 36: 533–36.

MacArthur, Robert H. 1960. "On the Relative Abundance of Species." *American Naturalist* 94: 25–36.

MacArthur, Robert H. 1962. "Review of Slobodkin's *Growth and Regulation of Animal Populations.*" *Ecology* 43: 579.

MacArthur, Robert H. 1966. "A Note on Mrs. Pielou's Comments." *Ecology* 47: 1074.

MacArthur, Robert H. 1969. "Species Packing, and What Interspecies Competition Minimizes." *Proceedings of the National Academy of Sciences (USA)* 64: 1369–71.

MacArthur, Robert H. 1970. "Species Packing and Competitive Equilibrium for Many Species." *Theoret. Pop. Biol.* 1: 1–11.

MacArthur, Robert H. 1972. *Geographical Ecology.* New York: Harper and Row.

MacArthur, Robert H., and Richard Levins. 1967. "The Limiting Similarity, Convergence, and Divergence of Coexisting Species." *American Naturalist* 101: 377–85.

MacArthur, Robert H., and Edward O. Wilson 1963. "An Equilibrium Theory of Insular Zoogeography." *Evolution* 17: 373–87.

MacArthur, Robert H., and Edward O. Wilson 1967. *The Theory of Island Biogeography.* Princeton: Princeton University Press.

MacFadyen, Amyan. 1975. "Some Thoughts on the Behavior of Ecologists." *Journal of Animal Ecology* 44: 351–63.

MacFarlane, A. I. 1981. "Dynamic Structure Theory: A Structural Approach to Social and Biological Systems." *Bull. Math. Biol.* 43: 579–91.

Malthus, Thomas. 1798 (1926). *Essay on Population.* Ed. James Bonar. London.

Margalef, Ramón D. 1963. "On Certain Unifying Principles in Ecology." *American Naturalist* 97: 357–74.

Margalef, Ramón D. 1968. *Perspectives in Ecological Theory.* Chicago: University of Chicago Press.

Marx, Leo. 1970. "American Institutions and Ecological Ideals." *Science* 170: 945–52.

Mason, Herbert L. 1947. "Evolution of Certain Floristic Associations in Western North America." *Ecological Monographs* 17: 201–10.

Matsuno, K. 1986. "From Physics to Biology and Back." Part I. *Revista di Biologia* 79: 269–85.

Matsuno, K., and S. N. Salthe. 1995. "Global Idealism/Local Materialism." *Biology and Philosophy* 10: 309–37.

Maturana, H. R., and F. J. Varela. 1980. *Autopoiesis and Cognition: The Realization of the Living.* Dordrecht: Reidel.

May, Robert M. 1972. "Will a Large Complex System Be Stable?" *Nature* 238: 413–14.

May, Robert M. 1973a. *Stability and Complexity in Model Ecosystems.* Princeton: Princeton University Press.

May, Robert M. 1973b. "Time-Delay versus Stability in Population Models with Two to Three Trophic Levels." *Ecology* 54: 315–25.

May, Robert M. 1974. "Biological Populations with Nonoverlapping Generations: Stable Points, Stable Cycles, and Chaos." *Science* 186: 645–47.

May, Robert M. 1975. "Patterns of Species Abundance and Diversity." In M. L. Cody and J. M. Diamond (eds.), *Ecology and Evolution of Communities.* Cambridge: Harvard University Press, pp. 81–120.

May, Robert M. 1976. "Models for Single Populations." In R. May (ed.), *Theoretical Ecology Principles and Applications.* Philadelphia: W. B. Saunders, pp. 4–25.

May, Robert M. 1979. "The Structure and Dynamics of Ecological Communities. *Symp. Br. Ecol. Soc.* 20: 385–407.

May, Robert M. 1980. "Mathematical Models in Whaling and Fisheries Management." In Y. F. Oster (ed.), *Some Mathematical Questions in Biology.* Volume 13. Providence, R.I.: American Mathematical Society, pp. 1–64.

May, Robert M. 1981a. "The Role of Theory in Ecology." *American Zoologist* 21: 903–10.

May, Robert M. 1981b. *Theoretical Ecology: Principles and Applications.* Second edition. Sunderland, Mass.: Sinauer.

May, Robert M., J. R. Beddington, C. W. Clark, S. J. Holt, and R. M. Laws. 1979. "Management of Multi-Species Fisheries." *Science* 205: 267–77.

May, Robert M., J. R. Beddington, J. W. Horwood, and J. G. Shepherd. 1978. "Exploiting Natural Populations in an Uncertain World." *Mathematical Biosciences* 42: 219–52.

May, Robert M., and G. F. Oster. 1976. "Bifurcations and Dynamic Complexity in Simple Ecological Models." *American Naturalist* 110: 573–99.

Maynard Smith, J. 1964. "Group Selection and Kind Selection: A Rejoinder." *Nature* 201: 1145–47.

Maynard Smith, J. 1976. "Group Selection." *Quarterly Review of Biology* 51: 277–83.

Maynard Smith, J. 1982. *Evolution and the Theory of Games.* Cambridge, Mass.: Cambridge University Press.

Mayr, Ernst. 1942. *Systematics and the Origin of Species: From the Viewpoint of a Zoologist.* New York: Columbia University Press.

Mayr, Ernst. 1959a. "Agassiz, Darwin, and Evolution." *Harvard Library Bulletin* 13: 165–94.

Mayr, Ernst. 1959b. "Review of Himmelfarb's 'Darwin and the Darwinian Revolution.'" *Scientific American* 201: 215.

Mayr, Ernst. 1959c. "Where Are We?" *Cold Spring Harbor Symposium of Quantitative Biology* 24: 409–40.

Mayr, Ernst. 1960. "The Emergence of Evolutionary Novelties." In S. Tax (ed.), *The Evolution of Life.* Chicago: University of Chicago Press, pp. 349–80.

Mayr, Ernst. 1961. "Cause and Effect in Biology." *Science* 134: 1501–6.

Mayr, Ernst. 1962. "Accident or Design, the Paradox of Evolution." In *The Evolution of Living Organisms: A Symposium of the Royal Society of Victoria, held in Melbourne, December, 1959.* Melbourne: Melbourne University Press, pp. 1–14.

Mayr, Ernst. 1963. *Animal Species and Evolution.* Cambridge: Harvard University Press.

Mayr, Ernst. 1966. (1975). Introduction to *On the Origin of Species.* Cambridge: Harvard University Press.

Mayr, Ernst. 1970. *Population, Species, and Evolution.* Cambridge: Harvard University Press.

Mayr, Ernst. 1975. "The Unity of the Genotype." *Biologisches Zentralblatt* 94: 377–88.

Mayr, Ernst. 1982a. "Adaptation and Selection." *Biologisches Zentralblatt* 101: 66–77.

Mayr, Ernst. 1982b. *The Growth of Biological Thought: Diversity, Evolution, and Inheritance.* Cambridge: Harvard University Press.

Mayr, Ernst. 1983. "How to Carry out the Adaptationist Program?" *American Naturalist* 121: 324–34.

Mayr, Ernst. 1987. "The Ontological Status of Species: Scientific Progress and Philosophical Terminology." *Biology and Philosophy* 2: 145–66.

Mayr, Ernst. 1988. *Toward a New Philosophy of Biology*. Cambridge: Harvard University Press.

Mayr, Ernst. 1991. *One Long Argument: Charles Darwin and the Genesis of Modern Evolutionary Thought*. Cambridge: Harvard University Press.

Mayr, Ernst. 1992. "The Idea of Teleology." *Journal of the History of Ideas* 53(1): 117–77.

McCoy, Earl D., H. R. Mushinsky, and D. S. Wilson. 1993. "Pattern in the Compass Orientation of Gopher Tortoise Burrows at Different Spatial Scales." *Global Ecology and Biogeography Letters* 3.

McEvoy, A. F. 1986. *The Fisherman's Problem: Ecology and Law in the California Fisheries*. Cambridge: Cambridge University Press.

McIntosh, Robert P. 1975. "H. A. Gleason: Individualistic Ecologist 1882–1975: His Contributions to Ecology." *Bulletin of the Torrey Botanical Club* 102: 253–73.

McIntosh, Robert P. 1976. "Ecology since 1900." In B. J. Taylor and T. J. White (eds.), *Issues and Ideas in America*. Norman: University of Oklahoma Press, pp. 353–72.

McIntosh, Robert P. 1980. "The Background and Some Current Problems for Theoretical Ecology." *Synthese* 43: 195–255.

McIntosh, Robert P. 1985. *The Background of Ecology: Concept and Theory*. New York: Cambridge University Press.

McKibben, Bill. 1989. *The End of Nature*. New York: Random House.

McNaughton, S. J. 1977. "Diversity and Stability of Ecological Communities: A Comment on the Role of Empiricism in Ecology." *American Naturalist* 111: 515–25.

Medawar, P. 1974. "A Geometric Model of Reduction and Emergence." In F. J. Ayala and T. Dobzhansky (eds.), *Studies in the Philosophy of Biology: Reduction and Related Problems*. Berkeley: University of California Press, pp. 57–63.

Meehl, P. E. 1983. "Subjectivity in Psychoanalytic Inference." In J. Earman (ed.), *Testing Scientific Theories*. Minneapolis: University of Minnesota Press, pp. 349–411.

Mendel, Gregor. 1866 (1965). *Experiments in Plant Hybridisation*. Trans. Ronald A. Fisher, ed. J. H. Bennett. London: Oliver and Boyd.

Merchant, Carolyn. 1980 (1990). *The Death of Nature: Women, Ecology and the Scientific Revolution*. San Francisco: Harper Collins.

Merriam, S. 1988. *Case Study Research in Education: A Qualitative Approach*. San Francisco: Jossey-Bass.

Merton, R. K. 1973. *The Sociology of Science*. Chicago: University of Chicago Press.

Mill, John Stuart. 1861 (1974). *Utilitarianism*. Ed. Mary Warnock. New York: Meridian Books.

Miller, J. G. 1978. *Living Systems*. New York: McGraw-Hill.

Mitchell, G. 1986. "Vampire Bat Control in Latin America." In G. Orians (ed.), *Ecological Knowledge and Environmental Problem Solving: Concepts and Case Studies*. Washington, D.C.: National Academy Press, pp. 151–64.

Mitchell, Rodger, Ramona A. Mayer, and Jerry Downhomer. 1976. "An Evaluation of Three Biome Programs." *Science* 192: 859.

Möbius, Karl A. 1877 (1881). "The Oyster and Oyster-Culture." Trans. H. J. Rice. In *Fish* and Fisheries, Annual Report of the Commission for the Year 1880. Volume 3, appendix H. Documents of the Senate of the United States for the Third Session of the Forty-sixth Congress and the Special Session of the Forty-seventh Congress, 1880–1881, pp. 721–24.

Montgomery, E. 1882. "Are We Cell-Aggregates?" Mind 7: 100–107.

Morton, E. S., M. S. Geitgey, and S. McGrath. 1978. "On Bluebird Responses to Apparent Female Adultery." American Naturalist 112: 968–71.

Mowry, B. 1985. "From Galen's Theory to William Harvey's Theory: A Case Study in the Rationality of Scientific Theory Change." Studies in History and Philosophy of Science 16: 49–82.

Muir, John. 1867 (1992). A Thousand Mile Walk to the Gulf. New York: Penguin.

Muller, H. J., et al. 1949. "Natural Selection and Adaptation." Proceedings of the American Philosophical Society 93: 459–519.

Munson, R. 1971. "Biological Adaptation." Philosophy and Science 38: 200–15.

Naess, Arne. 1973. "The Shallow and the Deep, Long-Range Ecology Movement: A Summary." Inquiry 16: 95–100.

Naess, Arne. 1981. "The Primacy of the Whole." In Holism and Ecology: Working Papers for the Project on Goals, Processes and Indicators of Development (GPID). Japan: United Nations University, pp. 1–10.

Naess, Arne. 1988. "Deep Ecology and Ultimate Premises." Ecologist 18: 128–31.

Naess, Arne. 1989 (1993). Ecology, Community, and Lifestyle: Outline of an Ecosophy. Trans. and ed. David Rothenberg. New York: Cambridge University Press.

Nagel, Ernest. 1961. The Structure of Science: Problems in the Logic of Scientific Explanation. New York: Harcourt, Brace and World.

Needham, J. 1936. Order and Life. New Haven: Yale University Press.

Needham, J. (1937). Integrative Levels: A Revaluation of the Idea of Progress. Oxford: Clarendon Press.

Needham, J. G. 1904. "Is the Course for College Entrance Requirements Best for Those Who Go No Further?" Science 19: 650–56.

Neill, W. E. 1974. "The Community Matrix and Interdependence of the Competition Coefficients." American Naturalist 108: 399–408.

Newell, R. 1986. Objectivity, Empiricism, and Truth. New York: Routledge and Kegan Paul.

Nicholson, A. J. 1933. "The Balance of Animal Populations." Journal of Animal Ecology 2 (suppl.): 132–78.

Nietzsche, Friedrich. 1887 (1989). On the Genealogy of Morals. Trans. Walter Kaufmann and R. J. Hollingdale. New York: Vintage.

Nietzsche, Friedrich. 1888 (1968). The Will to Power. Trans. Walter Kaufmann and R. J. Hollingdale. New York: Vintage.

Noble, I. R., and R. O. Slatyer. 1980. "The Use of Vital Attributes to Predict Successional Changes in Plant Communities Subject to Recurrent Disturbances." Vegetatio 43: 5–21.

Norse, E. A. 1990. Ancient Forests of the Pacific Northwest. Washington, D.C.: Island Press.

Novikoff, A. B. 1945. "The Concept of Integrative Levels and Biology." Science 101: 209–15.

Odum, Eugene P. 1962. "Relationships between Structure and Function in Ecosystems." *Japanese Journal of Ecology* 12: 108–18.

Odum, Eugene P. 1964. "The New Ecology." *BioScience* 14: 14–16.

Odum, Eugene P. 1969. "The Strategy of Ecosystem Development." *Science* 164: 262–70.

Odum, Eugene P. 1975. *Ecology: The Link between the Natural and the Social Sciences.* Second edition. New York: Holt, Rinehart and Winston.

Odum, Eugene P. 1977. "The Emergence of Ecology as a New Integrative Discipline." *Science* 195: 1289–93.

Odum, Eugene P. 1983. *Basic Ecology.* New York: W. B. Saunders.

Odum, Eugene P. 1986. "Introductory Review: Perspectives of Ecosystem Theory and Application." In Nicholas Polunin (ed.), *Ecosystem Theory and Application.* New York: John Wiley, pp. 1–11.

Odum, Eugene P. 1989 (1993). *Ecology and Our Endangered Life-Support Systems.* Sunderland, Mass.: Sinauer.

Odum, Eugene P., and L. J. Biever. 1984. "Resource Quality, Mutualism, and Energy Partitioning in Food Chains." *American Naturalist* 124: 360–76.

Odum, Eugene P., Gene A. Bramlet, Albert Ike, James R. Champlin, Joseph C. Zieman, and Herman H. Shugart. 1976. "Totality Indexes for Evaluating Environmental Impacts of Highway Alternatives." *Transportation Research Record No. 561.* Washington, D.C.: Transportation Research Board, National Academy of Sciences, pp. 57–67.

Odum, Eugene P., and J. L. Cooley. 1977. *Biological Evaluation of Environmental Impact.* Washington, D.C.: Council for Environmental Quality.

Odum, Eugene P., and Howard T. Odum. 1953. *Fundamentals of Ecology.* First edition. Philadelphia: W. B. Saunders.

Odum, Eugene P., and Howard T. Odum. 1955. "Trophic Structure and Productivity of a Windward Coral Reef Community on Eniwetok Atoll." *Ecological Monographs* 25: 291–320.

Odum, Eugene P., and Howard T. Odum. 1959. *Fundamentals of Ecology.* Second edition. Philadelphia: W. B. Saunders.

Odum, Eugene P., and Howard T. Odum. 1971. *Fundamentals of Ecology.* Third edition. Philadelphia: W. B. Saunders.

Odum, Howard T. 1957. "Trophic Structure and Productivity of Silver Springs, Florida." *Ecological Monographs* 27: 55–112.

Odum, Howard T. 1960. "Ecological Potential and Analogue Circuits for the Ecosystem." *American Scientist* 48: 1–8.

Odum, Howard T. 1983. *Systems Ecology: An Introduction.* New York: John Wiley.

Odum, Howard T., and R. C. Pinkerton. 1955. "Time's Speed Regulator: The Optimum Efficiency for Maximum Power Output in Physical and Biological Systems." *American Scientist* 43: 331–43.

Odum, Howard T., and Eugene P. Odum. 1976. *The Energy Basis for Man and Nature.* New York: McGraw-Hill.

Odum, Howard W. 1936. *Southern Regions of the United States.* Chapel Hill: University of North Carolina Press.

Odum, Howard W., and H. E. Moore. 1938. *American Regionalism.* New York: Holt.

Oelschlaeger, Max. 1991. *The Idea of Wilderness: From Prehistory to the Age of Ecology.* New Haven: Yale University Press.

Olding, A. 1978. "A Defence of Evolutionary Laws." *British Journal for the Philosophy of Science* 29: 131–43.

O'Neill, Robert V., D. L. DeAngelis, J. B. Waide, and T. F. H. Allen. 1986. *A Hierarchical Concept of Ecosystems.* Monographs in Population Biology 23. Princeton: Princeton University Press.

Onsager, L. 1931. "Reciprocal Relations in Irreversible Processes." *Physical Review A* 37: 405–26.

Oosting, Henry J. 1950. *The Study of Plant Communities: An Introduction to Plant Ecology.* San Francisco: W. H. Freeman.

Opler, P. A. 1978. "Insects of American Chestnut: Possible Importance and Conservation Concern." *Proceedings of the American Chestnut Symposium* 83–84.

Orians, Gordon H. 1962. "Natural Selection and Ecological Theory." *American Naturalist* 96: 257–63.

Orians, Gordon. 1973. "A Diversity of Textbooks: Ecology Comes of Age." *Science* 181: 1238–39.

Orians, Gordon H. 1986. *Ecological Knowledge and Environmental Problem Solving: Concepts and Case Studies.* Washington, D.C.: National Academy Press.

Orians, Gordon H., and O. T. Solbrig. 1977. "A Cost-Income Model of Leaves and Roots with Special Reference to Arid and Semiarid Areas. *American Naturalist* 111: 677–90.

Ortiz de Montellano, Bernard R. 1978. "Aztec Cannibalism: An Ecological Necessity?" *Science* 200: 611–17.

Oster, G. 1975. "Stochastic Behavior of Deterministic Models." In S. A. Levin (ed.), *Ecosystem Analysis and Prediction.* Philadelphia: SIAM-SIMS, pp. 24–38.

Pahl-Wostl, Claudia. 1995. *The Dynamic Nature of Ecosystems: Chaos and Order Entwined.* New York: John Wiley.

Paine, R. T. 1984. "Ecological Determinism in the Competition for Space." *Ecology* 65: 1339–48.

Paley, William. 1802 (1825). *Natural Theology. The Works of William Paley.* Volume 5. London: C. and J. Rivington.

Parker, G. G. 1989. "Are Currently Available Statistical Methods Adequate for Long-Term Studies?" In G. E. Likens (ed.), *Long Term Studies in Ecology: Approaches and Alternatives.* New York: Springer-Verlag, pp. 199–200.

Patrick, Ruth. 1949. "A Proposed Biological Measure of Stream Conditions Based on a Survey of Conestoga Basin, Lancaster County, Pennsylvania." *Proceedings of the Philadelphia Academy of Natural Sciences* 101: 277–341.

Patrick, Ruth. 1983. Editor's Comments. In *Diversity: Benchmark Papers in Ecology.* Volume 13. Stroudsburg, Pa.: Hutchinson Ross, pp. 8–13.

Patrick, Ruth, B. Crum, and J. Coles. 1969. "Temperature and Manganese as Determining Factors in the Presence of Diatom or Blue-Green Algal Floras in Streams." *Proceedings of the National Academy of Sciences (USA)* 64: 472–78.

Pattee, H. H. 1973. *Hierarchy Theory.* New York: Braziller.

Pattee, H. H. 1978. "The Complementarity Principle in Biological and Social Structures." *Journal of the Society of Biological Structures* 1: 191–200.

Pattee, H. H. 1979. "The Complementarity Principle and the Origin of Macromolecular Information." *BioSystems* 11: 217–26.

Patten, Bernard C. (ed.) 1971. *Systems Analysis and Simulation in Ecology*. Volume 1. New York: Academic Press.

Patten, Bernard C. 1959. "An Introduction to the Cybernetics of the Ecosystem: The Trophic-Dynamic Aspect." *Ecology* 40: 221–31.

Patten, Bernard C. 1975. "Ecosystem Linearization: An Evolutionary Design Problem." *American Naturalist* 109: 529–39.

Patten, Bernard C. 1993a. "Toward a More Holistic Ecology and Science: The Contribution of H. T. Odum." *Oecologia* 93: 597–602.

Patten, Bernard C. 1993b. "Discussion: Promoted Coexistence through Indirect Effects: Need for a New Ecology of Complexity." In H. Kawanabe, J. E. Cohen, and K. Iwasaki (eds.), *Mutualism and Community Organization*. Oxford: Oxford Science Publications, pp. 323–335.

Patten, Bernard C., and E. P. Odum. 1981. "The Cybernetic Nature of Ecosystems." *American Naturalist* 118: 886–95.

Peacocke, C. 1986. Reply. In J. Canfield (ed.), *The Philosophy of Wittgenstein*. Volume 10. New York: Garland, pp. 274–97.

Peirce, C. S. 1877. "The Fixation of Belief." *Popular Science Monthly* 12: 1–15.

Perrins, C. M. 1964. "Survival of Young Swifts in Relation to Brood-Size." *Nature* 201: 1147–48.

Peters, Robert H. 1991. *A Critique for Ecology*. Cambridge: Cambridge University Press.

Peterson, R. W. 1976. "The Impact Statement. Part II." *Science* 193: 193.

Phillips, John. 1931. "The Biotic Community." *Journal of Ecology* 19: 1–24.

Phillips, John. 1934. "Succession, Development, the Climax, and the Complex Organism: An Analysis of Concepts. Part I." *Journal of Ecology* 22: 554–71.

Phillips, John. 1935a. "Succession, Development, the Climax, and the Complex Organism: An Analysis of Concepts. Part II." *Journal of Ecology* 23: 210–43.

Phillips, John. 1935b. "Succession, Development, the Climax, and the Complex Organism: An Analysis of Concepts. Part III." *Journal of Ecology* 23: 488–508.

Piaget, J. 1970. *L'épistémologie génétique*. Paris: PUF.

Picardi, E. 1988. "Meaning and Rules." In J. Nyiri and B. Smith (eds.), *Practical Knowledge*. London: Croom Helm, pp. 90–121.

Pickett, Stewart T. A., Jurek Kolasa, and Clive G. Jones. 1994. *Ecological Understanding: The Nature of Theory and the Theory of Nature*. San Diego: Academic Press.

Pielou, E. C. 1969. *An Introduction to Mathematical Ecology*. New York: Wiley-Interscience.

Pimentel, D. 1966. "Complexity of Ecological Systems and Problems in Their Study and Management." In K. E. F. Watt (ed.), *Systems Analysis in Ecology*. New York: Academic Press, pp. 15–35.

Pimm, Stuart L. 1979. "Complexity and Stability: Another Look at MacArthur's Original Hypothesis." *Oikos* 33: 351–57.

Pimm, Stuart L. 1980. "Food Web Design and the Effects of Species Deletion." *Oikos* 35: 139–49.

Pimm, Stuart L. 1982. *Food Webs.* London: Chapman and Hall.

Pimm, Stuart L. 1984a. "The Complexity and Stability of Ecosystems." *Nature* 307: 321–26.

Pimm, Stuart L. 1984b. "Food Chains and Return Times." In D. R. Strong Jr., D. Simberloff, L. G. Abele, and A. B. Thistle (eds.), *Community Ecology: Conceptual Issues and the Evidence.* Princeton: Princeton University Press, pp. 397–412.

Pirsig, Robert M. 1974. *Zen and the Art of Motorcycle Maintenance.* New York: Bantam Books.

Plantinga, Alvin. 1974. *The Nature of Necessity.* Oxford: Oxford University Press.

Plato. (1982). *Phaedrus.* In Edith Hamilton and Huntington Cairns (eds.), *Plato: The Collected Dialogues.* Princeton: Princeton University Press.

Polanyi, Michael. 1959. *The Study of Man.* Chicago: University of Chicago Press.

Polanyi, Michael. 1964. *Personal Knowledge: Towards a Post-Critical Philosophy.* New York: Harper and Row.

Polanyi, Michael. 1968. "Life's Irreducible Structure." *Science* 160: 1308–12.

Ponyatovskaya, V. M. 1961. "On Two Trends in Phytocoenology." Trans. J. Major. *Vegetatio* 10: 373–85.

Poole, R. W. 1974. *An Introduction to Quantitative Ecology.* New York: McGraw-Hill.

Poole, R. W. 1977. "Periodic, Pseudoperiodic, and Chaotic Population Fluctuations." *Ecology* 58: 210–13.

Popper, Karl R. 1959 (1968). *The Logic of Scientific Discovery.* Third edition. London: Hutchinson.

Popper, Karl R. (1977). "The Bucket and the Searchlight: Two Theories of Knowledge." In Matthew Lipman (ed.), *Discovering Philosophy.* Englewood Cliffs, N.J.: Prentice-Hall, pp. 328–34.

Popper, Karl R. 1982. *The Open Universe: An Argument for Indeterminism.* Totowa, N.J.: Rowman and Littlefield.

Popper, Karl R. 1990. *A World of Propensities.* Bristol: Thoemmes.

Pound, Roscoe, and Frederic E. Clements. 1897. *The Phytogeography of Nebraska.* Part 1, *General Survey.* Second edition. Lincoln: University of Nebraska.

Preston, F. W. 1969. "Diversity and Stability in the Biological World." In G. M. Woodwell and H. H. Smith (eds.), *Diversity and Stability in Ecological Systems.* Brookhaven, N.Y.: Brookhaven National Laboratory, pp. 1–12.

Price, G. R. 1970. "Selection and Covariance." *Nature* 207: 520–21.

Price, G. R. 1972. "Extension of Covariance Selection Mathematics." *Ann. Human Genet. London* 35: 485–90.

Price, P. W., C. N. Slobodchikoff, and W. S. Gaud (eds.). 1984. *A New Ecology.* New York: John Wiley.

Price, Peter. 1996. *Biological Evolution.* Forth Worth: Saunders.

Prigogine, I. 1945. "Moderation et transformations irreversibles des systemes ouverts." *Bull. Cl. Sci. Acad. R. Belg.* Cinque E Ser. 31: 600–606.

Pyke, G. H. 1984. "Optimal Foraging Theory: A Critical Review." *Annual Review of Ecology and Systematics* 15: 523–75.

Quine, W. V. O. 1961. *From a Logical Point of View.* Cambridge: Harvard University Press.

Quinn, J. F., and A. E. Dunham. 1983. "On Hypothesis Testing in Ecology and Evolution." *American Naturalist* 122: 602–17.

Rabinow, Paul. 1996. *Essays on the Anthropology of Reason.* Princeton: Princeton University Press.

Ramade, F. 1992. "Lettre de la société française d'écologie." Octobre.

Randall, John H. Jr. 1960. *Aristotle.* New York: Columbia University Press.

Raup, H. M. 1942. "Trends in the Development of Geographic Botany." *Assoc. Amer. Geogr. Ann.* 32: 319–54.

Raven, C. E. 1942. *John Ray, Naturalist.* Cambridge.

Ray, John. 1686. *Historia Plantarum.* Volume 1. London.

Redfearn, Andrew, and Stuart L. Pimm. 1987. "Insect Outbreaks and Community Structure." In Pedro Barbosa and Jack C. Schultz (eds.), *Insect Outbreaks.* New York: Academic Press.

Regan, Tom. 1992. "The Thee Generation: Reflections on the Coming Revolution." In Stephen Satris (ed.), *Taking Sides: Clashing Views on Controversial Moral Issues.* Third edition. Guilford, Conn.: Dushkin.

Rehm, A., and H. J. Humm. 1973. "*Sphaeroma terebrans:* A Threat to the Mangroves of Southwestern Florida." *Science* 182: 173–74.

Reichenbach, H. 1956. *The Direction of Time.* Berkeley: University of California Press.

Reichenbach, Hans. 1951. "The Philosophical Significance of the Theory of Relativity." In Paul Arthur Schlipp (ed.), *Albert Einstein: Philosopher-Scientist.* New York: Tudor, pp. 269–311.

Reiners, W. A. 1986. "Complementary Models for Ecosystems." *American Naturalist* 127: 59–73.

Rejmanek, M., and P. Stary. 1979. "Connectance in Real Biotic Communities and Critical Values for Stability in Model Ecosystems." *Nature* 280: 311–13.

Rensch, B. 1959. *Evolution above the Species Level.* New York: Columbia University Press.

Rhoades, D. F. 1985. "Offensive-Defensive Interactions between Herbivores and Plants: Their Relevance in Herbivore Population Dynamics and Ecological Theory." *American Naturalist* 125: 205–38.

Risch, S. J., D. Andow, and M. Altieri. 1983. "Agroecosystem Diversity and Pest Control: Data, Tentative Conclusions and New Research Directions." *Environ. Entomol.* 12: 625–29.

Romanes, G. J. 1900. "The Darwinism of Darwin and of the Post-Darwinian Schools." In *Darwin, and after Darwin.* Volume 2. London: Longmans, Green.

Root, Richard B. 1967. "The Niche Exploitation Pattern of the Blue-Grey Gnatcatcher." *Ecological Monographs* 37: 317–50.

Root, Richard B. 1973. "Organization of a Plant Arthropod Association in Simple and

Diverse Habitats: The Fauna of Collards, *Brassicia oleracea.*" *Ecological Monographs* 43: 95–124.

Root, Richard B., and Stephen J. Chaplin. 1976. "The Life Styles of Tropical Milkweed Bugs, *Oncopeltus* (Hemiptera: Lygaeidae), Utilizing the Same Hosts." *Ecology* 57: 132–40.

Rorty, Richard. 1972. "Relations, Internal and External." In *The Encyclopedia of Philosophy,* Volume 7. New York: Macmillan and Free Press, pp. 125–33.

Rosen, R. 1985. "Information and Complexity." In R. E. Ulanowicz and T. Platt (eds.), *Ecosystem Theory for Biological Oceanography.* Ottawa: Canadian Bulletin of Fisheries and Aquatic Sciences 213, pp. 221–33.

Rosen, R. 1991. *Life Itself: A Comprehensive Inquiry into the Nature, Origin and Foundation of Life.* New York: Columbia University Press.

Rosenberg, Alexander. 1985. *The Structure of Biological Science.* New York: Cambridge University Press.

Ross, W. D. 1937. *Aristotle.* London.

Roszak, Theodore, Mary E. Gomes, and Allen D. Kanner (eds.). 1995. *Ecopsychology: Restoring the Earth, Healing the Mind.* San Francisco: Sierra Club Books.

Roughgarden, Jonathan. 1979. *Theory of Population Genetics and Evolutionary Ecology: An Introduction.* New York: Harper and Row.

Roughgarden, Jonathan. 1983. "Competition and Theory in Community Ecology." *American Naturalist* 122: 583–601.

Roux, W. 1881. *Kampf der Theile im Organismus.* Leipzig: Jena.

Rowe, J. S. 1961. "The Level-of-Integration Concept and Ecology." *Ecology* 42: 420–27.

Rudwick, M. J. S. 1964. "The Function of Zig-Zag Deflections in the Commissures of Fossil Brachiopods." *Palaeontology* 7: 135–71.

Ruse, Michael. 1971. "Narrative Explanation and the Theory of Evolution." *Canadian Journal of Philosophy* 1: 59–74.

Ruse, Michael (ed.). 1989. *What the Philosophy of Biology Is: Essays Dedicated to David Hull.* Dordrecht: Kluwer.

Russell, E. S. 1924. *Study of Living Things.* London: Methuen.

Sagoff, Mark. 1997. "Muddle or Muddle Through? Takings Jurisprudence Meets the Endangered Species Act." *William and Mary Law Review* 38: 825–993.

Sahlins, M. 1978. "Culture as Protein and Profit." *New York Review of Books* (November 23): 45–53.

Salmon, W. 1984. *Scientific Explanation and the Causal Structure of the World.* Princeton: Princeton University Press.

Salmon, W. 1989. "Four Decades of Scientific Explanations." In P. Kitcher and W. C. Salmon (eds.), *Scientific Explanation.* Minnesota Studies in the Philosophy of Science 13. Minneapolis: University of Minnesota Press, pp. 384–409.

Salt, George W. 1979. "A Comment on the Use of the Term 'Emergent Properties.'" *American Naturalist* 113(1): 145–48.

Salthe, S. N. 1985. *Evolving Hierarchical Systems: Their Structure and Representation.* New York: Columbia University Press.

Salthe, S. N. 1993. *Development and Evolution: Complexity and Change in Biology.* Cambridge: MIT Press.

Salwasser, H. 1986. "Conserving a Regional Spotted Owl Population." In G. Orians (ed.), *Ecological Knowledge and Environmental Problem Solving.* Washington, D.C.: National Academy Press, pp. 227–47.

Sattler, R. 1986. *Biophilosophy: Analytic and Holistic Perspectives.* New York: Springer-Verlag.

Schaefer, M. B. 1957. "A Study of the Dynamics of the Fishery for Yellowfin Tuna in the Eastern Tropical Pacific Ocean." *Inter. Amer. Trop. Tuna Comm. Bull.* 2: 247–85.

Schaffer, W. M., and P. F. Elson. 1975. "The Adaptive Significance of Variations in Life History among Local Populations of Atlantic Salmon in North America." *Ecology* 56: 577–90.

Schaffner, Kenneth F. 1967. "Approaches to Reduction." *Philosophy of Science* 34: 137–47.

Schaffner, Kenneth F. 1969. "The Watson-Crick Model and Reductionism." *British Journal for the Philosophy of Science* 20: 325–48.

Schaffner, Kenneth F. 1986. "Reduction in Biology: Prospects and Problems." In E. Sober (ed.), *Conceptual Issues in Evolutionary Biology.* Cambridge: MIT Press, pp. 428–45.

Schaffner, Kenneth F. 1993. "Theory Structure, Reduction, and Disciplinary Integration in Biology." *Biology and Philosophy* 8: 319–47.

Scheiner, Samuel M., André J. Hudson, and Mark A. VanderMeulen. 1993. "An Epistemology for Ecology." *Ecological Society of America Bulletin* 74: 17–21.

Schindler, D. W. 1976. "The Impact Statement Boondoggle." *Science* 192: 509.

Schneider, E. D. 1987. "Thermodynamics, Ecological Succession, and Natural Selection: A Common Thread." In B. H. Weber, D. J. DePew, and J. D. Smith (eds.), *Entropy, Information, and Evolution.* Cambridge: MIT Press, pp. 107–38.

Schoener, Thomas W. 1972. "Mathematical Ecology and Its Place among the Sciences." *Science* 178: 389–91.

Schoener, Thomas W. 1974a. "Some Methods for Calculating Competition Coefficients from Resource-Utilization Spectra." *American Naturalist* 108: 332–40.

Schoener, Thomas W. 1974b. "Competition and the Form of Habitat Shift." *Theor. Pop. Biol.* 6: 265–307.

Schoener, Thomas W. 1975. "Presence and Absence of Habitat Shift in Some Widespread Lizard Species." *Ecological Monographs* 45: 233–58.

Schoener, Thomas W. 1982. "The Controversy over Interspecific Competition." *American Scientist* 70: 586–95.

Schoener, Thomas W. 1983a. "Field Experiments on Interspecific Competition." *American Naturalist* 122: 240–85.

Schoener, Thomas W. 1983b. "Simple Models of Optimal Feeding-Territory Size: A Reconciliation." *American Naturalist* 121: 608–29.

Schoener, Thomas W. 1985. "Kinds of Ecological Communities: Ecology Becomes Pluralistic." In J. M. Diamond and T. Case (eds.), *Ecological Communities.* New York: Harper and Row.

Schoener, Thomas W. 1986. "Mechanistic Approaches to Ecology: A New Reductionism?" *American Zoologist* 26: 81–106.

Sears, Paul B. 1964. "Ecology: A Subversive Subject." *BioScience* 14: 11–13.

Shapely, D. 1973. "Congress Picks up Technology Gauntlet." *Science* 180: 392.

Shapere, D. 1974. "On the Relations between Compositional and Evolutionary Theories." In F. J. Ayala and T. Dobzhansky (eds.), *Studies in the Philosophy of Biology*. Berkeley: University of California Press, pp. 187–204.

Shea, B. T. 1977. "Eskimo Craniofacial Morphology, Cold Stress and the Maxillary Sinus." *Am. J. Phys. Anthrop.* 47: 289–300.

Sheail, John. 1987. *Seventy-five Years in Ecology*. Oxford: Blackwell.

Shelford, Victor E. 1913. *Animal Communities in Temperate America*. Chicago: University of Chicago Press.

Shelford, Victor E. 1931. "Some Concepts of Bioecology." *Ecology* 12: 455–67.

Shrader-Frechette, Kristin S. 1985. *Science Policy, Ethics, and Economic Methodology: Some Problems of Technology Assessment and Environmental Impact Analysis*. Boston: Reidel.

Shrader-Frechette, Kristin S. 1991. *Risk and Rationality: Philosophical Foundations for Populist Reforms*. Berkeley: University of California Press.

Shrader-Frechette, Kristin S., and Earl D. McCoy. 1993. *Method in Ecology: Strategies for Conservation*. Cambridge: Cambridge University Press.

Shrader-Frechette, Kristin S., and Earl D. McCoy. 1994. "Applied Ecology and the Logic of Case Studies." *Philosophy of Science* 61: 228–49.

Shugart, H. H. 1984. *A Theory of Forest Dynamics*. New York: Springer-Verlag.

Shvarts, S. S. 1977. *The Evolutionary Ecology of Animals*. New York: Consultants Bureau.

Simberloff, Daniel S. 1976. "Experimental Zoogeography of Islands: Effects of Island Size." *Ecology* 57: 629–48.

Simberloff, Daniel S. 1980. "A Succession of Paradigms in Ecology: Essentialism to Materialism and Probabilism." *Synthese* 43: 3–39.

Simberloff, Daniel S. 1981. "The Sick Science of Ecology: Symptoms, Diagnosis, and Prescription." *Eidema* 1: 49–54.

Simberloff, Daniel S. 1983. "Competition Theory, Hypothesis Testing, and Other Community-Ecological Buzzwords." *American Naturalist* 122: 626–35.

Simberloff, Daniel S., and L. G. Abele. 1976. "Island Biogeography Theory and Conservation Practice." *Science* 191: 285–86.

Simberloff, Daniel S., and Edward O. Wilson. 1970. "A Two-Year Record of Colonization." *Ecology* 51: 934–37.

Simon, H. A. 1962. "The Architecture of Complexity." *Proceedings of the American Philosophical Society* 106: 467–82.

Simon, H. A. 1969. *The Sciences of the Artificial*. Cambridge: MIT Press.

Simon, H. A. 1973. "The Organization of Complex Systems." In H. H. Pattee (ed.), *Hierarchy Theory*. New York: Braziller, pp. 3–27.

Simpson, G. G. 1961. *The Principles of Animal Taxonomy*. New York: Columbia University Press.

Simpson, G. G. 1964. *This View of Life: The World of an Evolutionist.* New York: Harcourt, Brace and World.

Skellam, J. G. 1955. "The Mathematical Approach to Population Dynamics." In J. B. Cragg and N. W. Pirie (eds.), *The Numbers of Man and Animals.* Edinburgh: Oliver and Boyd, pp. 31–46.

Skyrms, B. 1980. *Causal Necessity: A Pragmatic Investigation of the Necessity of Laws.* New Haven: Yale University Press.

Slatkin, M., and J. Maynard Smith. 1979. "Models of Coevolution." *Quarterly Review of Biology* 54: 233–63.

Slobodkin, L. B. 1972. "On the Inconstancy of Ecological Efficiency and the Form of Ecological Theories." *Transactions of the Connecticut Academy of Arts and Sciences* 44: 293–305.

Slobodkin, Lawrence B. 1975. "Comments from a Biologist to a Mathematician." In S. A. Levin (ed.), *Ecosystem Analysis and Prediction.* Proceedings of SIAM-SIMS Conference on Ecosystems, Alta, Utah, July 1–5, pp. 318–29.

Smith, B. 1988. "Knowing How versus Knowing That." In J. Nyiri and B. Smith (eds.), *Practical Knowledge.* London: Croom Helm, pp. 1–16.

Smith, F. E. 1975. "Ecosystems and Evolution." *Bulletin of the Ecological Society of America* 56: 2–6.

Smuts, Jan Christian. 1926. *Holism and Evolution.* New York: Macmillan.

Snyder, Gary. 1974. *Turtle Island.* New York: New Directions.

Sober, Eliot. 1981. "Revisability, a Priori Truth, and Evaluation." *Australasian Journal of Philosophy* 59: 68–85.

Sober, Eliot. 1987. "Parsimony, Likelihood, and the Principle of the Common Cause." *Philosophy of Science* 54: 465–69.

Sober, Eliot. 1988. "The Principle of the Common Cause." In J. Fetzer (ed.), *Probability and Causality: Essays in Honor of W. C. Salmon.* Boston: Reidel, pp. 211–28.

Sober, Eliot, and Richard C. Lewontin. 1982. "Artifact, Cause, and Genic Selection." *Philosophy and Science* 49: 157–80.

Sokal, P., and P. Sneath 1963. *Principles of Numerical Taxonomy.* San Francisco: Freeman.

Somenzi, V. 1987. "Storicità unicità complessità." *Prometeo* 19:64–69.

Soulé, Michael E. 1995. "What Is Conservation Biology?" *BioScience* 35: 727–34.

Spiller, David A. 1986. "Consumptive Competition Coefficients: An Experimental Analysis with Spiders." *American Naturalist* 127: 604–14.

Stauffer, Robert C. 1957. "Haeckel, Darwin, and Ecology." *Quarterly Review of Biology* 32: 138–44.

Steingraber, Sandra. 1997. *Living Downstream: An Ecologist Looks at Cancer and the Environment.* New York: Vintage Books.

Stent, G. S. 1978. *The Coming of the Golden Age: A View of the End of Progress.* Garden City, N.Y.: Natural History.

Stern, J. T. 1970. "The Meaning of 'Adaptation' and Its Relation to the Phenomenon of Natural Selection. *Evolutionary Biology* 4: 39–66.

Steward, Jane C., and Robert F. Murphy. 1977. *Evolution and Ecology: Essays of Social Transformation.* Chicago: University of Illinois Press.

Storer, N. W. 1973. Introduction to *The Sociology of Science,* by R. K. Merton. Chicago: University of Chicago Press.

Strong, D. R., and D. A. Levin. 1979. "Species Richness of Plant Parasites and Growth Form of Their Hosts." *American Naturalist* 114: 1–22.

Strong, D. R., D. Simberloff, L. G. Abele, and A. B. Thistle. 1984a. *Ecological Communities: Conceptual Issues and the Evidence.* Princeton: Princeton University Press.

Strong, D. R., J. H. Lawton, and T. R. E. Southwood. 1984b. *Insects on Plants.* Oxford: Blackwell.

Sutherland, J. P. 1974. "Multiple Stable Points in Natural Communities." *American Naturalist* 108: 959–73.

Sweeney, B. W., and R. L. Vannote. 1978. "Size Variation and the Distribution of Hemimetabolous Aquatic Insects: Two Thermal Equilibrium Hypotheses." *Science* 200: 444–46.

Tanner, J. T. 1966. "Effects of Population Density on the Growth Rates of Animal Populations." *Ecology* 47: 433–45.

Tanner, J. T. 1975. "The Stability and the Intrinsic Growth Rates of Prey and Predator Populations." *Ecology* 56: 855–67.

Tansley, Arthur G. 1920. "The Classification of Vegetation and the Concept of Development." *Journal of Ecology* 8: 118–44.

Tansley, Arthur G. 1929. "Succession: The Concept and Its Values." *Proceedings of the International Congress of Plant Sciences, 1926,* pp. 677–86.

Tansley, Arthur G. 1935. "The Use and Abuse of Vegetational Concepts and Terms." *Ecology* 16: 284–307.

Taylor, P. J. 1985. "Construction and Turnover of Multispecies Communities: A Critique of Approaches to Ecological Complexity." Ph.D. diss., Harvard University, Cambridge.

Teilhard de Chardin, Pierre. 1955 (1969). *The Future of Man.* Trans. Norman Denny. New York: Harper and Row.

Tennyson, Alfred. 1850 (1987). "In Memoriam A. H. H." In Christopher Ricks (ed.), *The Poems of Tennyson.* Volume 2. Berkeley: University of California Press, pp. 304–459.

Theophrastus. (1916). *Enquiry into Plants.* Trans. Arthur Hort. Cambridge: Cambridge University Press.

Theophrastus. (1927). *Causes of Plants.* Trans. R. E. Dengler. Philadelphia.

Thienemann, August. 1939. "Grundzuge einer allgemeinen Okologie." *Archives of Hydrobiology* 35: 267–85.

Thomas, J. W., E. D. Forsman, J. B. Lint, E. C. Meslow, B. R. Moon, and J. Verner. 1990. *A Conservation Strategy for the Northern Spotted Owl.* Portland, Ore.: USDA, Forest Service; USDI, Bureau of Land Management; USDI, Fish and Wildlife Service; USDI, National Park Service.

Thompson, D'Arcy W. 1942. *Growth and Form.* New York: Macmillan.

Thoreau, Henry David. 1860 (1980a). "The Succession of Forest Trees." In *The Natural History Essays.* Salt Lake City: Peregrine Smith.

Thoreau, Henry David. 1862 (1980b). "Walking." In *The Natural History Essays.* Salt Lake City: Peregrine Smith.

Tilman, David. 1976. "Ecological Competition between Algae: Experimental Confirmation of Resource-Based Competition Theory." *Science* 192: 463–65.

Tilman, David. 1977. "Resource Competition between Planktonic Algae: An Experimental and Theoretical Approach." *Ecology* 58: 338–48.

Tilman, David. 1986. "A Consumer-Resource Approach to Community Structure." *American Zoologist* 26: 5–22.

Tilman, David, and John A. Downing. 1994. "Biodiversity and Stability in Grasslands." *Nature* 367: 363–65.

Tiwari, J. L., and J. E. Hobbie. 1976a. "Random Differential Equations as Models of Ecosystems, Part I: Monte Carlo Simulation Approach." *Mathematical Biosciences* 28: 25–44.

Tiwari, J. L., and J. E. Hobbie. 1976b. "Random Differential Equations as Models of Ecosystems, Part II: Initial Conditions and Parameter Specifications in Terms of Maximum Entropy Distributions." *Mathematical Biosciences* 31: 37–53.

Tobey, Ronald C. 1981. *Saving the Prairies: The Life Cycle of the Founding School of American Plant Ecology, 1895–1955.* Berkeley: University of California Press.

Totafurno, J., C. J. Lumsden, and L. E. H. Trainor. 1980. "Structure and Function in Biological Hierarchies: An Ising Model Approach." *Journal of Theoretical Biology* 85: 171–98.

Traub, R. 1980. "Some Adaptive Modifications in Fleas." In R. Traub and H. Starcke (eds.), *Fleas.* Rotterdam: A. A. Baldema, pp. 33–67.

Tribus, M., and E. C. McIrvine. 1971. "Energy and Information." *Scientific American* 225: 179–88.

Turelli, M. 1981. "Niche Overlap and Invasion of Competitors in Random Environments, Part I: Models without Demographic Stochasticity." *Theor. Pop. Biol.* 20: 1–56.

Turresson, Gote. 1923. "The Scope and Import of Genecology." *Hereditas* 4: 171–76.

Tylor, E. B. 1871. *Primitive Culture: Researches into the Development of Mythology, Philosophy, Religion, Language, Art and Custom.* London.

Ulanowicz, Robert E. 1980. "An Hypothesis on the Development of Natural Communities." *Journal of Theoretical Biology* 85: 223–45.

Ulanowicz, Robert E. 1986. *Growth and Development: Ecosystems Phenomenology.* New York: Springer-Verlag.

Ulanowicz, Robert E. 1990. "Aristotelean Causalities in Ecosystem Development." *Oikos* 57: 42–48.

Ulanowicz, Robert E. 1995a. "Beyond the Material and the Mechanical: Occam's Razor Is a Double-Edged Blade." *Zygon* 30: 249–66.

Ulanowicz, Robert E. 1995b. "*Utricularia's* Secret: The Advantages of Positive Feedback in Oligotrophic Environments." *Ecological Modelling* 79: 49–57.

Ulanowicz, Robert E. 1996. "The Propensities of Evolving Systems." In E. L. Khalil and K. E. Boulding (eds.), *Evolution, Order and Complexity.* New York: Routledge, pp. 217–33.

Ulanowicz, Robert E. 1997. *Ecology, the Ascendent Perspective.* New York: Columbia University Press.

Ulanowicz, Robert E. 1998. "Theoretical and Philosophical Considerations Why Ecosystems May Exhibit a Propensity to Increase in Ascendency." In F. Mueller (ed.), *Eco-Targets, Goal Functions and Orientors.* Berlin: Springer-Verlag, pp. 177–92.

Ulanowicz, Robert E. 1999. "Life after Newton: An Ecological Metaphysic." *BioSystems* 50: 127–42.

Ulanowicz, Robert E., and D. Baird. 1999. "Nutrient Controls on Ecosystem Dynamics: The Chesapeake Mesohaline Community." *Journal of Marine Systematics* 19: 159–72.

Ulanowicz, Robert E., and B. M. Hannon. 1987. "Life and the Production of Entropy." *Proceedings of the Royal Society of London: Biological Sciences* 32: 181–92.

Ulanowicz, Robert E., and J. Norden. 1990. "Symmetrical Overhead in Flow Networks." *International Journal of Systems Science* 21(2): 429–37.

U.S. Congress. 1990. *Report of the Interagency Scientific Committee to Address the Conservation of the Northern Spotted Owl.* Senate Hearings. Washington, D.C.: U.S. Government Printing Office, pp. 101–850.

Vandermeer, John. 1981. *Elementary Mathematical Ecology.* New York: John Wiley.

Van Der Steen, W., and H. Kamminga. 1991. "Laws and Natural History in Biology." *British Journal for the Philosophy of Science* 42: 445–67.

Van Valen, L. 1976. "Ecological Species, Multispecies, and Oaks." *Taxon* 25: 233–39.

Vogel, Steven. 1996. *Against Nature: The Concept of Nature in Critical Theory.* Albany: State University of New York Press.

Voltaire. 1759 (1999). *Candide.* Trans. Daniel Gordon. Boston: St. Martin's Press.

Volterra, Vito. 1926. "Fluctuations in the Abundance of a Species Considered Mathematically." *Nature* 118: 558–60.

Voorhees, D. W. (ed.) 1983. *Concise Dictionary of American Science.* New York: Scribner's.

Wade, M. J. 1978. "A Critical Review of the Models of Group Selection." *Quarterly Review of Biology* 53: 101–13.

Wade, M. J. 1980. "Kin Selection: Its Components." *Science* 210: 665–67.

Wallace, A. R. 1899. *Darwinism.* London: Macmillan.

Walters, C. J. 1975. "Dynamic Models and Evolutionary Strategies." In S. A. Levin (ed.), *Ecosystem Analysis and Prediction.* Philadelphia: Society for Industrial and Applied Mathematics, pp. 68–82.

Wangersky, P. J. 1970. "An Ecology of Probability." *Ecology* 51: 940.

Warren, Karen J. 1990. "The Power and the Promise of Ecological Feminism." *Environmental Ethics* 12: 125–46.

Watson, James, and Francis Crick. 1953a. "Molecular Structure of Nucleic Acids: A Structure for Deoxyribose Nucleic Acid." *Nature* 171: 737.

Watson, James, and Francis Crick. 1953b. "Genetical Implications of the Structure of Deoxyribose Nucleic Acid." *Nature* 171: 964.

Watt, K. E. F. (ed.). 1966a. *Systems Analysis in Ecology.* New York: Academic Press.

Watt, K. E. F. 1966b. "The Nature of Systems Analysis." In K. E. F. Watt (ed.), *Systems Analysis in Ecology.* New York: Academic Press, pp. 1–14.

Watt, K. E. F. 1968. *Ecology and Resource Management.* New York: McGraw-Hill.

Watt, K. E. F. 1975. "Critique and Comparison of Biome Ecosystem Modeling." In B. C.

Patten (ed.), *Systems Analysis and Simulation in Ecology*. Volume 5. New York: Academic Press, pp. 139–52.

Webster, J. R. 1979. "Hierarchical Organization of Ecosystems." In E. Halfon (ed.), *Theoretical Systems Ecology*. New York: Academic Press, pp. 119–29.

Webster's New Collegiate Dictionary. 1981. Springfield, Mass.: Merriam Co.

Weiss, P. A. 1967. "1 + 1 ≠ 2 (One plus One Does Not Equal Two)." In G. C. Quarton, T. Meinechuck, and F. O. Schmitt (eds.), *The Neurosciences, a Study Program*. New York: Rockefeller University Press, pp. 801–21.

Werner, Earl E., and Mittelbach, Gary G. 1981. "Optimal Foraging: Field Tests of Diet Choice and Habitat Switching." *American Zoologist* 21: 813–29.

Wertheimer, M. 1945. *Productive Thinking*. New York: Macmillan.

Westfall, Richard S. 1983. *The Construction of Modern Science: Mechanism and Mechanics*. New York: Cambridge University Press.

Westfall, Richard S. 1993. *The Life of Isaac Newton*. Cambridge: Cambridge University Press.

White, Gilbert. 1789 (1981). *The Illustrated Natural History of Selborne*. London: Macmillan.

White, Lynn Jr. 1967. "The Historical Roots of Our Ecologic Crisis." *Science* 155: 1203–7.

Whitehead, Alfred N. 1925 (1967). *Science and the Modern World*. New York: Free Press.

Whitehead, Alfred N. 1938 (1966). *Modes of Thought*. New York: Free Press.

Whitham, T. G., A. G. Williams, and A. M. Robinson. 1984. "The Variation Principle: Individual Plants as Temporal and Spatial Mosaics of Resistance to Rapidly Evolving Pests." In P. W. Price, C. N. Slobodchikoff, and W. S. Gaud (eds.), *A New Ecology*. New York: John Wiley, pp. 15–52.

Whitman, Walt. 1855 (1965). "Song of the Open Road." In *Leaves of Grass*. New York: New York University Press.

Whittaker, Robert H. 1953. "A Consideration of the Climax Theory: The Climax as a Population and Pattern." *Ecological Monographs* 23: 41–78.

Whittaker, Robert H. 1956. "Vegetation of the Great Smoky Mountains." *Ecological Monographs* 26: 1–80.

Whittaker, Robert H. 1967. "Gradient Analysis of Vegetation." *Biological Review* 42: 207–64.

Whittaker, Robert H. 1975. *Communities and Ecosystems*. Second edition. London: Macmillan.

Whittaker, Robert H., Simon A. Levin, and Richard B. Root. 1973. "Niche, Habitat and Ecotope." *American Naturalist* 107: 321–38.

Whittaker, Robert H., Simon A. Levin, and Richard B. Root. 1975. "On the Reasons for Distinguishing *Niche, Habitat*, and *Ecotope*." *American Naturalist* 109: 479–82.

Whittaker, Robert H., and G. M. Woodwell. 1971. "Evolution of Natural Communities." In J. A. Wiens (ed.), *Ecosystem Structure and Function*. Corvallis: Oregon State University Press, pp. 137–59.

Whyte, L. L., A. G. Wilson, and D. Wilson (eds.). 1969. *Hierarchical Structures*. New York: Elsevier.

Wilbur, H. M. 1977. "Interactions of Food Level and Population Density in *Rana sylvatica.*" *Ecology* 58: 206–9.

Wilcove, D. S. 1990. "Of Owls and Ancient Forests." In E. A. Norse (ed.), *Ancient Forests of the Pacific Northwest.* Washington, D.C.: Island Press, pp. 76–83.

Williams, Bernard. 1985. *Ethics and the Limits of Philosophy.* Cambridge: Harvard University Press.

Williams, G. C. 1966. *Adaptation and Natural Selection: A Critique of Some Current Evolutionary Thought.* Princeton: Princeton University Press.

Williams, Terry Tempest. 1995. *Desert Quartet.* New York: Pantheon.

Williamson, M. 1983. "Variations in Population Density and Extinction." *Nature* 303: 201.

Wilson, David S. 1980. *Natural Selection of Populations and Communities.* Menlo Park, Calif.: Benjamin Cummings.

Wilson, Edward O. 1978. *On Human Nature.* Cambridge: Harvard University Press.

Wilson, Edward O. 1992. *The Diversity of Life.* New York: W. W. Norton.

Wimsatt, W. C. 1974. "Complexity and Organization." In K. F. Schaffner and R. S. Cohen (eds.), *Boston Studies in the Philosophy of Science* 20: 67–86.

Wimsatt, W. C. 1976a. "Reductive Explanation: A Functional Account." In R. S. Cohen, C. A. Hooker, A. C. Michalos, and J. W. Van Evra (eds.), *PSA 1975.* Dordrecht: Reidel, pp. 671–710.

Wimsatt, W. C. 1976b. "Reductionism, Levels of Organization, and the Mind-Body Problem." In G. G. Globus, G. Maxwell, and I. Savodnik (eds.), *Consciousness and the Brain: A Scientific and Philosophical Inquiry.* New York: Plenum Press, pp. 205–67.

Wimsatt, W. C. 1980a. "Reductionist Research Strategies and Their Biases in the Units of Selection Controversy." In T. Nickles (ed.), *Scientific Discovery: Case Studies.* Dordrecht: Reidel.

Wimsatt, W. C. 1980b. "Randomness and Perceived Randomness in Evolutionary Biology." *Synthese* 49: 287–329.

Wimsatt, W. C. 1986. "Reductive Explanation: A Functional Account." In E. Sober (ed.), *Conceptual Issues in Evolutionary Biology.* Cambridge: MIT Press, pp. 477–508.

Wimsatt, W. C. 1994. "The Ontology of Complex Systems: Levels of Organization, Perspectives, and Causal Thickets." In J. Matthen and R. X. Ware (eds.), *Biology and Society: Reflections on Methodology. Canadian Journal of Philosophy,* Supplemental volume 20, pp. 207–74.

Wisdom, J. 1965. *Paradox and Discovery.* Oxford: Blackwell.

Wittgenstein, Ludwig. (1969). *On Certainty.* Oxford: Blackwell.

Wittgenstein, Ludwig. (1973). *Philosophical Investigations.* Trans. G. E. M. Anscombe. Oxford: Blackwell.

Wittgenstein, Ludwig. (1979). "Remarks on Frazer's *The Golden Bough.*" Trans. J. Beversluis. In C. Luckhardt (ed.), *Wittgenstein.* Ithaca: Cornell University Press, pp. 61–81.

Woodger, J. H. 1932. "Some Apparently Unavoidable Characteristics of Natural Scientific Theory." *Proceedings of the Aristotelian Society* 32: 95–120.

Woodwell, G. M., and H. H. Smith (eds.). 1969. *Diversity and Stability in Ecological*

Systems. Brookhaven Symposium of Biology 22. Springfield, Va.: U.S. Department of Commerce.

Worster, Donald. 1977 (1994). *Nature's Economy: A History of Ecological Ideas.* Second edition. New York: Cambridge University Press.

Wright, S. 1931. "Evolution in Mendelian Populations." *Genetics* 16: 97–159.

Wright, S. 1949. "Adaptation and Selection." In G. Jepsen, E. Mayr, and G. G. Simpson (eds.), *Genetics, Paleontology, and Evolution.* Princeton: Princeton University Press, pp. 365–89.

Wundt, W. M. 1912. *Introduction to Psychology.* New York: Macmillan.

Wynne-Edwards, V. C. 1962. *Animal Dispersion in Relation to Social Behaviour.* Edinburgh: Oliver and Boyd.

Wynne-Edwards, V. C. 1977. "Society versus the Individual in Animal Evolution." In B. Stonehouse and C. Perrins (eds.), *Evolutionary Ecology.* Baltimore: University Park Press.

Yin, R. K. 1984. *Case Study Research: Design and Methods.* Beverly Hills: Sage.

Zadeh, L. A. 1965. "Fuzzy Sets." *Information and Control* 8: 338.

Zeller, Eduard. 1931 (1963). *Outlines of the History of Greek Philosophy.* Thirteenth edition. Trans. L. R. Palmer, ed. Wilhelm Nestle. New York: Humanities Press.

Zirkle, C. 1959. "Species before Darwin." *Proceedings of the American Philosophical Society* 103.

Zucker, W. V. 1983. "Tannins: Does Structure Determine Function? An Ecological Perspective." *American Naturalist* 121: 335–65.

INDEX

adaptationism, 241–42, 263–76, 277–87, 316

Alexander, Samuel, 31, 174

Allee, W. C., 215, 290

Allen, T. F. H., 177, 178

Anaxagoras (*also* Anaxagorian), 206

Andrewartha, H. G., 288, 289, 295

anthropocentrism, 4, 5, 6

Aristotle (*also* Aristotelian), 6, 19 (n. 9), 31, 82, 86, 206, 228, 247, 248–52, 253–54, 255, 256

Augustine, 31

autecology (*see also* ecology, merological), 11, 171

Bacon, Francis (*also* Baconian), 82, 83, 86, 133, 141, 151, 152, 172, 258

Barash, D. P., 270–71

Bergandi, Donato, 177

Bergson, Henri, 31

Berkeley, George, 133

biocenosis (*biocœnosis*, biocönose), 62, 102, 110, 113

biogeocoenosis, 9

Birch, L. C., 288, 289, 295

Bisol, P. M., 269, 270

Blitz, David, 171–72

Bookchin, Murray, 6

boundaries, ecological, 23–24, 27

Bramwell, Anna, 18 (n. 3)

Brown, James, 243

Buffon, Georges de, 7, 247, 249, 254–56, 258, 261

Burden-Sanderson, John, 9

Campbell, D. T., 157, 160, 169 (n. 1)

Campbell, John, 240

Carnap, Rudolp, 19 (n. 10)

Carson, Rachel, 5–6

case studies, method of, 139, 153–69

Chamberlin, Thomas, 139

chaos. *See* nature, indeterminism of

Clements, Frederic (*also* Clementsian), xi, 18, 19 (n. 15), 24–26, 27–28, 55–56, 57, 59, 60, 61, 66, 69, 70, 71–74, 77, 82, 85, 104, 108, 137, 176, 177, 206, 209, 215, 219, 220, 291, 293

climax community (*see also* succession, ecological), 24, 35, 60, 66, 69, 71

Cole, Lamont, 292, 293

Collins, James, 243

community, ecological, 16, 24, 65, 67–69, 85, 101–4, 108, 110, 111–14, 119–23, 124–31, 175, 176, 221–23, 311–12

—as a complex organism, 61–63, 68, 69, 70, 209

—contrasted with biotic association, 28

—definition of, 212

conservation biology, 18 (n. 4)

constructivism, social, 1, 12, 13, 14

Copernicus, Nicholas, 173

Costa, R., 269, 270

Cowles, Henry, 26, 290

Crick, Francis, 174

Crumley, Carole, 30

Curtis, John, 28

Cuvier, Georges, 247, 249, 257–58, 259

cybernetics, 83, 214, 227

Darling, Frank Frasier, 31

Darwin, Charles (*also* Darwinian), 7, 9, 32, 80, 82, 99, 106, 151, 219, 224, 226, 233–38, 239, 240, 243, 245, 247, 259–62, 266, 267, 272–74, 275–76, 277, 282, 283–84, 285, 287, 289, 290, 291, 299

Dawkins, Richard, 238, 241

Dayton, Paul, 3
Deleuze, Gilles, 13
Democritus, 29
Dennett, Daniel, 241, 242, 245 (n. 1)
Derrida, Jacques, 13
Descartes, René (*also* Cartesian), 13, 14,
  18, 31, 83, 133, 172, 179 (n. 1), 219, 245
determinism, 18, 29, 71, 73, 79–80, 84–85,
  172
diversity, biological, 16, 102, 105–7,
  119–23
Dobzhansky, T., 281, 284, 291
Downing, John, 108
Duhl, Leonard, 15

ecofeminism, 6
ecological
—process, 30–32
—worldview (*see also* ecology, definition
  of word), 2, 3
ecology (*see also* protoecology)
—arcadian, 10
—community, 102, 104, 153–69, 175, 181,
  183–93, 225
—and conservation, 3
—deep, 6
—definition of word, 2, 7, 9, 195, 196
—dialectical, 177, 178
—and economics, 201–2
—ecosystem (*also* systems), 10, 11,
  18 (n. 5), 138, 176, 177, 189, 304–19
—and environmentalism, 3
—and ethics, 2, 18, 320
—holological, 11, 175–77, 178
—imperial, 10
—merological, 11, 174–75, 178
—and politics, 5–6, 202–3, 320
—population, 10, 11, 18 (n. 5), 153–69,
  175, 183–92, 225
—Romantic (literary), 4–5
—as a science of synthesis, 15, 320
—scientific, 3, 6–11: definition of, 9–11,
  14; history of, 6–9; mathematics in, 75,

88–94, 138, 147–51, 181, 198, 315, 317;
  subdisciplines of, 4, 181, 183–86
—scope of, 1, 11, 14–15, 320
—social, 6
—as subversive, 5, 245
ecosystem, 9, 10, 16, 23, 26–27, 30, 63–67,
  68, 85, 99, 106, 175, 177, 194, 196–98,
  208–9, 306–9, 310, 313, 316–19
—human, 14, 68
—as machine, 26, 176
ecotope, 29, 30, 117, 118
Edelson, M., 164
Edson, Michael, 175
Egerton, Frank, 107
Einstein, Albert, 14, 19 (n. 13), 79–80, 164
Eldredge, Niles, 154, 241, 245 (n. 1)
Elger, Frank, 137
Elton, Charles, 72, 77, 105, 108, 109, 124,
  126, 138, 292
emergent property (emergentism), 17,
  171–75, 177, 178, 194–96, 205, 213,
  215–17, 227–31, 307, 308
Emerson, Alfred, 293, 299
Emerson, Ralph Waldo, 4
Empedocles, 251
empiricism, 17, 133, 134
—inductive, 138–39
—naïve, 134
entities
—ecological, 23–27, 29
—metaphysics of, 22–23, 171
environmentalism (*see also* ecology, and
  politics), 3
ethics
—and ecology, 2, 18, 320
—environmental, 3, 6, 16
Evernden, Neil, 13

feminism, ecological, 6
Fenchel, Tom, 3–4, 18 (n. 6)
Ferré, Frederick, 22–23, 31, 33, 81
Feyerabend, Paul, 152
Forbes, Stephen, 137

Foucault, Michel, 13
Frankfurt School, 13
Franklin, Rosalind, 174

Gaia hypothesis, 28
Galileo, 83, 172
Gause, George F., 105, 137, 292
Geisteswissenschaften (humanities), 15, 320
genetics, 17, 99, 119, 161, 239–41, 243, 302
Glacken, Clarence, 105–6, 107
Gleason, Henry, 25–26, 27, 28, 73, 82, 104, 137, 187
Gleick, James, 109
Glymour, C., 164
Golley, Frank, 176
Gould, Stephen J., 241–42, 245 (n. 1), 277, 278, 280–81, 283–87, 305, 310
Grene, Marjorie, 29
Grünbaum, A., 164

Haeckel, Ernst, 7–9, 14, 19 (n. 10), 26, 243, 290
Hagen, Joel, 19 (n. 12), 139, 291
Harner, Michael, 266
Harvey, William, 172, 252, 282
Hegel, G. W. F. (*also* Hegelianism), 29, 31, 206
Heisenberg, Werner, 14, 19 (n. 13), 77
Hempel, Carl, 174, 175, 178
Heraclitus, 31
Herodotus, 107
hierarchies, 86, 175, 177–79, 226–31
—ecological, 29–30, 183–85, 306
—nested and nonnested (control), 29–30, 178, 228–31
hierarchy theory, 87, 177, 313–15
Hobbes, Thomas, 13, 29, 83, 172
holism, 2, 17, 27, 62, 63, 77–79, 82, 171, 172, 175, 177, 179, 194–203, 204–17, 218, 220, 221, 284, 287, 305
Hull, David, 139, 235
Hume, David, 64, 133, 135, 136, 234

Hutchinson, G. Evelyn, 11, 19 (n. 12), 74, 82, 215, 297
hypothetico-deductive method, 134–37, 152, 158, 168

industrialism, 1, 4, 320
interconnectedness, ontological, 2, 23, 225
irrationalism, 133, 134

Jeffers, Robinson, 5

Kant, Immanuel (*also* Kantianism), 21–22, 134, 141, 142, 179 (n. 1)
Kauffman, Stuart, 109
Kekulé, Friedrich, 135
Kepler, Johannes, 4, 172
Kerner von Marilaun, Anton, 102
Kodric-Brown, Astrid, 243
Kuhn, Thomas (*also* Kuhnian), 28, 165, 166
Kulesza, George, 115, 116, 117

Lack, David, 293, 297, 298
Lamarck, Jean-Baptiste, 7, 247, 256, 257, 258–59, 261
Langton, Christopher, 109
Laszlo, Ervin, 173
Leibniz, Gottfried, 133, 234, 242
L'Enfant, Pierre, 103
Leopold, Aldo, 5
Levin, Simon, 104
Levins, Richard, 177, 231
Levy, H., 64
Lewontin, Richard, 75, 177, 224, 241–42, 275, 277, 278, 280–81, 283–87, 305, 310
Lindeman, Raymond, 176, 210
Linnaeus, Carolus, 7, 14, 33, 107–8, 247, 253–54, 255, 256
Locke, John, 4, 133
Loehle, Craig, 244
Lopez, Barry, 5
Lotka, Alfred, 137, 138, 150, 292
Lovejoy, Arthur, 105, 173, 179 (n. 1)

Lucretius, 64
Lyotard, Jean-François, 13

MacArthur, Robert, 75, 109, 120, 121, 123, 124, 126, 133, 138, 186
MacFadyen, Amyan, 76, 137
Malthus, Thomas, 7, 82, 106, 235–36
Margalef, Ramón, 215, 229
Marsh, George Perkins, 3
Marx, Karl (*also* Marxian, Marxism), 6, 79, 219, 224
Marx, Leo, 3
materialism
—dialectical, 177, 178, 219, 221
—mechanistic (*also* mechanism, mechanistic worldview), 3, 4, 5, 15, 18, 26, 33, 171, 172, 181–93, 218, 219, 223, 320
May, Robert, 75, 125, 137, 231
Mayr, Ernst, 76, 77, 239–40, 242, 260–61, 294–96
McCoy, Earl, 82, 139
McIntosh, Robert, 71, 72, 79, 102, 290
McNaughton, S. J., 122
mechanism. *See* materialism, mechanistic
Mendel, Gregor (*also* Mendelian), 74, 224, 239, 291, 298
Mill, John Stuart, 19 (n. 14), 133
Mittelbach, Gary, 139
Möbius, Karl, 102, 110
Muir, John, 4, 5
mutualism (symbiosis), 311, 313–14, 315

Naess, Arne, 6, 81, 179
Nagel, Ernest, 181–82, 208
natural history, 6–7, 9, 27, 106, 195, 226
naturalism, 2, 6, 11, 12
natural selection, 7, 9, 32, 235–38, 241–42, 301, 307, 308, 309
natural theology, 233–35
nature (*also* natural systems), 1, 4, 11–14
—ancient Greek concept of, 72, 79–80, 107, 244, 248–52
—balance (equilibrium) of, 12, 15, 66, 107, 108

—Christian concept of, 106, 107, 244, 252–53
—definition of word, 11–12, 13, 18 (n. 1)
—indeterminism (*also* stochasticity, chaos) of, 12, 14, 25–26, 27, 28, 32, 33, 71, 75–77, 79–80, 82, 85, 87, 109, 219, 225, 238–39
—teleology (design) in, 4, 15, 16, 25, 28, 87, 107, 108, 233–35, 238–39
Naturwissenschaften (natural sciences), 15, 320
Newton, Isaac (*also* Newtonian), 14, 15, 18, 19 (n. 13), 31, 73, 81, 83–84, 96–99, 172, 223, 224
niche, 16, 104–5, 115–18
Nietzsche, Friedrich (*also* Nietzschean), xi, 13, 31, 239
nominalism, 86, 87
nonanthropocentricism, 2

objectivity, 1–2, 13, 14, 163
Odum, Eugene, 77, 79, 82, 95, 108, 121, 176, 177, 204–17
Odum, Howard, 82, 121, 176, 230
Oppenheim, Paul, 174, 175, 178
organicism, 70, 171, 172, 176, 206
Orians, Gordon, 243, 288, 294, 295

Pahl-Wostl, Claudia, 32
Paley, William, 18–19 (n. 7), 233, 238
Parmenides, 31
Patrick, Ruth, 106
Patten, Bernard, 77–78, 176
Pechmann, Joseph, 244
Peters, Robert, 138–39, 153
Phillips, John, 26, 55, 56, 57, 59, 60, 61, 62, 63, 64, 66, 67, 70, 215
Pimm, Stuart, 109, 125, 126
Pirsig, Robert, 135–36
Plato (*also* Platonic), 5, 6, 28, 31, 105, 107, 206, 224, 247, 248–49
Pliny the Elder, 7
pluralism, 139, 192, 218, 241, 283
Plutarch, 7